Feature Papers in Compounds

Feature Papers in Compounds

Editor

Juan C. Mejuto

MDPI • Basel • Beijing • Wuhan • Barcelona • Belgrade • Manchester • Tokyo • Cluj • Tianjin

Editor
Juan C. Mejuto
University of Vigo
Spain

Editorial Office
MDPI
St. Alban-Anlage 66
4052 Basel, Switzerland

This is a reprint of articles from the Special Issue published online in the open access journal *Compounds* (ISSN 2673-6918) (available at: https://www.mdpi.com/journal/compounds/special_issues/feature_papers_compounds).

For citation purposes, cite each article independently as indicated on the article page online and as indicated below:

LastName, A.A.; LastName, B.B.; LastName, C.C. Article Title. *Journal Name* **Year**, *Volume Number*, Page Range.

ISBN 978-3-0365-5987-2 (Hbk)
ISBN 978-3-0365-5988-9 (PDF)

Contents

About the Editor

Juan C. Mejuto

Juan C. Mejuto (PhD) studied chemistry at the Universidad de Santiago de Compostela from 1987 to 1992 and received his PhD in chemistry, focusing on organic reactivity in biomimetic media and the effects of catalytic processes on reactions of biological interest in colloidal aggregates, from the same university in 1996. His study of these systems examined not only their restricted kinetic implications but also their thermodynamic properties and their internal dynamics. Currently, Dr. Mejuto is a full professor in the Physical Chemistry Department of the University of Vigo, Ourense Campus (since 2009), where he previously worked as an assistant professor (1996–2009). He is member of the Argo-environmental and Food Chemistry Research Group at the Ourense Campus. His research interest comprise physical organic and physical inorganic chemistry, reactivity mechanisms in homogeneous and micro-heterogeneous media, and the stability of self-assembly aggregates. He has also carried out research on supramolecular chemistry, particularly with respect to cyclodextrins, crown ethers and cryptands, and their interactions with colloidal aggregates, from the perspectives of both their internal dynamics and the effects that these chemical systems have upon chemical reactivity. He also works on the modeling of complex systems, particularly colloidal systems, using different computational tools. In collaboration with different research groups, he has applied mathematical models and machine learning tools to the prediction of different environmental phenomena. Finally, his interests also include food chemistry.

Preface to "Feature Papers in Compounds"

"Feature Papers in Compounds" comprises a collection of open access, high-quality, peer-reviewed research and review articles published in Compounds by members of the Editorial Board and the authors invited by the Editorial Office and Editor-in-Chief.

Compounds is an interdisciplinary journal that was established in 2021 with the aim of showcasing the achievements of the scientific community through the modeling of molecules and transformation of matter according to the needs of modern societies. Compounds intends to inform its readers and stimulate interest and discussion regarding the development of the scientific basis of this discipline. The journal regularly publishes research articles, communications, letters, short notes, and reviews contributing to the enrichment of this expanding field of chemical research.

We thank the authors who have chosen to publish their research in Compounds, as well as the editorial committee and the editorial office who have supported this endeavor over the past two years.

Juan C. Mejuto
Editor

compounds

Editorial

Feature Papers in *Compounds*

Juan C. Mejuto

Physical Chemistry Department, Faculty of Sciences, University of Vigo, E32004 Ourense, Spain; xmejuto@uvigo.es

Citation: Mejuto, J.C. Feature Papers in *Compounds*. *Compounds* 2022, 2, 237–239. https://doi.org/10.3390/compounds2040019

Received: 28 September 2022
Accepted: 28 September 2022
Published: 1 October 2022

Publisher's Note: MDPI stays neutral with regard to jurisdictional claims in published maps and institutional affiliations.

1. Introduction and Scope

Nearly two years ago, when *Compounds* was founded, the journal was introduced as an interdisciplinary tool for the scientific community to present their scientific results in an open access format so that their findings are disseminated quickly and efficiently [1]. In this sense, practically at the same time, a Special Issue was launched under the title of "Feature Papers in Compounds" to make up a collection of high-quality articles published in open access by members of the Editorial Board and authors invited by the Editorial Office as per the Editor-in-Chief [2].

2. Contributions

This Special Issue, as we have already indicated, is made up of 14 manuscripts, of which 4 are review articles, 8 are research articles and 2 are communications, whose thematic variety is clear proof of the interdisciplinary spirit that we seek in *Compounds*.

In this way, contributions related to the synthesis [3–6] and characterization [6,7] of the compounds'—with a special interest in natural products [8–12]—chemical reactivity [13,14] were incorporated. Likewise, there have been contributions in the field of crystalline structures [15] and computational modeling in quantum chemistry and molecular dynamics [16].

In the synthesis section, Damkaci et al. [3] addressed the development of a simple, efficient and profitable synthesis of a library of thiazolidinedione compounds (glitazones) whose interest lies in the fact that they are compounds with important applications in the pharmaceutical industry, because they exhibit antidiabetic, anti-inflammatory and anticancer properties. Microwave irradiation together with reduction under pressurized hydrogen gas using palladium hydroxide was used for the synthesis. Thus, a fast and simple synthetic route is presented that, in addition, obtains satisfactory yields. Further, Aubert et al. [4] presents the synthesis of on-symmetrical atropisomeric polyhalogenated $4,4'$-bipyridines and the subsequent functionalization using cross-coupling reactions. Moreover, a review article on the preparation of various carbon compounds from glycerol (by-product of the transesterification processes of fats and oils) together with their different applications is included [5].

Along with the previous contribution, we must comment on the one made by Peosi et al. [6], where they present the synthesis of four thiosemicarbazones derived from naphthaldehyde and anthraldehyde and their copper complexes with biological properties (particularly their antileukemic properties). These compounds were characterized using different instrumental techniques, their stability under physiological conditions was verified and they were subjected to in vitro biological tests against a histiocytic lymphoma cell line, finding interesting results.

Kim et al. [7] evaluated the effects of structural defects caused by impurities in silicon carbide, characterizing them with various instrumental techniques.

Within this Special Issue, we should also point out that where studies on natural products are concerned, we must point out studies on the determination by chromatographic techniques of volatile organic compounds present in *Dactylorhiza* (*D. viridis D. romana*, *D. saccifera* and *D. sambucina*) [8], *Himantoglossum* (*H. hircinum*, *H. adriaticum*, *H.

robertianuma) [9,10], *Barlia robertiana* [10], *Orchis* (*O. anthropophra, O. purpurea, O. italica, O. masculua, O. pauciflora, O. quadrpunctata, O. provincialis* and *O. pallens*) [11]. Likewise, the volatile fraction of avocado oil (*Persea americana*, Greek variety "Zutano") was evaluated [12].

Regarding chemical reactivity studies, the Special Issue includes a study on the effect of electrolytes on the dediazoniation of aryldiazonium ions in acidic mixtures [13] and a review article on electrophilic iodination of organic compounds using elemental iodine or iodides [14].

Therefore, a review article presents several illustrative crystal structures from various crystal structure databases (CSD and ICSD), proving that imide nitrogen is not the only case where nitrogen can act as an electrophilic agent. It is shown that covalently bound nitrogen presents a σ-hole or even a π-hole, with the ability to establish attractive compromises with negative sites that allow for both intramolecular and intermolecular interactions [15].

To close this Special Issue, Malyshkina and Novikov [16] compiled the most popular modern programs for quantum chemistry and molecular dynamics calculations (classical, ab initio and QM/MM). The authors place special emphasis on those with applicability to nanotubes, surfaces and films, polymers, and crystalline solids.

3. Conclusions and Outlook

In conclusion, we must state that this Special Issue compiles interesting contributions that illustrate the transversal vocation of *Compounds*, as well as its interdisciplinary nature within pure and applied chemistry. At the end of the process, we must indicate that 14 contributions have been published that have accumulated 13,000 views and 15 citations, which demonstrates the interest that *Compounds* is beginning to arouse in the scientific community. This success allows us to face the challenge of a new Special Issue under the title "Feature Papers in Compounds (2022–2023)" [17], in which we hope to incorporate new and valuable contributions of interest to *Compounds* readers.

Conflicts of Interest: The author declares no conflict of interest.

References

1. Mejuto, J.C. Introducing Compounds: An Interdisciplinary Open Access Journal. *Compounds* **2021**, *1*, 1–2. [CrossRef]
2. Compounds Special Issue "Feature Papers in Compounds". Available online: https://www.mdpi.com/journal/compounds/special_issues/feature_papers_compounds (accessed on 26 September 2022).
3. Damkaci, F.; Szymaniak, A.A.; Biasini, J.P.; Cotroneo, R. Synthesis of Thiazolidinedione Compound Library. *Compounds* **2022**, *2*, 182–190. [CrossRef]
4. Aubert, E.; Wenger, E.; Peluso, P.; Mamane, V. Convenient Access to Functionalized Non-Symmetrical Atropisomeric 4,4′-Bipyridines. *Compounds* **2021**, *1*, 58–74. [CrossRef]
5. Batista, M.; Carvalho, S.; Carvalho, R.; Pinto, M.L.; Pires, J. Waste-Glycerol as a Precursor for Carbon Materials: An Overview. *Compounds* **2022**, *2*, 222–236. [CrossRef]
6. Pelosi, G.; Pinelli, S.; Bisceglie, F. DNA and BSA Interaction Studies and Antileukemic Evaluation of Polyaromatic Thiosemicarbazones and Their Copper Complexes. *Compounds* **2022**, *2*, 144–162. [CrossRef]
7. Kim, S.-K.; Jung, E.Y.; Lee, M.-H. Defect-Induced Luminescence Quenching of 4H-SiC Single Crystal Grown by PVT Method through a Control of Incorporated Impurity Concentration. *Compounds* **2022**, *2*, 68–79. [CrossRef]
8. Mecca, M.; Racioppi, R.; Romano, V.A.; Viggiani, L.; Lorenz, R.; D'Auria, M. Volatile Organic Compounds in Dactylorhiza Species. *Compounds* **2022**, *2*, 121–130. [CrossRef]
9. Mecca, M.; Racioppi, R.; Romano, V.A.; Viggiani, L.; Lorenz, R.; D'Auria, M. The Scent of Himantoglossum Species Found in Basilicata (Southern Italy). *Compounds* **2021**, *1*, 164–173. [CrossRef]
10. Romano, V.A.; Rosati, L.; Fascetti, S.; Cittadini, A.M.R.; Racioppi, R.; Lorenz, R.; D'Auria, M. Spatial and Temporal Variability of the Floral Scent Emitted by Barlia robertiana (Loisel.) Greuter, a Mediterranean Food-Deceptive Orchid. *Compounds* **2022**, *2*, 37–53. [CrossRef]
11. Mecca, M.; Racioppi, R.; Romano, V.A.; Viggiani, L.; Lorenz, R.; D'Auria, M. Volatile Organic Compounds from *Orchis* Species Found in Basilicata (Southern Italy). *Compounds* **2021**, *1*, 83–93. [CrossRef]
12. Xagoraris, M.; Galani, E.; Valasi, L.; Kaparakou, E.H.; Revelou, P.-K.; Tarantilis, P.A.; Pappas, C.S. Estimation of Avocado Oil (Persea americana Mill., Greek "Zutano" Variety) Volatile Fraction over Ripening by Classical and Ultrasound Extraction Using HS-SPME–GC–MS. *Compounds* **2022**, *2*, 25–36. [CrossRef]

13. Losada-Barreiro, S.; Bravo-Díaz, C. Effects of Electrolytes on the Dediazoniation of Aryldiazonium Ions in Acidic MeOH/H$_2$O Mixtures. *Compounds* **2022**, *2*, 54–67. [CrossRef]
14. Ajvazi, N.; Stavber, S. Electrophilic Iodination of Organic Compounds Using Elemental Iodine or Iodides: Recent Advances 2008–2021: Part I. *Compounds* **2022**, *2*, 3–24. [CrossRef]
15. Varadwaj, P.R.; Varadwaj, A.; Marques, H.M.; Yamashita, K. The Nitrogen Bond, or the Nitrogen-Centered Pnictogen Bond: The Covalently Bound Nitrogen Atom in Molecular Entities and Crystals as a Pnictogen Bond Donor. *Compounds* **2022**, *2*, 80–110. [CrossRef]
16. Malyshkina, M.V.; Novikov, A.S. Modern Software for Computer Modeling in Quantum Chemistry and Molecular Dynamics. *Compounds* **2021**, *1*, 134–144. [CrossRef]
17. Compounds Special Issue "Feature Papers in Compounds (2022–2023)". Available online: https://www.mdpi.com/journal/compounds/special_issues/871D2T3QU1 (accessed on 26 September 2022).

 *compounds*

Communication

Synthesis of Thiazolidinedione Compound Library

Fehmi Damkaci *, Adam A. Szymaniak, Jason P. Biasini and Ryan Cotroneo

Department of Oswego, SUNY Oswego, Oswego, NY 13126, USA; szymania@oswego.edu (A.A.S.);
biasini@oswego.edu (J.P.B.); cotroneo@oswego.edu (R.C.)
* Correspondence: fehmi.damkaci@oswego.edu; Tel.: +1-315-312-2698

Abstract: Thiazolidinediones (TZDs), also known as Glitazones, have anti-diabetic, anti-inflammatory and anti-cancer properties. A simple, efficient and cost-effective synthesis of a thiazolidinedione compound library was developed. The synthesis is facilitated by microwave irradiation in three of the four steps followed by reduction under pressurized hydrogen gas using palladium hydroxide. All reactions, except one, were completed within an hour and provided desired products in moderate to good yields after a simple work-up.

Keywords: thiazolidinedione; microwave synthesis; compound library; rosiglitazone; Knoevangel condensation

1. Introduction

The occurrence of diabetes and other diseases contracted around the globe has vastly increased over the past few decades driven by the global rise in the prevalence of obesity. Worldwide there is a projected increase in the frequency of diabetes from 285 million in 2010 to 439 million in 2030. Estimates in developing countries show marked increases, particularly in areas where populations are rapidly adopting Western lifestyles [1]. The increase in the occurrence of childhood obesity has led to the development of type II diabetes in children, and young adults, particularly those in high susceptible ethnic groups [1]. For this reason, the availability of drugs and therapeutics aimed at diabetes must increase, most specifically type II diabetes mellitus, which accounts for approximately 90–95% of all diagnosed cases of diabetes [2,3]. Along with diabetes treatments comes the necessity to treat other associated cardiovascular diseases such as hypertension, atherosclerosis, dyslipidemia, coagulation abnormalities, heart disease and many more [4].

Thiazolidinediones (TZDs), also known as Glitazones, are a class of insulin-sensitizing agents, which are used in the oral therapy of type II diabetes mellitus [5–7]. TZDs were introduced in the late 1990s and have been widely used since due to their clinical advantages of treating insulin resistance and sustaining glycemic control [8,9]. The first thiazolidinedione drug approved by the FDA, called troglitazone, was withdrawn from the market within three years due to severe liver damage in some patients [10,11]. The only thiazolidinedione drugs currently in use and on the market are rosiglitazone and pioglitazone (Figure 1) [12].

After its release, it was found that rosiglitazone was associated with an increased risk of myocardial infarction, and in November of 2011, the FDA began restricting access, granting the drug only to patients with no cardiovascular risk, and whose diabetes is not well controlled with other medications [8,13,14]. However, in November of 2013, the FDA removed the restriction due to recent findings of no risk of heart failure from the use of rosiglitazone [15,16].

The mechanism in which TZDs work is relatively well known. Treatment of type II diabetes is achieved from the thiazolidinedione ring binding to, and activating the peroxisome proliferators-activated receptor γ (PPAR-γ), which promotes glucose utilization, primarily in adipose tissue [17,18]. Peroxisome proliferator-activated receptor γ is a nuclear receptor

Citation: Damkaci, F.; Szymaniak, A.A.; Biasini, J.P.; Cotroneo, R. Synthesis of Thiazolidinedione Compound Library. *Compounds* **2022**, *2*, 182–190. https://doi.org/10.3390/compounds2030013

Academic Editor: Juan Mejuto

Received: 3 April 2022
Accepted: 21 June 2022
Published: 5 July 2022

Publisher's Note: MDPI stays neutral with regard to jurisdictional claims in published maps and institutional affiliations.

that modulates the transcription of insulin-responsive genes involved in the control of glucose lipid metabolism, as well as the gene involved in inflammatory responses [19,20]. Thiazolidinediones were also shown to have significant anti-inflammatory effects, which would be beneficial in patients that suffer from both diabetes and atherosclerosis [11,21]. Furthermore, the discovery of anti-cancer properties, and the suggestion that TZDs may improve cognitive abilities in patients with Alzheimer's disease and dementia, add other possibilities for the potential uses of thiazolidinediones [15,22,23].

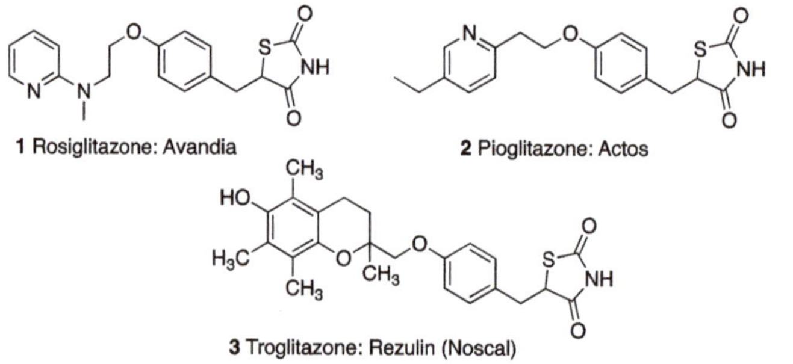

Figure 1. Previous and current TZD drugs on the U.S. market.

Given the side effects of previously marketed TZDs, and increasing potential effectiveness in many diseases, the necessity for newly developed TZD compounds is high. The goal of this study was to develop an efficient synthesis of a thiazolidinedione compound library over four steps, three of which were accomplished in less than an hour using microwave heating. Given that all thiazolidinediones previously and currently on the market rely solely upon the addition of substituents to the methylene of the TZD ring (Scheme 1), this project focused on the effects of adding substituents to the nitrogen of the TZD ring as well. This was accomplished through microwave-assisted N-benzylation reactions and subsequent reduction of the olefin for decreased rigidity. It is known that N-alkylation of TZDs lowers the antidiabetic activity, however, we are also interested in a new mode of action with TZD derivatives. Unlike conventional heating, microwave radiation causes a uniform increase in temperature throughout the sample, which allows for shorter reaction times, increased yields, and less side product formation [24].

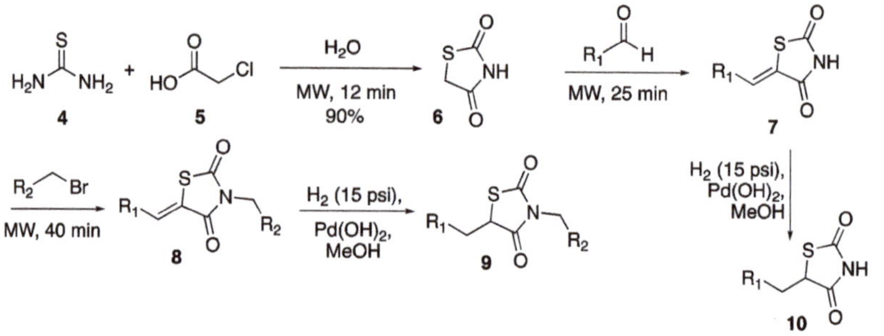

Scheme 1. Synthesis of TZD library.

2. Materials and Methods

2.1. General Information

Reactions using microwave irradiation were performed using a Milestone Start S dual-move microwave synthesizer (Milestone, Sorisole, Italy) and contained in a Synthware pressure vial. Reactions under H_2 gas were carried out using a Parr pressure apparatus.

All chemicals were purchased from Acros Organics (Morris Plains, NJ, USA), Aldrich (Springfield, MO, USA) and Alfa Aesar (Tewksbury, MA, USA), and were used without further purification.

^{1}H and ^{13}C NMR spectra were recorded on Varian UNITY I Nova 300 MHz, Bruker Ultrashield 300 MHz and Bruker Ascend 500 MHz. Dimethyl Sulfoxide-d_6 and Chloroform-d were used as the reference point in ^{1}H and ^{13}C NMR spectra (2.50, 39.5 and 7.24, 77.23, respectively). Coupling constants (J values) are given in hertz (Hz). Spin multiplicities are indicated by the following symbols: s (singlet), d (doublet), t (triplet), q (quartet), p (pentet), sx (sextet), sp (septet), o (octet), br (broad), dd (doublet of doublets), td (triplet of dublets), m (multiplet).

High-resolution mass spectra were collected using JEOL AccuTOF mass spectrometer (JEOL, Tokyo, Japan).

IR spectra were collected using neat samples on a Nicolet iS5 infrared spectrometer (ThermoFisher Scientific, Waltham, MA, USA). Band positions are given in reciprocal centimeters (cm^{-1}) and relative intensities are listed as s (strong), m (medium), w (weak) or br (broad).

Melting points were taken in soft glass capillary tubes using an uncalibrated Mel-Temp II capillary melting point apparatus (Barnstead International, Dubuque, IA, USA).

Of the total, 38 compounds were tested for antimicrobial activity towards E. coli, Bacillus subtilis, Staphyloccus aueus, Salmonella typhimurium, Sinomonas atrocyanea and Rhodococcus erythropolis using the paper diffusion method. The selection of compounds for testing was made based on grouping compounds according to their functional groups.

Supplementary Material includes all new compound spectral data.

2.2. Experimental Procedure for Thiazolidine-2,4-dione (6)

A mixture of thiourea (**4**, 3.34 g, 43.4 mmol) and monochloroacetic acid (**5**, 4.16 g, 44.0 mmol) in 8 mL of water was added to a 15 mL pressure vial equipped with a stir bar. The reaction mixture was allowed to stir for 1 h at room temperature and microwave irradiated at 110 °C and 350 W for 12 min. (2 min. ramp, 10 min. sustain). The resulting solution was cooled and stirred at room temperature for 1 h. The precipitate was recrystallized from water to produce the product as a white crystalline solid (4.57 g) in 90% yield.

2.3. General Experimental Procedure for Compound 7

A mixture of substituted aryl aldehyde (1.00 mmol), thiazolidine-2,4-dione (**6**, 1.50 mmol), silica gel (200 mg), 5 drops (~0.25 mL) of both acetic acid and piperidine in 2 mL toluene were added to a Synthware pressure vial equipped with a stir bar. The mixture was microwave irradiated for 25 min at 110 °C and 300 W (5 min. ramp at 500 W, 20 min. sustain). The resulting mixture was diluted with 4 mL of water and precipitated on ice for 15 min. Silica gel was removed by vacuum filtration and washed with hot methanol and the filtrate was concentrated under reduced pressure. The resulting solid was recrystallized using ethanol and dried in vacuo to give the products as colored solids in 35–75% yield.

2.4. General Experimental Procedure for Compound 8

A mixture of monosubstituted thiazolidine-2,4-dione (1.00 mmol), substituted benzyl bromide (1.00 mmol) potassium hydroxide (100 mg, 1.78 mmol), tertbutylammonium hydrogen sulfate (110 mg, 0.324 mmol) in 2 mL of water and 3 mL of toluene were added to a Synthware pressure vial equipped with a stir bar. The reaction mixture was microwave irradiated at 85 °C and 250 W for 45 min while pausing every 2 min and the reaction vial shaken to obtain sufficient agitation (5 min ramp, 40 min sustain). The resulting reaction mixture was diluted with 5 mL of water, extracted twice with 30 mL of ethyl acetate, washed with 20 mL of water, and dried with magnesium sulfate. The magnesium sulfate was filtered, the filtrate concentrated under reduced pressure, rinsed with 20 mL of ethanol and the solid dried in vacuo resulting in the products as colored, textured solids in 27–97% yield.

2.5. General Experimental Procedure for Compounds **9** and **10**

A mixture of disubstituted thiazolidine-2,4-dione (100 mg) and 20% palladium hydroxide on activated carbon (120 mg, 0.855 mmol) in 20 mL of methanol was added to a 30 psi pressure vial and shaken by a pressurized reaction apparatus at 15 psi under hydrogen atmosphere for 15 h. The resulting mixture was filtered using celite, dried with silica gel and concentrated under reduced pressure to give the products as solids in 43–98% yield.

3. Results and Discussion

Overall, 76 thiazolidinedione compounds were synthesized by the synthetic pathway shown in Scheme 1, with the utilization of microwave irradiation. The synthesis of thiazolidinedione-2,4-dione (**6**) was accomplished by following an established literature procedure using water as the solvent and reagent [22–24]. The synthesis of TZD ring **6** was easily scaled-up to provide four grams of the product without any reduction in the yield. Next, Knoevangel condensation of an aldehyde with thiazolidine-2,4-dione **6** was performed resulting in the formation (Scheme 1) of 1-(benzylidene)-3-thiazolidine-2,4-dione (**7**), by following a modified literature procedure (at a lower temperature and using different work-up procedure) (Table 1) [22]. Ten different aldehydes were chosen to react with TZD **6** based on their electronic (electron-rich and -poor), steric (ortho-substituted), and hydrogen bond donor (containing OH group) properties. Overall, the reaction provided the desired coupled derivatives of **7** in moderate to good yields. In general, electron-poor aldehyde derivatives (Table 1, **7E–G**) led to slightly better results compared to electron-rich derivatives (Table 1, **7B–D**). Both hydroxyl-containing aldehydes worked under the conditions, however, 2-hydroxybenzaldehyde (Table 1, **7J**) resulted in a lesser yield compared to 4-hydroxybenzaldehyde (Table 1, **7I**).

Table 1. Knoevangel condensation of thiazolidine-2,4-dione.

Compound	R_1	Yield (%) [1]
7A	4-iPr	75
7B	4-OMe	55
7C	4-Me	52
7D	3-OMe-4-OH	60
7E	4-Cl	75
7F	2-Cl	71
7G	4-Br	70
7H	4-F	57
7I	4-OH	72
7J	2-OH	51

[1] Isolated yields.

Seven out of ten derivatives of TZD **7** were successfully carried further in derivatization efforts for *N*-benzylation with various benzyl bromides resulting in the formation of thiazolidine-2,4-dione derivative **8** (Table 2).

Table 2. *N*-benzylated derivatives of Compound **7**.

Compound	R_1	R_2	Yield (%) [1]
8A-1	4-iPr	H	83
8A-2	4-iPr	4-Me	67
8A-3	4-iPr	3-OMe	76
8A-4	4-iPr	2-Cl	82
8A-5	4-iPr	3-NO2	71
8A-6	4-iPr	4-(2-CN-Ph)-Ph	86
8B-1	4-OMe	H	82
8B-2	4-OMe	4-Me	55
8B-3	4-OMe	3-OMe	58
8C-1	4-Me	H	97
8C-2	4-Me	4-Me	87
8C-3	4-Me	3-OMe	77
8C-4	4-Me	2-Cl	56
8C-5	4-Me	4-(2-CN-Ph)-Ph	37
8D-1	3-OMe-4-OH	H	76
8D-2	3-OMe-4-OH	4-Me	81
8D-3	3-OMe-4-OH	3-OMe	70
8D-4	3-OMe-4-OH	2-Cl	57
8E-1	4-Cl	H	71
8E-2	4-Cl	4-Me	57
8E-3	4-Cl	3-OMe	75
8E-4	4-Cl	2-Cl	72
8E-5	4-Cl	3-NO2	64
8E-6	4-Cl	4-(2-CN-Ph)-Ph	46
8F-1	2-Cl	H	86
8F-2	2-Cl	4-Me	75
8F-3	2-Cl	2-Cl	86
8F-4	2-Cl	3-NO2	85
8F-5	2-Cl	4-(2-CN-Ph)-Ph	67
8G-1	4-Br	H	63
8G-2	4-Br	4-Me	55
8G-3	4-Br	3-OMe	65
8G-4	4-Br	2-Cl	41

[1] Isolated yields.

TZD **7H–J** in Table 1 gave extremely poor yields in reacting with benzyl bromide derivatives, and it was difficult to purify their reaction mixtures. Thus, these results do not appear in Table 2. *N*-benzylation of the TZD ring with benzyl bromides proved to be the critical step due to the fact that excessive agitation of the reaction mixture was required in order to achieve acceptable yields. We were able to synthesize 33 variations of compound **8** by using a phase transfer catalyst in a biphasic reaction mixture under microwave heating. Except for a few (Table 2, **8C-5**, **8E-6**, and **8G-4**), most of the derivatives were obtained in good to very good yields. The reaction did not appear to depend upon the electronic or steric nature of the benzyl bromide derivatives, since it provided mixed results with various derivatives. For example, sterically hindered and electron-poor 2-chlorobenzyl bromide provided good yields in the synthesis of derivatives **8A-4**, **8E-4**, and **8F-3** (71–86% yield) while providing moderate yields in the synthesis of derivatives **8C-4**, **8D-4**, and **8G-4** (41–56% yield). *N*-benzylation using phase transfer catalysis either provided pure products after a simple work-up or provided a complex mixture of products and starting materials with very low yield.

Finally, in order to test the importance of rigidity around C1 of the TZD ring, the benzylidene double bond of compound **8** was reduced to give racemic mixtures of fully functionalized compound **9** (Table 3). The reduction reaction was initially performed using magnesium in methanol by following a literature procedure [22], which resulted in no products even after several optimization efforts of reaction conditions. However, reduction of the olefin using palladium hydroxide in methanol under hydrogen pressure led to the desired product with very good yields. In addition, the same procedure was tested in the reduction of select derivatives of **7**, which successfully provided compounds **10** in very good yields (Table 4).

Table 3. Olefin reduction of Compound **8** derivatives.

Compound	R$_1$	R$_2$	Yield (%) [1]
9A-1	4-iPr	H	98
9A-2	4-iPr	4-Me	87
9A-3	4-iPr	3-OMe	82
9A-4	4-iPr	2-Cl	91
9A-5	4-iPr	3-NH2	71
9A-6	4-iPr	4-(2-CN-Ph)-Ph	95
9B-1	4-OMe	H	66
9B-2	4-OMe	4-Me	49
9B-3	4-OMe	3-OMe	83
9C-1	4-Me	H	88
9C-2	4-Me	4-Me	82
9C-3	4-Me	3-OMe	49
9C-4	4-Me	2-Cl	64
9D-1	3-OMe-4-OH	H	83
9D-2	3-OMe-4-OH	4-Me	74
9D-3	3-OMe-4-OH	3-Me	93
9D-4	3-OMe-4-OH	2-Cl	43
9E-1	4-Cl	H	83
9E-2	4-Cl	4-Me	87
9E-3	4-Cl	3-OMe	82
9E-4	4-Cl	2-Cl	76
9F-1	2-Cl	H	66
9F-2	2-Cl	4-Me	93
9F-3	2-Cl	2-Cl	76
9F-4	2-Cl	3-NH2	61
9F-5	2-Cl	4-(2-CN-Ph)-Ph	72
9G-1	4-Br	H	85
9G-2	4-Br	4-Me	70
9G-3	4-Br	2-Cl	90

[1] Isolated yields.

Table 4. Olefin reduction of 1-(sub. benzylidene)-3-thiazolidine-2,4-dione (**7**).

Compound	R_1	Yield (%) [1]
10A	4-iPr	95%
10B	4-OMe	80%
10C	4-Me	88%
10E	4-Cl	89%

[1] Isolated yields.

Of the 76 compounds synthesized, 39 were tested for antimicrobial activity against *E. coli*, *Bacillus subtilis*, *Staphylococcus aureus*, *Salmonella typhimurium*, *Sinomonas atrocyanea* and *Rhodococcus erythropolis* (Table 5). Compounds **7A** and **7E** were found to have antimicrobial properties toward *Bacillus subtilis*, *Staphylococcus aureus*, *Sinomonas atrocyanea* and *Rhodococcus erythropolis*. Microorganisms were chosen based on what was available to hand at the time of testing with no further reasoning. Further activity tests will be conducted on the synthesized derivatives in the future.

Table 5. The results of antimicrobial testing using paper disk method [25] [1].

Microorganism	7A	7B	7C	7E	7H	7J	8D2	9A1	9A5	9D4	10B	10C
E. Coli K-12												
Bacillus subtillis JCM1465	17			13	11T							
Staphylococcus aureas	17			16	15T							
Salmonella typhimurium												
Sinomonas atrocyanea JCM1329	16			11				9T	8T		11T	8T
Rhodococcus erthropolis JCM3201	23	8T	8T	16T	13T	10T	8T			9T	9T	

[1] The results were given in millimeter and T means turbid. The compounds were soluble in acetone (generally 10 mg/mL, or 5 mg/mL if solubility was a problem) and 5 μL were soaked in 6 mm paper disks, dried and tested. Results **7A**, **7C**, **7E**, **7G**, **7H**, **7I**, **8A-1–6**, **8C-1–3**, **8D-1–4**, **8E-1**, **8E-3**, **8G-1**, **9A-1**, **9A-3**, **9A-5**, **9C-1** and **9D-3** were soluble in 10 mL of acetone. Results **7B**, **7D**, **8C-4**, **8E-2**, **8E-5**, **8E-6**, **9A-5**, **9C-2**, **9C-3** and **9D-4** were soluble in 5 mL of acetone. The rest, **9A-2**, **9A-4**, **9A-6**, **9D-1**, **9E-1**, **9E-2**, **9E-4**, **10A** and **10E** were not tested due to low solubility in acetone.

4. Conclusions

In conclusion, we developed a simple, efficient, and cost-effective synthesis of a thiazolidinedione compound library with the use of microwave irradiation and phase transfer catalysis. It should be noted all products were obtained with a simple work-up without column chromatography. Future studies will entail expansion of the compound library, improvements in yield and further biological evaluation of the compounds.

Supplementary Materials: https://www.mdpi.com/article/10.3390/compounds2030013/s1, Table S1: ^{1}H and ^{13}C-NMR spectral data for compound **7** derivatives; Table S2: ^{1}H and ^{13}C-NMR spectral data for compound **8** derivatives; Table S3: ^{1}H and ^{13}C-NMR spectral data for compound **9** derivatives; Table S4: ^{1}H and ^{13}C-NMR spectral data for compound **10** derivatives.

Author Contributions: Conceptualization, F.D.; methodology development, F.D., A.A.S. and J.P.B.; performing reactions and formal analysis, A.A.S., J.P.B. and R.C.; writing—original draft preparation, J.P.B.; writing—review and editing, F.D.; supervision and project administration, F.D. All authors have read and agreed to the published version of the manuscript.

Funding: This research received no external funding.

Institutional Review Board Statement: Not applicable.

Informed Consent Statement: Not applicable.

Data Availability Statement: Not applicable.

Acknowledgments: The authors gratefully acknowledge SUNY at Oswego for providing the Student Scholarly Creativity and Activity Grant and the Department of Chemistry for providing additional funds. Special thanks to Michael Knopp (The University of Maine at Presque Isle) and to Wataru Kitagawa of AIST Hokkaido for performing the antimicrobial cultures.

Conflicts of Interest: The authors declare no conflict of interest.

References

1. Forouhi, N.G.; Wareham, N.J. Epidemiology of diabetes. *Medicine* **2010**, *38*, 602–606. [CrossRef]
2. Shaw, J.E.; Sicree, R.A.; Zimmet, P.Z. Global estimates of the prevalence of diabetes for 2010 and 2030. *Diabetes Res. Clin. Pract.* **2010**, *87*, 4–14. [CrossRef] [PubMed]
3. Statistics Report | Data & Statistics | Diabetes | CDC. Available online: http://www.cdc.gov/diabetes/data/statistics/2014 statisticsreport.html (accessed on 7 July 2015).
4. Sowers, J.R.; Epstein, M.; Frohlich, E.D. Diabetes, hypertension, and cardiovascular disease: An update. *Hypertension* **2001**, *37*, 1053–1059. [CrossRef] [PubMed]
5. Chinthala, Y.; Kumar Domatti, A.; Sarfaraz, A.; Singh, S.P.; Kumar Arigari, N.; Gupta, N.; Satya, S.K.V.N.; Kotesh Kumar, J.; Khan, F.; Tiwari, A.K.; et al. Synthesis, biological evaluation and molecular modeling studies of some novel thiazolidinediones with triazole ring. *Eur. J. Med. Chem.* **2013**, *70*, 308–314. [CrossRef]
6. Millioni, R.; Puricelli, L.; Iori, E.; Arrigoni, G.; Tessari, P. The effects of rosiglitazone and high glucose on protein expression in endothelial cells. *J. Proteome Res.* **2010**, *9*, 578–584. [CrossRef]
7. Flowers, E.; Aouizerat, B.E.; Abbasi, F.; Lamendola, C.; Grove, K.M.; Fukuoka, Y.; Reaven, G.M. Circulating MicroRNA-320a and MicroRNA-486 Predict Thiazolidinedione Response: Moving Towards Precision Health for Diabetes Prevention. *Metabolism* **2015**, *64*, 1051–1059. [CrossRef]
8. Hsu, J.C.; Ross-Degnan, D.; Wagner, A.K.; Zhang, F.; Lu, C.Y. How Did Multiple FDA Actions Affect the Utilization and Reimbursed Costs of Thiazolidinediones in US Medicaid? *Clin. Ther.* **2015**, *37*, 1420–1432. [CrossRef]
9. Zhu, Z.-N.; Jiang, Y.-F.; Ding, T. Risk of fracture with thiazolidinediones: An updated meta-analysis of randomized clinical trials. *Bone* **2014**, *68*, 115–123. [CrossRef]
10. Helal, M.A.; Darwish, K.M.; Hammad, M.A. Homology modeling and explicit membrane molecular dynamics simulation to delineate the mode of binding of thiazolidinediones into FFAR1 and the mechanism of receptor activation. *Bioorg. Med. Chem. Lett.* **2014**, *24*, 5330–5336. [CrossRef]
11. Mohanty, P.; Aljada, A.; Ghanim, H.; Hofmeyer, D.; Tripathy, D.; Syed, T.; Al-Hadad, W.; Dhindsa, S.; Dandona, P. Evidence for a Potent Antiinflammatory Effect of Rosiglitazone. *J. Clin. Endocrinol. Metab.* **2004**, *89*, 2728–2735. [CrossRef]
12. Keil, A.M.; Frederick, D.M.; Jacinto, E.Y.; Kennedy, E.L.; Zauhar, R.J.; West, N.M.; Tchao, R.; Harvison, P.J. Cytotoxicity of Thiazolidinedione-, Oxazolidinedione- and Pyrrolidinedione-Ring Containing Compounds in HepG2 Cells. *Toxicol. In Vitro* **2015**, *29*, 1887–1896. [CrossRef]
13. Hoffmann, B.R.; El-Mansy, M.F.; Sem, D.S.; Greene, A.S. Chemical Proteomics-Based Analysis of Off-Target Binding Profiles for Rosiglitazone and Pioglitazone: Clues for Assessing Potential for Cardiotoxicity. *J. Med. Chem.* **2012**, *55*, 8260–8271. [CrossRef]
14. Park, K.W.; Kang, J.; Park, J.J.; Yang, H.-M.; Kwon, Y.-W.; Lee, H.-Y.; Kang, H.-J.; Koo, B.-K.; Oh, B.-H.; Park, Y.-B.; et al. Thiazolidinedione usage is associated with decreased response to clopidogrel in DM patients. *Int. J. Cardiol.* **2013**, *168*, 608–610. [CrossRef]
15. Mughal, A.; Kumar, D.; Vikram, A. Effects of Thiazolidinediones on metabolism and cancer: Relative influence of PPARgamma and IGF-1 signaling. *Eur. J. Pharmacol.* **2015**, *768*, 217–225. [CrossRef]
16. Roussel, R.; Hadjadj, S.; Pasquet, B.; Wilson, P.W.F.; Smith, S.C., Jr.; Goto, S.; Tubach, F.; Marre, M.; Porath, A.; Krempf, M.; et al. Thiazolidinedione use is not associated with worse cardiovascular outcomes: A study in 28,332 high risk patients with diabetes in routine clinical practice. *Int. J. Cardiol.* **2013**, *167*, 1380–1384. [CrossRef]
17. Sotiriou, A.; Blaauw, R.H.; Meijer, C.; Gijsbers, L.H.; van der Burg, B.; Vervoort, J.; Rietjens, I.M.C.M. Correlation between activation of PPARγ and resistin downregulation in a mouse adipocyte cell line by a series of thiazolidinediones. *Toxicol. Vitr.* **2013**, *27*, 1425–1432. [CrossRef]
18. He, J.; Xu, C.; Kuang, J.; Liu, Q.; Jiang, H.; Mo, L.; Geng, B.; Xu, G. Thiazolidinediones attenuate lipolysis and ameliorate dexamethasone-induced insulin resistance. *Metabolism* **2015**, *64*, 826–836. [CrossRef]
19. Sinagra, T.; Tamburella, A.; Urso, V.; Siarkos, I.; Drago, F.; Bucolo, C.; Salomone, S. Reversible inhibition of vasoconstriction by thiazolidinediones related to PI3K/Akt inhibition in vascular smooth muscle cells. *Biochem. Pharmacol.* **2013**, *85*, 551–559. [CrossRef]
20. Lecca, D.; Nevin, D.K.; Mulas, G.; Casu, M.A.; Diana, A.; Rossi, D.; Sacchetti, G.; Carta, A.R. Neuroprotective and anti-inflammatory properties of a novel non-thiazolidinedione PPARγ agonist in vitro and in MPTP-treated mice. *Neuroscience* **2015**, *302*, 23–35. [CrossRef]
21. Molero, J.C.; Lee, S.; Leizerman, I.; Chajut, A.; Cooper, A.; Walder, K. Effects of rosiglitazone on intramyocellular lipid accumulation in Psammomys obesus. *Biochim. Biophys. Acta* **2010**, *1802*, 235–239. [CrossRef]

22. Gaonkar, S.L.; Shimizu, H. Microwave-assisted synthesis of the antihyperglycemic drug rosiglitazone. *Tetrahedron* **2010**, *66*, 3314–3317. [CrossRef]
23. Seaquist, E.R.; Miller, M.E.; Fonseca, V.; Ismail-Beigi, F.; Launer, L.J.; Punthakee, Z.; Sood, A. Effect of Thiazolidinediones and Insulin on Cognitive Outcomes in ACCORD-MIND. *J. Diabetes Complicat.* **2013**, *27*, 485–491. [CrossRef] [PubMed]
24. Lidström, P.; Tierney, J.; Wathey, B.; Westman, J. Microwave Assisted Organic Synthesis—A Review. *Tetrahedron* **2001**, *57*, 9225–9283. [CrossRef]
25. Hudzicki, J. *Kirby-Bauer Disk Diffusion Susceptibility Test Protocol*; American Society for Microbiology: Washington, DC, USA, 2009. Available online: https://asm.org/getattachment/2594ce26-bd44-47f6-8287-0657aa9185ad/Kirby-Bauer-Disk-Diffusion-Susceptibility-Test-Protocol-pdf.pdf (accessed on 17 May 2022).

Article

Convenient Access to Functionalized Non-Symmetrical Atropisomeric 4,4′-Bipyridines

Emmanuel Aubert [1], Emmanuel Wenger [1], Paola Peluso [2,*] and Victor Mamane [3,*]

[1] CNRS, CRM2, Université de Lorraine, F-54000 Nancy, France; emmanuel.aubert@univ-lorraine.fr (E.A.); emmanuel.wenger@univ-lorraine.fr (E.W.)

[2] Institute of Biomolecular Chemistry ICB, CNR, Secondary Branch of Sassari, Traversa La Crucca 3, Regione Baldinca, Li Punti, 07100 Sassari, Italy

[3] Strasbourg Institute of Chemistry, UMR CNRS 7177, Team LASYROC, 1 Rue Blaise Pascal, University of Strasbourg, 67008 Strasbourg, France

* Correspondence: paola.peluso@cnr.it (P.P.); vmamane@unistra.fr (V.M.); Tel.: +39-079-2841218 (P.P.); +33-3-68851612 (V.M.)

Abstract: Non-symmetrical chiral 4,4′-bipyridines have recently found interest in organocatalysis and medicinal chemistry. In this regard, the development of efficient methods for their synthesis is highly desirable. Herein, a series of non-symmetrical atropisomeric polyhalogenated 4,4′-bipyridines were prepared and further functionalized by using cross-coupling reactions. The desymmetrization step is based on the *N*-oxidation of one of the two pyridine rings of the 4,4′-bipyridine skeleton. The main advantage of this methodology is the possible post-functionalization of the pyridine *N*-oxide, allowing selective introduction of chlorine, bromine or cyano groups in 2- and 2′-postions of the chiral atropisomeric 4,4′-bipyridines. The crystal packing in the solid state of some newly prepared derivatives was analyzed and revealed the importance of halogen bonds in intermolecular interactions.

Keywords: atropisomerism; 4,4′-bipyridine; pyridine *N*-oxidation; halogenation; halogen bond; cyanation; Finkelstein reaction; Suzuki coupling

Citation: Aubert, E.; Wenger, E.; Peluso, P.; Mamane, V. Convenient Access to Functionalized Non-Symmetrical Atropisomeric 4,4′-Bipyridines. *Compounds* **2021**, *1*, 58–74. https://doi.org/10.3390/compounds1020006

Academic Editor: Juan Mejuto

Received: 19 April 2021
Accepted: 31 May 2021
Published: 1 July 2021

Publisher's Note: MDPI stays neutral with regard to jurisdictional claims in published maps and institutional affiliations.

1. Introduction

4,4′-Bipyridines are useful ligands for the design of coordination polymers and metal-organic frameworks (MOFs) [1], and are key components in the preparation of violo-gens [2,3]. In contrast to 2,2′-bipyridines for which a large number of chiral derivatives were developed [4–6], chiral 4,4′-bipyridines were much less explored. In the latter deriva-tives, the chirality can be brought by specific functions on the pyridine rings [7,8] or by atropisomery [9,10]. Indeed, atropisomeric 4,4′-bipyridines are the particular case where rotation around the pyridyl-pyridyl bond is blocked by the presence of three or four sub-stituents. They were first used for the preparation of metallo-supramolecular squares [9] and some years later for building chiral MOFs [11]. These last years, our groups were involved in the development of halogenated chiral 4,4′-bipyridines and in the study of their performances as halogen- [12,13] and chalcogen [14,15] bond donors in different applications such as organocatalysis [16] and medicinal chemistry [17].

Particularly, we have shown that homocoupling reactions represent straightforward ways for the synthesis of symmetrical chiral 4,4′-bipyridines [10,18]. However, these methods are not relevant for the synthesis of the non-symmetrical derivatives; therefore, desymmetrization processes have to be employed. Very limited methods for the desym-metrization of the 4,4′-bipyridine framework to non-symmetrical chiral 4,4′-bipyridines were reported. During their work on the 3,3′-dilithiation of octachloro-4,4′-bipyridine **1**, Foulger and Wakefield observed that quenching with dichlorodiphenylsilane and di-π-cyclopentadienyltitanium dichloride generated 4,4′-bipyridines **2** and **3**, respectively [19] (Figure 1a). Later, we showed that 3-mono- and 3,5′-dilithiation of 2,2′-dibromo-5,5′-dichloro-4,4′-bipyridine **4** and subsequent electrophilic trapping furnished 4,4′-bipyridines

5 and **6**, respectively [20] (Figure 1b). Finally, very recently, we used lithiation and cross-coupling reactions in 2-position of 2,2′-diiodo-3,3′,5,5′-tetrachloro-4,4′-bipyridine **7** as non-selective routes to 4,4′-bipyridines **8** and **9** [17] (Figure 1c). It is worth mentioning that in all these examples the yields were low to moderate.

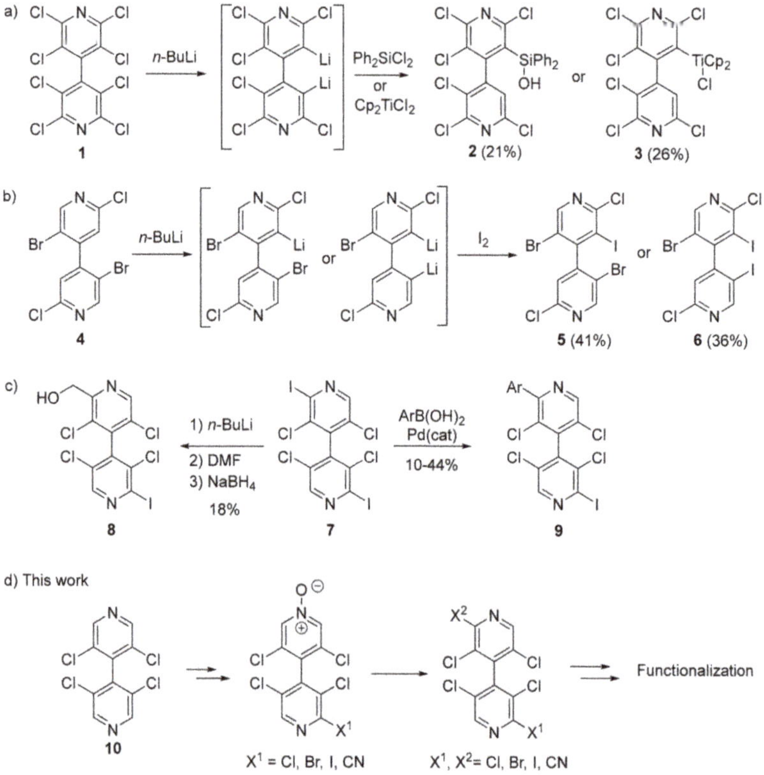

Figure 1. Non-symmetrical chiral 4,4′-bipyridines obtained by desymmetrization of parent 4,4′-bipyridines in the literature (**a–c**) and in the present work (**d**).

Herein, the *N*-oxidation as a straightforward and operatively easy method was used to desymmetrize 3,3′,5,5′-tetrachloro-4,4′-bipyridine **10**. The advantage of the reported methodology relies on the chemical transformations allowed by the pyridine N-O function, such as halogenation and cyanation [21]. The halogenated 4,4′-bipyridines were further functionalized through metal-catalyzed coupling reactions (Figure 1d). Moreover, X-ray diffraction analysis of selected compounds revealed very interesting solid-state packing features.

2. Materials and Methods

2.1. General Information

Proton (^{1}H NMR) and carbon (^{13}C NMR) nuclear magnetic resonance spectra were recorded on a Bruker Avance III instrument operating at 300, 400, or 500 MHz (Bruker Corporation, Billerica, MA, USA). The chemical shifts are given in parts per million (ppm) on the delta scale. The solvent peak was used as reference values for ^{1}H NMR (CDCl$_3$ = 7.26 ppm) and for ^{13}C NMR (CDCl$_3$ = 77.16 ppm). Data are presented as follows: Chemical shift, multiplicity (s = singlet, d = doublet, t = triplet, q = quartet, quint = quintet, m = multiplet, b = broad), integration, and coupling constants (J/Hz). High-resolution mass spectra (HRMS) data were recorded on a micrOTOF spectrometer (Bruker Corporation,

Billerica, MA, USA) equipped with an orthogonal electrospray interface (ESI). Analytical thin layer chromatography (TLC plates from Merck KGaA, Darmstadt, Germany) was carried out on silica gel 60 F254 plates with visualization by ultraviolet light. Reagents and solvents were purified using standard means. Tetrahydrofuran (THF) was distilled from sodium metal/benzophenone and freshly used. Dry dichloromethane was obtained by passing through activated alumina under a positive pressure of argon using GlassTechnology GTS100 devices. Dry dioxane (over molecular sieve) was purchased from Aldrich, triethylamine and diisopropylamine were distilled over CaH_2 and stored over KOH under an argon atmosphere. Anhydrous reactions were carried out in flame-dried glassware and under an argon atmosphere. All other chemicals were used as received.

2.2. Syntheses

2.2.1. 3,3′,5,5′-Tetrachloro-4,4′-Bipyridine 10

To a solution of diisopropylamine (27.5 mmol, 3.85 mL) in THF (200 mL) at −40 °C was added a solution of *n*-BuLi (1.6 M in hexanes, 27.5 mmol, 17.2 mL) and the mixture was stirred for 20 min. A solution of 3,5-dichloropyridine **11** (50 mmol, 7.4 g) in THF (100 mL) was added during 1 h while maintaining the temperature close to −40 °C. The temperature of the cloudy solution was slowly raised to −15 °C (during 1h) to give a homogeneous dark red solution. The temperature was lowered to −78 °C then a solution of I_2 (30 mmol, 7.62 g) in THF (50 mL) was slowly added. After 10 min at −78 °C, the temperature was raised to room temperature and the reaction was quenched by the addition of aqueous saturated solution of $Na_2S_2O_3$ (100 mL). Water was added (200 mL) and the mixture was extracted three times with ethyl acetate (3 × 200 mL). The organic phases were combined, washed with brine (100 mL) and dried over $MgSO_4$. After concentration, the crude compound was purified by chromatography on silica gel (cyclohexane/ethyl acetate 95/5) to give 4,4′-bipyridine **10** (3.64 g) and compound **12** (2.7 g). Compound **12** was diluted with ethyl acetate (200 mL) and I_2 (18 mmol, 4.57g) was added and the mixture was stirred overnight at room temperature in an open flask. The reaction was quenched by the addition of aqueous saturated solution of $Na_2S_2O_3$ (25 mL). Water was added (50 mL) and the organic phases washed with brine (100 mL) and dried over anhydrous $MgSO_4$. After concentration, the crude compound was purified by chromatography on silica gel (cyclohexane/ethyl acetate 95/5) to give 4,4′-bipyridine **10** (2.4 g). A total mass of 6.04 g of **10** was obtained which corresponds to an overall yield of 82%. The NMR data for **10** are in complete agreement with the literature [22].

3,3′,5,5′-Tetrachloro-1,4-dihydro-4,4′-bipyridine (**12**). Yellow solid, mp 155–157 °C. ^{1}H NMR (500 MHz, CD_3COCD_3) δ 8.57 (s, 1H), 8.56 (s, 1H), 7.55 (s, 1H, NH), 6.56 (s, 1H), 6.55 (s, 1H), 5.87 (s, 1H). ^{13}C NMR (126 MHz, CD_3COCD_3) δ 150.7, 148.3, 142.5, 134.3, 134.0, 127.7, 102.8, 50.6. HRMS (ESI-TOF) [M + H]$^+$ *m/z*: Calcd. for $C_{10}H_6Cl_5N_2$ 294.9285, found: 294.9327 (Figure S1).

2.2.2. 3,3′,5,5′-Tetrachloro-[4,4′-Bipyridine] 1-Oxide 13

4,4′-Bipyridine **10** (12.38 mmol, 3.64 g) was dissolved in CH_2Cl_2 (62 mL) at room temperature. *m*-CPBA (*m*-chloro-perbenzoic acid) (77% purity, 12.38 mmol, 2.8 g) was added and the mixture was stirred for 24 h. After dilution with CH_2Cl_2 (140 mL), the mixture was washed with NaOH 1M (2 × 100 mL) and with brine. The organic phase was dried over $MgSO_4$, concentrated and purified by chromatography on silica gel (CH_2Cl_2/methanol 99/1) to give respectively the starting compound **10** (850 mg, 23%), compound **13** (2.0 g, 54% Yield) and the bis *N*-oxide **14** (925 mg, 23%).

3,3′,5,5′-Tetrachloro-[4,4′-bipyridine] 1-oxide (**13**). White solid, mp 218–220 °C. ^{1}H NMR (500 MHz, $CDCl_3$) δ 8.68 (s, 2H), 8.31 (s, 2H). ^{13}C NMR (126 MHz, $CDCl_3$) δ 148.0, 139.0, 138.3, 132.9, 132., 130.2. HRMS (ESI-TOF) [M + H]$^+$ *m/z*: Calcd. for $C_{10}H_5Cl_4N_2O$ 308.9150, found: 308.9153 (Figure S2).

3,3′,5,5′-Tetrachloro-[4,4′-bipyridine] 1,1′-dioxide (**14**). White solid, mp 270–272 °C. ^{1}H NMR (500 MHz, $CDCl_3$) δ 8.31 (s, 4H). ^{13}C NMR (126 MHz, $CDCl_3$) δ 138.3, 133.7, 128.9.

HRMS (ESI-TOF) $[M + H]^+$ *m/z*: Calcd. for $C_{10}H_5Cl_4N_2O_2$ 324.9100, found: 324.9095 (Figure S3).

2.2.3. 2-Bromo-3,3′,5,5′-Tetrachloro-4,4′-Bipyridine **15**

3,3′,5,5′-Tetrachloro-[4,4′-bipyridine] 1-oxide **13** (6.37 mmol, 1.91 g) was dissolved in dibromoethane (15 mL) and triethylamine (12.73 mmol, 1.7 mL) was added. The mixture was cooled to −40 °C and oxalyl bromide (12.73 mmol, 1.8 mL) was slowly added while maintaining the temperature below −35 °C. After stirring 1h at −40 °C, methanol (1 mL) was added and temperature was raised to room temperature. A saturated solution of ammonium chloride (20 mL) was added and the mixture was extracted with CH_2Cl_2 (3 × 50 mL). The organic phase was dried over $MgSO_4$, filtered and concentrated under vacuum. The crude compound was purified by chromatography on silica gel (pentane/CH_2Cl_2 7/3 to 2/1) to give pure 4,4′-bipyridine **15** (2.18 g, 94% yield). White solid, mp 100–102 °C. ^{1}H NMR (500 MHz, CDCl$_3$) δ 8.68 (s, 2H), 8.48 (s, 1H). ^{13}C NMR (126 MHz, CDCl$_3$) δ 148.1, 147.2, 142.1, 140.7, 140.4, 132.9, 130.9, 130.6. HRMS (ESI-TOF) $[M + H]^+$ *m/z*: Calcd. for $C_{10}H_4BrCl_4N_2$ 370.8306, found: 370.8302 (Figure S4).

2.2.4. 3,3′,5,5′-Tetrachloro-[4,4′-Bipyridine]-2-Carbonitrile **16**

3,3′,5,5′-Tetrachloro-[4,4′-bipyridine] 1-oxide **13** (1 mmol, 310 mg) was dissolved in acetonitrile (2 mL) and triethylamine (0.5 mL). Trimethylsilylcyanide (3 mmol, 375 µL) was slowly added and the mixture was refluxed at 100 °C for 24 h. The temperature lowered to 0 °C, a solution of NaOH (5 M, 20 mL) was slowly added and the mixture was extracted with CH_2Cl_2 (3 × 30 mL). The organic phase was dried over $MgSO_4$, filtered and concentrated under vacuum. The crude compound was purified by chromatography on silica gel (pentane/ethyl acetate 7/3) to give 4,4′-bipyridine **16** (267 mg, 83% yield). White solid, mp 99–101 °C. ^{1}H NMR (500 MHz, CDCl$_3$) δ 8.76 (s, 1H), 8.71 (s, 2H). ^{13}C NMR (126 MHz, CDCl$_3$) δ 149.1, 148.2, 141.9, 138.7, 135.40, 135.35, 132.3, 130.9. HRMS (ESI-TOF) $[M + H]^+$ *m/z*: Calcd. for $C_{11}H_3Cl_4N_3$ 317.9154, found: 317.9151 (Figure S5).

2.2.5. 2-Bromo-3,3′,5,5′-Tetrachloro-[4,4′-Bipyridine] 1-Oxide **17**

2-Bromo-3,3′,5,5′-tetrachloro-4,4′-bipyridine **15** (5.36 mmol, 2.0 g) was dissolved in CH_2Cl_2 (28 mL) at room temperature. *m*-CPBA (77% purity, 11 mmol, 2.47 g) was added and the mixture was stirred for 60 h. After dilution with CH_2Cl_2 (100 mL), the mixture was washed with NaOH 1M (2 × 80 mL) and with brine. The organic phase was dried over $MgSO_4$, concentrated and purified by chromatography on silica gel (CH_2Cl_2/methanol 99/1) to give compound **17** (1.72 g, 83%). White solid, mp 212–214 °C. ^{1}H NMR (500 MHz, CDCl$_3$) δ 8.48 (s, 1H), 8.31 (s, 2H). ^{13}C NMR (126 MHz, CDCl$_3$) δ 147.2, 141.1, 140.8, 138.4, 133.6, 132.6, 131.3, 130.4. HRMS (ESI-TOF) $[M + H]^+$ *m/z*: Calcd. for $C_{10}H_3BrCl_4N_2O$ 386.8256, found: 386.8265 (Figure S6).

2.2.6. 3,3′,5,5′-Tetrachloro-2′-Cyano-[4,4′-Bipyridine] 1-Oxide **18**

Procedure 2.2.5. was used starting from 3,3′,5,5′-tetrachloro-[4,4′-bipyridine]-2-carbonitrile **16** (0.5 mmol, 160 mg) and *m*-CPBA (77% purity, 1 mmol, 224 mg) in CH_2Cl_2 (2.5 mL). After purification by chromatography on silica gel (CH_2Cl_2/methanol 98/2), compound **18** was obtained (158 mg, 94%). White solid, mp 193–195 °C. ^{1}H NMR (500 MHz, CDCl$_3$) δ 8.78 (s, 1H), 8.32 (s, 2H). ^{13}C NMR (126 MHz, CDCl$_3$) δ 149.2, 141.0, 138.5, 136.2, 136.1, 132.7, 132.3, 128.4, 113.9. HRMS (ESI-TOF) $[M + H]^+$ *m/z*: Calcd. for $C_{11}H_3Cl_4N_3O$ 333.9103, found: 333.9091 (Figure S7).

2.2.7. 2-Bromo-2′,3,3′,5,5′-Pentachloro-4,4′-Bipyridine **19**

Procedure 2.2.3. was used starting from 2-bromo-3,3′,5,5′-tetrachloro-[4,4′-bipyridine] 1-oxide **17** (0.179 mmol, 70 mg), oxalyl chloride (0.358 mmol, 31 µL), triethylamine (0.358 mmol, 50 µL) in CH_2Cl_2 (0.5 mL). After purification by chromatography on silica gel (pentane/ethyl acetate 97/3), compound **19** was obtained (51 mg, 70%). White solid, mp

119–121 °C. ^{1}H NMR (500 MHz, CDCl$_3$) δ 8.50 (s, 1H), 8.49 (s, 1H). ^{13}C NMR (126 MHz, CDCl$_3$) δ 148.7, 147.3, 146.9, 142.9, 142.2, 140.8, 132.6, 130.3, 129.9, 129.8. HRMS (ESI-TOF) [M + H]$^+$ *m/z*: Calcd. for C$_{10}$H$_2$BrCl$_5$N$_2$ 404.7917, found: 404.7921 (Figure S8).

2.2.8. 2′,3,3′,5,5′-Pentachloro-[4,4′-Bipyridine]-2-Carbonitrile 20

Procedure 2.2.3. was used starting from 3,3′,5,5′-tetrachloro-[4,4′-bipyridine]-2-carbonitrile **16** (0.179 mmol, 60 mg), oxalyl chloride (0.358 mmol, 31 µL), triethylamine (0.358 mmol, 50 µL) in CH$_2$Cl$_2$ (0.5 mL). After purification by chromatography on silica gel (pentane/ethyl acetate 95/5), compound **20** was obtained (45 mg, 71%). Colorless syrup. ^{1}H NMR (500 MHz, CDCl$_3$) δ 8.78 (s, 1H), 8.53 (s, 1H). ^{13}C NMR (126 MHz, CDCl$_3$) δ 149.2, 149.0, 147.1, 142.1, 141.2, 135.1, 132.4, 129.9, 129.7, 113.9. HRMS (ESI-TOF) [M + H]$^+$ *m/z*: Calcd. for C$_{11}$H$_2$Cl$_5$N$_3$ 351.8764, found: 351.8774 (Figure S9).

2.2.9. 3,3′,5,5′-Tetrachloro-2′-Iodo-[4,4′-Bipyridine] 1-Oxide 21

In a dry tube were placed 2-bromo-3,3′,5,5′-tetrachloro-[4,4′-bipyridine] 1-oxide **17** (4.11 mmol, 1.6 g), NaI (8.22 mmol, 1.23 g), CuI (0.411 mmol, 78 mg) and (rac)-*trans*-cyclohexane-1,2-diamine (0.822 mmol, 117 mg). The tube was evacuated and filled with argon before addition of degassed dioxane (15 mL). The tube was sealed with a screw cap and heated at 120 °C for 60 h. After cooling to room temperature, NH$_4$OH (20 mL) and H$_2$O (40 mL) were added then the product was extracted with dichloromethane (3 × 50 mL). After drying over MgSO$_4$, filtration and concentration, the crude compound was purified by chromatography on silica gel (CH$_2$Cl$_2$/methanol 99/1) to give compound **21** (1.29 g, 72%). White solid, mp 244–246 °C. ^{1}H NMR (500 MHz, CDCl$_3$) δ 8.47 (s, 1H), 8.30 (s, 2H). ^{13}C NMR (126 MHz, CDCl$_3$) δ 147.8, 139.2, 138.3, 138.0, 132.6, 131.8, 130.7, 119.4. HRMS (ESI-TOF) [M + H]$^+$ *m/z*: Calcd. for C$_{10}$H$_3$Cl$_4$IN$_2$O 434.8117, found: 434.8109 (Figure S10).

2.2.10. 2,3,3′,5,5′-Pentachloro-2′-Iodo-4,4′-Bipyridine 22

Procedure 2.2.3. was used starting from 3,3′,5,5′-tetrachloro-2′-iodo-[4,4′-bipyridine] 1-oxide **21** (0.179 mmol, 78 mg), oxalyl chloride (0.358 mmol, 31 mg), triethylamine (0.358 mmol, 50 µL) in CH$_2$Cl$_2$ (0.5 mL). After purification by chromatography on silica gel (pentane/ethyl acetate 97/3), compound **22** was obtained (65 mg, 80%). White solid, mp 140–142 °C. ^{1}H NMR (500 MHz, CDCl$_3$) δ 8.48 (s, 1H), 8.49 (s, 1H). ^{13}C NMR (126 MHz, CDCl$_3$) δ 148.7, 147.9, 146.9, 143.1, 140.4, 137.0, 130.8, 129.8, 129.7, 119.5. HRMS (ESI-TOF) [M + H]$^+$ *m/z*: Calcd. for C$_{10}$H$_2$Cl$_5$IN$_2$ 452,7778, found: 452.7749 (Figure S11).

2.2.11. 2′-Bromo-3,3′,5,5′-Tetrachloro-[4,4′-Bipyridine]-2-Carbonitrile 23

Procedure 2.2.3. was used starting from 3,3′,5,5′-tetrachloro-2′-cyano-[4,4′-bipyridine] 1-oxide **18** (0.297 mmol, 100 mg), oxalyl bromide (0.595 mmol, 85 µL), triethylamine (0.595 mmol, 80 µL) in CH$_2$Br$_2$ (1 mL). After purification by chromatography on silica gel (pentane/ethyl acetate 95/5), compound **23** was obtained (107 mg, 90%). White solid, mp 131–133 °C. ^{1}H NMR (500 MHz, CDCl$_3$) δ 8.77 (s, 1H), 8.52 (s, 1H). ^{13}C NMR (126 MHz, CDCl$_3$) δ 149.2, 147.4, 142.2, 141.0, 140.7, 135.06, 135.05, 132.5, 132.4, 130.2, 113.9. HRMS (ESI-TOF) [M + H]$^+$ *m/z*: Calcd. for C$_{11}$H$_2$BrCl$_4$N$_3$ 395.8259, found: 395.8277 (Figure S12).

2.2.12. 2-Bromo-3,3′,5,5′-Tetrachloro-2′-Iodo-4,4′-Bipyridine 24

Procedure 2.2.3. was used starting from 3,3′,5,5′-tetrachloro-2′-iodo-[4,4′-bipyridine] 1-oxide **21** (1.74 mmol, 760 mg), oxalyl bromide (3.49 mmol, 496 µL), triethylamine (3.49 mmol, 463 µL) in CH$_2$Br$_2$ (4.2 mL). After purification by chromatography on silica gel (pentane/CH$_2$Cl$_2$ 7/3), compound **24** was obtained (710 mg, 81%). White solid, mp 165–167 °C. ^{1}H NMR (500 MHz, CDCl$_3$) δ 8.414 (s, 1H), 8.410 (s, 1H). ^{13}C NMR (126 MHz, CDCl$_3$) δ 147.9, 147.3, 142.6, 140.8, 140.5, 137.0, 132.5, 130.8, 130.2, 119.5. HRMS (ESI-TOF) [M + H]$^+$ *m/z*: Calcd. for C$_{10}$H$_2$BrCl$_4$IN$_2$ 496.7273, found: 496.7242 (Figure S13).

2.2.13. 3,3′,5,5′-Tetrachloro-2′-Iodo-[4,4′-Bipyridine]-2-Carbonitrile 25

Procedure 2.2.4. was used starting from 3,3′,5,5′-tetrachloro-2′-iodo-[4,4′-bipyridine] 1-oxide **21** (0.395 mmol, 172 mg), acetonitrile (1.5 mL), triethylamine (0.4 mL) and trimethylsilylcyanide (1.18 mmol, 150 µL). After purification by chromatography on silica gel (pentane/ethyl acetate 95/5), compound **25** was obtained (110 mg, 62%). White solid, mp 144–146 °C. ^{1}H NMR (500 MHz, CDCl$_3$) δ 8.77 (s, 1H), 8.51 (s, 1H). ^{13}C NMR (126 MHz, CDCl$_3$) δ 149.2, 148.0, 142.4, 138.9, 137.0, 135.1, 135.0, 132.4, 130.7, 119.7, 114.0. HRMS (ESI-TOF) [M + H]$^+$ *m/z*: Calcd. for C$_{11}$H$_2$Cl$_4$IN$_3$ 443.8120, found: 443.8121 (Figure S14).

2.2.14. 2-Bromo-3,3′,5,5′-Tetrachloro-2′-Phenyl-4,4′-Bipyridine 26

In a dry tube were placed 2-bromo-3,3′,5,5′-tetrachloro-2′-iodo-4,4′-bipyridine **24** (0.3 mmol, 151.5 mg), Pd(PPh$_3$)$_4$ (0.015 mmol, 17.3 mg), phenylboronic acid (0.3 mmol, 36.6 mg) and K$_3$PO$_4$ (0.3 mmol, 64 mg). The tube was evacuated and filled with argon before addition of degassed mixture of toluene/water/ethanol (6:1:1, 4 mL). The tube was sealed with a screw cap and heated at 100 °C for 15 h. The cooled reaction mixture was diluted with water (10 mL) and extracted with CH$_2$Cl$_2$ (3 × 15 mL). After drying over MgSO$_4$, filtration and concentration, the crude compound was purified by chromatography on silica gel (pentane/ethyl acetate 98/2) to give compound **26** (122 mg, 89%). Colorless syrup. ^{1}H NMR (500 MHz, CDCl$_3$) δ 8.76 (s, 1H), 8.49 (s, 1H), 7.78–7.25 (m, 2H), 7.52–7.46 (m, 3H). ^{13}C NMR (126 MHz, CDCl$_3$) δ 156.2, 147.5, 147.2, 143.1, 141.7, 140.7, 137.0, 132.8, 130.6, 129.6, 129.5, 129.1, 128.7, 128.4. HRMS (ESI-TOF) [M + H]$^+$ *m/z*: Calcd. for C$_{16}$H$_7$BrCl$_4$N$_2$ 446.8619, found: 446.8625 (Figure S15).

2.2.15. 2-Bromo-2′-(4-((Tert-Butyldimethylsilyl)oxy)phenyl)-3,3′,5,5′-Tetrachloro-4,4′-Bipyridine 27

Procedure 2.2.14. was used starting from 2-bromo-3,3′,5,5′-tetrachloro-2′-iodo-4,4′-bipyridine **24** (0.3 mmol, 151.5 mg), Pd(PPh$_3$)$_4$ (0.015 mmol, 17.3 mg), 4-((tert-butyldimethylsilyl)oxy)phenyl boronic acid (0.3 mmol, 75.7 mg) and K$_3$PO$_4$ (0.3 mmol, 64 mg). After purification by chromatography on silica gel (pentane/ethyl acetate 98/2), compound **27** was obtained (137 mg, 79%). Colorless syrup. ^{1}H NMR (500 MHz, CDCl$_3$) δ 8.72 (s, 1H), 8.49 (s, 1H), 7.69 (d, J = 8.5 Hz, 2H), 6.94 (d, J = 8.5 Hz, 2H), 1.00 (s, 9H), 0.24 (s, 6H). ^{13}C NMR (126 MHz, CDCl$_3$) δ 157.1, 155.8, 147.3, 147.2, 143.3, 141.6, 140.7, 132.9, 131.1, 130.7, 130.0, 128.5, 128.4, 120.0, 25.8, 18.4, −4.2. HRMS (ESI-TOF) [M + H]$^+$ *m/z*: Calcd. for C$_{22}$H$_{22}$BrCl$_4$N$_2$OSi 576.9433, found: 576.9428 (Figure S16).

2.2.16. 2″-(4-((Tert-Butyldimethylsilyl)oxy)phenyl)-3′,3″,5′,5″-Tetrachloro-4,2′:4′,4″-Terpyridine 31

Procedure 2.2.14. was used starting from 2-bromo-3,3′,5,5′-tetrachloro-2′-iodo-4,4′-bipyridine **24** (0.143 mmol, 82.5 mg), Pd(PPh$_3$)$_4$ (0.007 mmol, 8.3 mg), 4-pyridine boronic acid (0.215 mmol, 31 mg) and K$_3$PO$_4$ (0.215 mmol, 46 mg). After purification by chromatography on silica gel (pentane/ethyl acetate 7/3), compound **31** was obtained (60 mg, 73%). Colorless syrup. ^{1}H NMR (500 MHz, CDCl$_3$) δ 8.80 (s, 1H), 8.78 (d, J = 5.5 Hz, 2H), 8.74 (s, 1H), 8.76–8.66 (m, 4H), 6.95 (d, J = 8.5 Hz, 2H), 1.00 (s, 9H), 0.24 (s, 6H). ^{13}C NMR (126 MHz, CDCl$_3$) δ 157.1, 155.9, 153.2, 150.1, 147.9, 147.3, 144.7, 142.9, 141.8, 131.09, 131.05, 130.1, 129.5, 128.7, 128.6, 124.0, 120.0, 25.8, 18.4, −4.2. HRMS (ESI-TOF) [M + H]$^+$ *m/z*: Calcd. for C$_{27}$H$_{26}$Cl$_4$N$_3$OSi 576.0594, found: 576.0589 (Figure S17).

2.2.17. 4-(3′,3″,5′,5″-Tetrachloro-[4,2′:4′,4″-Terpyridin]-2″-yl)phenol 32

2″-(4-((Tert-butyldimethylsilyl)oxy)phenyl)-3′,3″,5′,5″-tetrachloro-4,2′:4′,4″-terpyridine **31** (0.078 mmol, 45 mg) was dissolved in THF (2 mL) and the solution was cooled to 0 °C. A tetra-butylammonium fluoride (TBAF) solution (1M in THF, 0.117 mmol, 117 µL) was slowly added and the mixture was stirred at 0 °C for 30 min. After addition of a saturated solution of NH$_4$Cl (1 mL), the temperature was raised to ambient. Water was added (5 mL) and the mixture was extracted with CH$_2$Cl$_2$ (3 × 10 mL). The organic phase was washed with brine, dried over $MgSO_4$, filtered and concentrated. After filtration on silica gel (ethyl acetate), compound **32** was obtained (36 mg, 100%). White solid, mp 254–256 °C. ^{1}H NMR (500 MHz, DMSO-*d6*) δ 9.998 (broad s, 1H), 9.11 (s, 1H), 8.99 (s, 1H), 8.77 (d, J = 6.0 Hz, 2H), 7.73 (d, J = 6.0 Hz, 2H), 7.62 (d, J = 9.0 Hz, 2H), 6.90 (d, J = 9.0 Hz, 2H). ^{13}C NMR (126 MHz, DMSO-*d6*) δ 158.8, 155.4, 153.1, 150.0, 148.2, 147.5, 143.9, 142.0, 141.1, 131.1, 130.2, 128.4, 127.6, 127.4, 127.2, 115.1. HRMS (ESI-TOF) [M + H]$^+$ *m/z*: Calcd. for $C_{21}H_{12}Cl_4N_3O$ 461.9729, found: 461.9701 (Figure S18).

2.3. Single Crystal X-ray Diffraction

2.3.1. Crystallizations and Analysis

Suitable crystals of compounds **13**, **18** and **24** were obtained by slow evaporation of a dichloromethane/hexane (1:1) solution. The solid-state structures were determined by single crystal X-ray diffraction at low temperature (T = 100 K).

2.3.2. Crystal Data

CCDC 2077648, 2077649 and 2077650 contains the supplementary crystallographic data for this paper. These data can be obtained free of charge via http://www.ccdc.cam.ac.uk/conts/retrieving.html (accessed on 14 April 2021) (or from the CCDC, 12 Union Road, Cambridge CB2 1EZ, UK; Fax: +44 1223 336033; E-mail: deposit@ccdc.cam.ac.uk).

Crystal Data for **24** $C_{10}H_2BrCl_4IN_2$ (M = 498.75 g/mol): Monoclinic, space group $P2_1/c$ (no. 14), a = 7.9297(2) Å, b = 15.6388(4) Å, c = 11.4938(3) Å, β =105.632(2)°, V = 1372.64(6) Å^3, Z = 4, T = 100(2) K, μ(MoKα) = 6.002 mm^{-1}, D_{calc} = 2.413 g/cm^3, 51767 reflections measured (2.254° $\leq \Theta \leq$ 36.150°), 6552 unique (R_{int} = 0.0298), which were used in all calculations. The final R_1 was 0.0239 (I > 2σ(I)) and wR2 was 0.0548 (all data). CCDC 2077649.

Crystal Data for **18** $C_{11}H_3Cl_4N_3$ (M = 334.96 g/mol): Monoclinic, space group $P2_1/c$ (no. 14), a = 7.2662(2) Å, b = 12.0325(4) Å, c = 14.4319(5) Å, β = 93.491(3)°, V = 1259.45(7) Å^3, Z = 4, T = 100(2) K, μ(MoKα) = 0.931 mm^{-1}, D_{calc} = 1.767 g/cm^3, 48102 reflections measured (2.205° $\leq \Theta \leq$ 33.667°), 5000 unique (R_{int} = 0.0424), which were used in all calculations. The final R_1 was 0.0323 (I > 2σ(I)) and wR2 was 0.0732 (all data). CCDC 2077650.

Crystal Data for **13** $C_{10}H_4Cl_4N_2$ (M = 309.95 g/mol): Monoclinic, space group $P2_1/c$ (no. 14), a = 7.2697(2) Å, b = 11.5462(3) Å, c = 14.0808(4) Å, β = 97.804(3)°, V = 1170.96(6) Å^3, Z = 4, T = 100(2) K, μ(MoKα) = 0.991 mm^{-1}, D_{calc} = 1.758 g/cm^3, 48887 reflections measured (2.290° $\leq \Theta \leq$ 37.162°), 6021 unique (R_{int} = 0.0444), which were used in all calculations. The final R_1 was 0.0384 (I > 2σ(I)) and wR2 was 0.0840 (all data). CCDC 2077648.

2.4. Isolated Molecule Calculations

Molecular structures of **24**, **18** and **13** were optimized with Gaussian09 software at DFT level of theory using the B3LYP functional completed with D3 dispersion correction and the Def2TZVPP basis set. Frequency calculations were performed in order to check that true energy minimum were obtained. Electrostatic maps (Figure 2, Figures S31 and S32) were drawn using the AIMAll software, and locations of ESP extrema $V_{S,max}$ were searched using MultiWFN program (Table S4). Integrated atomic charges (AIMAll) are depicted on Figures S33–S35.

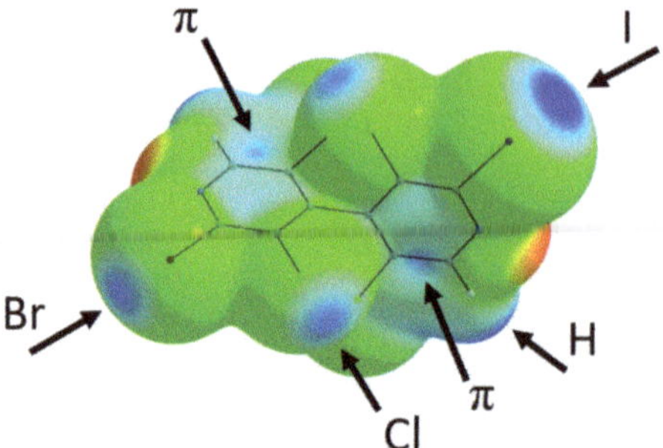

Figure 2. Electrostatic potential mapped on the ρ = 0.002 a.u. electron density isosurface of **24**. Coloring from red = −0.05 a.u. to blue = +0.05 a.u. Black arrows point to some representative electropositive σ- and π-holes.

3. Results

3.1. Improved Synthesis of 3,3′,5,5′-Tetrachloro-4,4′-Bipyridine **10**

The starting 4,4′-bipyridine **10** was already prepared by us with 44% yield by dimerization of 3,5-dichlopyridine **11** [22]. When the synthesis was reproduced on large scale (50 mmol), the yield was improved to 50%. Moreover, during the chromatographic purification, the more polar non-rearomatized compound **12** was also isolated with 40% yield. Compound **12** could be oxidized with I_2 in ethyl acetate at room temperature for 12 h, furnishing **10** with 90% yield. Overall, the yield of **10** in the dimerization process could be increased to 82% yield (Scheme 1).

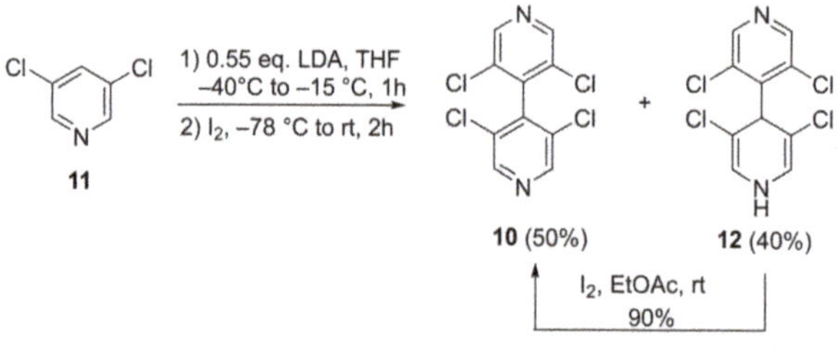

Scheme 1. Improved synthesis of 3,3′,5,5′-tetrachloro-4,4′-bipyridine **10**.

3.2. Desymmetrization of 3,3′,5,5′-Tetrachloro-4,4′-Bipyridine **10**

The mono *N*-oxidation of 4,4′-bipyridine **10** was performed using 1 equivalent of *m*-CPBA in CH_2Cl_2 (0.2 M) at room temperature. After treatment using NaOH 1M to remove the benzoic acid, the crude mixture was purified by chromatography on silica gel to give the expected mono-oxidized 4,4′-bipyridine **13** with 54% yield. Unreacted 4,4′-bipyridine **10** (23%) as well as doubly oxidized 4,4′-bipyridine **14** (23%) were also obtained after the purification. Based on the recovery of the starting 4,4′-bipyridine **10**, the yield of **14** was of 75%. The *N*-oxide **13** could be further transformed to the pentahalogenated 4,4′-bipyridine **15** in very good yields by using oxalyl bromide with Et_3N in CH_2Br_2 at low temperature [23], and to the cyano derivative **16** by using $TMSCN/Et_3N$ in refluxing dioxane [24] (Scheme 2).

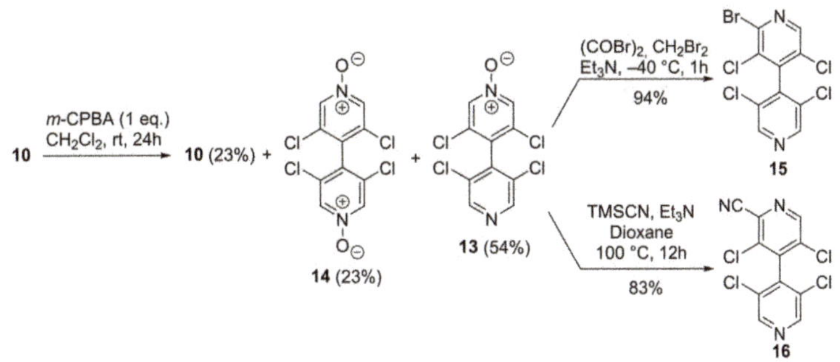

Scheme 2. Mono-oxidation of 4,4′-bipyridine **10** and its functionalization.

3.3. New Chiral Non-Symmetrical 4,4′-Bipyridines Based on the 3,3′,5,5′-Tetrachloro-4,4′-Bipyridine Core

4,4′-Bipyridines **15** and **16** were submitted to a new sequence of *N*-oxidation/halogenation in order to access to novel chiral non-symmetrical 4,4′-bipyridines. Thus, regioselective *N*-oxidation of the less substituted ring of **15** and **16** furnished, respectively, 4,4′-bipyridine *N*-oxides **17** and **18** in high yields. Chlorination of *N*-oxides **17** and **18** using oxalyl chloride with Et$_3$N in CH$_2$Cl$_2$ at low temperature [23] afforded 4,4′-bipyridines **19** and **20** in good yields. Alternatively, *N*-oxide **17** was transformed by a copper-catalyzed Finkelstein reaction [25] to the iodinated 4,4′-bipyridine *N*-oxide **21** which was further submitted to the chlorination to provide 4,4′-bipyridine **22**. *N*-oxides **18** and **21** were also brominated as described in Scheme 2 to give two other chiral non-symmetrical 4,4′-bipyridines **23** and **24** in good yields (Scheme 3). Finally, 4,4′-bipyridine **21** was transformed in good yield to the iodinated 4,4′-bipyridine **25** by using TMSCN/Et$_3$N in refluxing dioxane.

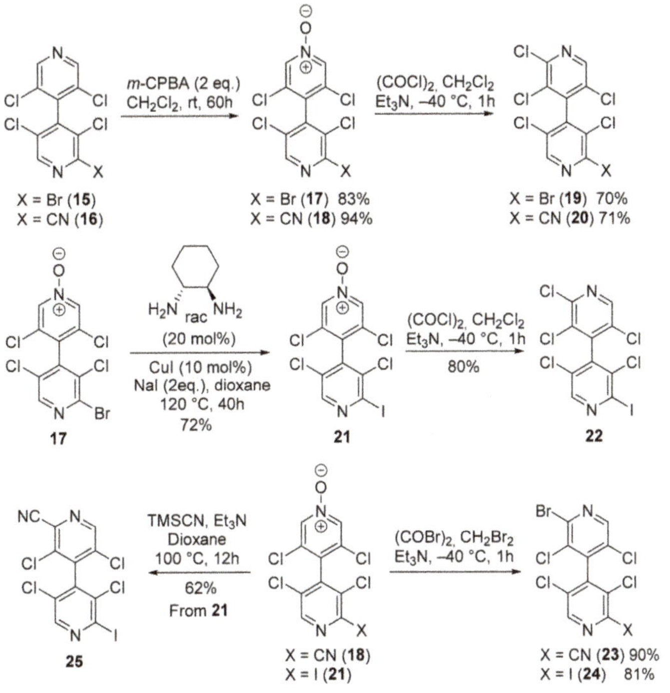

Scheme 3. Synthesis of six new chiral non-symmetrical 4,4′-bipyridines.

3.4. Cross-Coupling Reactions with 4,4′-Bipyridine **24**

4,4′-Bipyridine **24** bearing two different halogens (Br and I) in 2 and 2′-positions was chosen to perform selective cross-coupling reactions. In particular, 4,4′-bipyridines **28** and **30**, which were recently described as good transthyretin fibrillogenesis inhibitors [17], could be obtained in excellent yields and with very high purity by successive Suzuki and Finkelstein reactions, followed by silyl group deprotection in the case of **30**. Moreover, their enantiomers could be separated by HPLC on chiral stationary phase [26,27]. High selectivity for mono-coupling products **26** and **27** was obtained by using the Suzuki coupling [28] with one equivalent of arylboronic acid, letting the C–Br bond untouched for a further coupling reaction such as the copper-catalyzed Finkelstein Br/I exchange [18]. Moreover, a new Suzuki cross-coupling of 4,4′-bipyridine **27** with 4-pyridylboronic acid delivered compound **31**, which, after silyl group deprotection, furnished the chiral 4,4′-bipyridine **32**, bearing two different functional groups (Scheme 4).

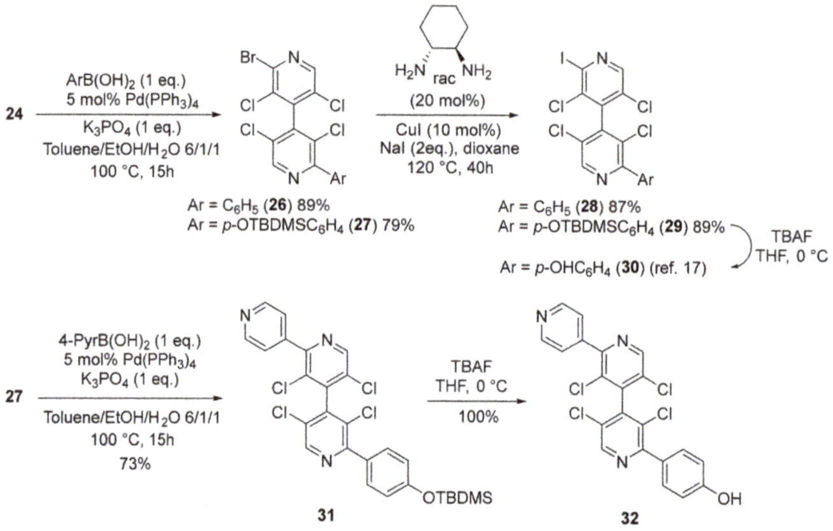

Scheme 4. Cross-coupling reactions with 4,4′-bipyridine **24**.

3.5. X-ray Diffraction Analysis

One non-symmetrical 4,4′-bipyridine (**24**) and two 4,4′-bipyridine *N*-oxides (**13** and **18**) solid state structures were determined by single X-ray diffraction.

2-Bromo-3,3′,5,5′-tetrachloro-2′-iodo-4,4′-bipyridine **24** crystallizes in $P2_1/c$ space group with one molecule in the asymmetric unit. A positional disorder (corresponding to two different orientations of the molecule) exchanges the Br and I halogen atom positions, the major component having a population of 0.7441(13). According to the energy packing analysis performed with CrystalExplorer software [29], the solid-state structure of **24** is primarily determined by the formation of intermolecular interactions with a predominance of stabilizing dispersion over electrostatic contributions (Table S1, entries 1–6). These contributions involve first the formation of strongly bound dimers centered on an inversion center where $\pi\cdots\pi$ stacking occurs (N7-pyridine interplane distance of 3.272 Å) (Table S1, entry 1 and Figure S19), with a complementary electrostatic interaction between the electropositive C12-H12 group (integrated atomic charges of Q(C12) = +0.59 and Q(H12) = +0.06) and the electronegative crown of Br1 (Q = −0.02), as evidenced from the electrostatic potential mapped on the isodensity molecular surface (Figure 2). In this dimer, the electronegative region about N7 is also near the electropositive area of Cl1, although the σ-hole of the halogen is not optimally oriented (C3-Cl1···N7 = 127.59°). The geometry of the next two intermolecular interactions (x, −y + 3/2, z − 1/2 and 1 + x, −y + 3/2, z + 1/2, Table S1, entries 2–3) can also be understood from electrostatic point of view, since electronegative regions (respectively Cl3 with Q(Cl3) = −0.17 and N1 with Q(N1) = −1.12) are close to electropositive area (respectively C6 with Q(C6) = +0.59 and C12 with Q(C12) = +0.59) (Figures S20 and S21).

The most remarkable feature of the molecular packing is the formation of a four-membered ring involving four I/Br atoms engaged in type-II halogen··· halogen bonds (Figure 3). Although the I/Br positional disorder can induce bias in the modeled inter-atomic distances, these latter show a clear interpenetration of the van der Waals spheres and an interaction geometry typical of type-II hal··· hal bonds [30,31] (I1··· Br1 = 3.564(2) Å RR = 0.93; C2-I1··· Br1 = 170.22(7)°; I1··· Br1-C8 = 114.6(1)°; Br1··· I1 = 3.716(3) Å RR = 0.97; C8- Br1··· I1 = 170.4(1)°; Br1··· I1-C2 = 81.45(6)°)). Noticeably, the shortest type-II hal··· hal bond is obtained when the σ-hole of the I atom points toward the crown of the Br atom; in-deed, the σ-hole is more electropositive for iodine than for bromine (Figure 2 and Table S4), whereas the electronegative crown is more pronounced for the latter. From the inter-molecular energy decomposition (Table S1), these interactions appear with predominant electrostatic contribution (entry 7: −1 + x, y, −1 + z), but also with major dispersion contribution along with significant electrostatic contribution (entry 4: 1 − x, 1 − y, 1 − z). This is due to the fact that in the former dimer the molecules are well separated (interacting only through the halogen atoms), whereas in the latter a close proximity of polarizable halogen atoms and aromatic rings is observed.

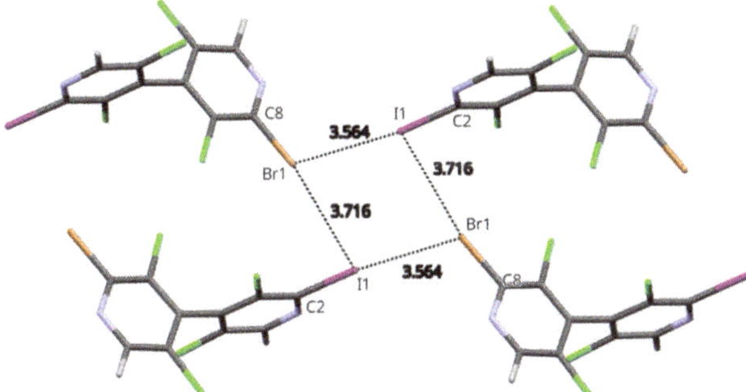

Figure 3. Four-membered ring involving four I/Br atoms (x, y, z; −1 + x, y, −1 + z; −x, 1 − y, −z; 1 − x, 1 − y, 1 − z) engaged in type-II halogen··· halogen bonds in the crystal structure of **24**. Representative distances are given in Å. Only the major component disorder is shown.

In order to rationalize the observed positional disorder that involves the iodine and bromine atoms, DFT calculations were undertaken on the isolated tetramer centered about the type-II hal··· hal bond's four-membered ring (Figure S36). From the results (Table S5), it appears that the most stable configuration is obtained for the observed largest component disorder (labelled IBrIBr), where the halogen atoms alternate in the sequence Hal1=I, Hal2=Br, Hal3=I, Hal4=Br, as depicted on Figure 3. However, the other configurations are not excessively less stabilizing, with the most unfavorable case (IIII) being only 8.92 kJ/mol higher in energy. Boltzmann populations calculated, taking into account all the possible configurations, finally lead to an equivalent average disorder population of 0.675, in remarkable qualitative agreement with the refined population parameter (0.7441(13)). This may indicate that the observed dissymmetry in the disorder population (i.e., different from 50/50) results from the preferential interaction of the σ-hole of iodine atom toward the crown of bromine at the short distance, at the expense of the reverse situation (interaction of bromine σ-hole toward the crown of iodine).

N-oxides **13** and **18** are isostructural, crystallizing in $P2_1/c$ space group with similar unit cell parameters. Indeed, they present similar crystal packing, with notable differences only about the substituent in 2 position (-H in **13**, -CN in **18**) (Figures S22 and S23). The analysis of intermolecular interaction energies (Tables S2 and S3) shows that in both structures the main interaction (entry 1) corresponds to the formation of a strong cyclic R2,2(8) hydrogen bond motif about an inversion center, involving the oxygen atom of the N-O as the acceptor (Figure 4; **18**: C12-H12··· O13 = 2.17 Å; 160°; **13**: 2.07 Å; 162°).

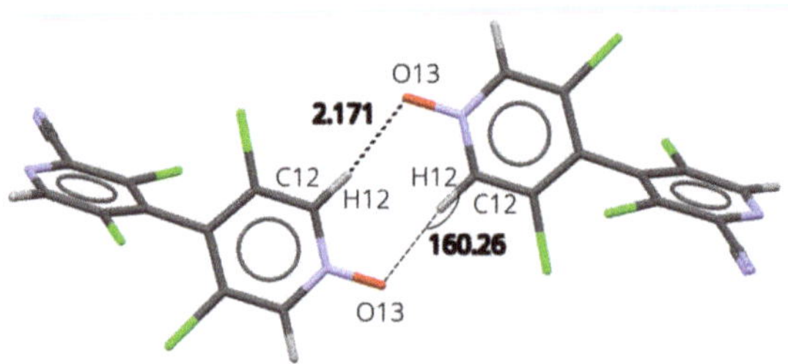

Figure 4. Packing in the solid-state structure of **18**. Interaction between x, y, z, and −1 + x, 1−y, −z molecules. Representative distance is given in Å and angle in °.

The main differences between the two structures concern the next two most intense interactions involving 2 − x, 1 − y,1 − z and 1 − x, 1 − y,1 − z molecules. The first dimer is strongly stabilizing in **18**, with the cyano groups of the two neighboring molecules arranged in a head-to-tail manner (Figure S24), associated with a large electrostatic component (Table S2, entry 2); on the contrary, in **13** the corresponding molecules does not present any contact below van der Waals limit and dispersion is the main stabilizing contribution to the interaction (Table S3, entry 2 and Figure S25). The second dimer (with 1 − x, 1 − y,1 − z, Tables S2 and S3, entry 3) is twice more stabilizing in **18** than in **13** and involves a contact at van der Waals limit between the Cl1 and C14 atoms of the cyano group, the crown of the halogen atom pointing in the direction of the positively charged carbon atom (C3-Cl1··· C14 = 3.442 Å, 110.04°; Q(C14) = 0.94) (Figure S26); in such a way, both electrostatic and dispersion components are more stabilizing in **18**, while in **13** a C3-Cl1··· Cl1 contact at van der Waals limit (3.490 Å, 78.83°) implies the crown of both halogen atoms in a unfavorable disposition (Figure S27).

The next interaction (Tables S2 and S3, entry 5; x, −y + 3/2, z − 1/2) evidences also some differences between the structures, although to a lesser extent. In both structures this dimer displays a halogen bond between the σ-hole of Cl3 and the O13 oxygen atom of the N-oxide group (**18**: C9-Cl3··· O13 = 3.152 Å RR = 0.96, 156.54°; **13**: 3.021 Å, RR = 0.92, 157.94°; Figures S28 and S29). However, in **18** a second halogen bond participates in the stabilization, the σ-hole of Cl2 being properly oriented toward the nitrogen atom of the cyano group (C5-Cl2··· N15 = 3.128 Å, RR = 0.95, 156.31°).

The last significant interaction implies 1 − x, y + 1/2, −z + 1/2 molecule (Tables S2 and S3, entry 6) and is slightly more stabilizing for **13** than for **18** due to a more negative electrostatic contribution. In this latter Cl2 atom presents three contacts at distances smaller than van der Waals limits with this neighbor, namely with N7 (3.179 Å, RR = 0.96), C8 (3.380 Å, RR = 0.98) and C12 (3.426 Å, RR = 0.99), but with an orientation that does not involves the halogen σ-hole (C5-Cl2··· N7 = 109.10°) (Figure S30). The electrostatic stabilization may then result from the head-to-tail relative orientation of the two molecules which bear a dipole moment parallel to their long N-N axis in the case of **13** (1.87 D), whereas in **18** the molecular dipole moment (3.42 D) is almost parallel to the –CN group. In this situation, the neighboring molecules have almost orthogonal dipole moments and thus a reduced overall electrostatic stabilization.

4. Discussion

The synthesis of 3,3′,5,5′-tetrachloro-4,4′-bipyridine **10** [22] was greatly improved by adjusting the amount of LDA to 0.55 equivalent with regard to 3,5-dichloropyridine

11. Moreover, a large quantity (44%) of dihydropyridine **12** was also isolated during the reaction, confirming the proposed mechanism for the dimerization process [22]. Indeed, after deprotonation of half equivalent of **11**, the nucleophilic lithiated **11-Li** added in 4-position of the remaining half equivalent of **11** to give **12-Li**, which after oxidation with iodine and hydrolysis delivered **10** and non-oxidized **12** (Figure 5A). After purification, dihydropyridine **12** was very stable in the solid state with no noticeable oxidation after three months on the bench at ambient temperature. In acetone solution, a slow oxidation occurred with almost complete formation of 4,4′-bipyridine **10** after 10 days, as shown by ^{1}H NMR (Figure 5B).

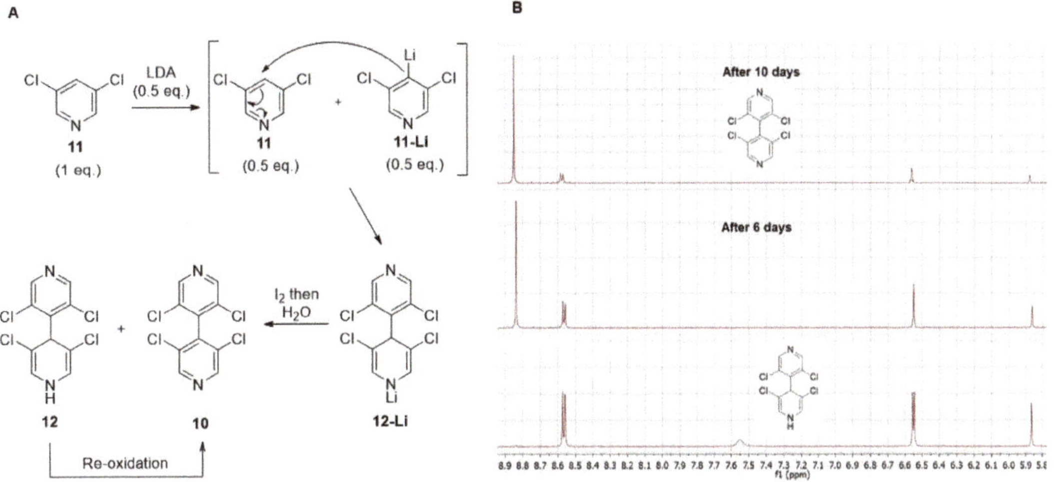

Figure 5. (**A**) Proposed mechanism for the formation of 4,4′-bipyrine **10**; (**B**) ^{1}H NMR spectra showing oxidation of dihydropyridine **12** in acetone over time.

The desymmetrization route of 4,4′-bipyridines used in this study is based on the *N*-oxidation of one of the two pyridine nitrogens in the presence of one equivalent of m-CPBA. *N*-oxidation of electron-deficient pyridines generally requires harsher conditions that necessitate prior *N*-activation with trifluoroacetic acid anhydride [32,33]. As expected, this first *N*-oxidation is not selective; however, the chemical nature of the three compounds of the reaction mixture (starting material with free nitrogens, mono *N*-oxide and bis *N*-oxide) allowed for simple purification with easy recovery of unreacted starting material. The *N*-oxidation of the electron-deficient pentasubstituted 4,4′-bipyridines **15** and **16** was highly selective with no formation of the bis *N*-oxide. The high selectivity of the reaction is due to both electronic and steric factors. Indeed, the trisubstituted pyridine ring is electronically impoverished by the three electron-withdrawing groups (Cl, Cl and Br or CN) and the *N*-pyridine is sterically hindered by the substituent in 2-position (Br or CN).

The choice of the pyridine *N*-oxides in our synthesis was also guided by the numerous known methodologies for their functionalization [21]. Halogenation and cyanation of 4,4′-bipyridine *N*-oxides used in this work are examples of such reactions which efficiently provided 4,4′-bipyridines functionalized in 2-position with Cl, Br and CN. A general mechanism can be proposed for these transformations (Figure 6). After activation of the *N*-oxide by (COCl)$_2$, (COBr)$_2$ or TMSCN, the intermediate may evolve following two different paths. In path a, the generated nucleophile during the first step (X$^-$ = Cl$^-$, Br$^-$ or CN$^-$) attacks the carbon in 2-position followed by base-assisted elimination of RO$^-$ to give the functionalized pyridine. In path b, the deprotonation of the acidic hydrogen in 2-position first occurs to generate a carbene intermediate which is trapped by the nucleophile with concomitant elimination of RO$^-$.

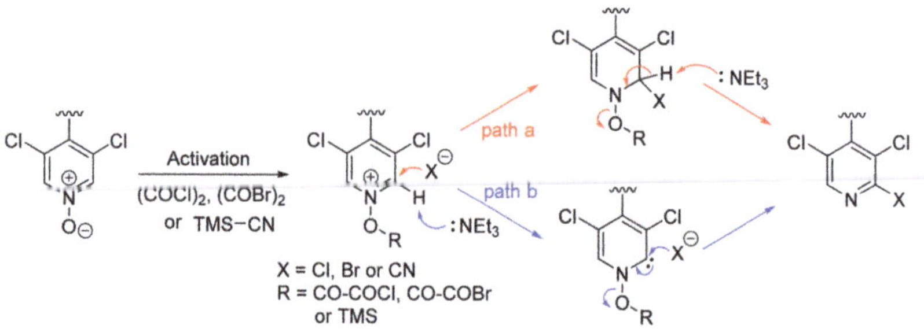

Figure 6. Proposed mechanism for the functionalization of pyridine *N*-oxides.

The installed functions in 2,2′-positions represent a potential entry to new functional groups by nucleophilic substitution of the chlorine, hydrolysis of the cyano group and cross-coupling reaction with the C-Br bond. In the latter case, we have used the copper-catalyzed Finkelstein reaction to exchange bromine by iodine, and the Suzuki reaction to introduce aryl groups in 2-positions. In this regard, 4,4′-bipyridine **24** possessing both bromine and iodine in 2,2′-positions is a compound of choice in order to selectively introduce two different aryl groups on the 3,3′,5,5′-tetrachloro-4,4′-bipyridine scaffold. This was highlighted by the synthesis of 4,4′-bipyridine **32** whose supramolecular arrangement in solution and in the solid state is currently under investigation in our laboratory.

5. Conclusions

A new methodology for the synthesis of chiral non-symmetrical 4,4′-bipyridines was developed. It is based on the successive functionalization of 2- and 2′-positions of 4,4′-bipyridines by using the sequence *N*-oxidation then halogenation or cyanation. Starting from 3,3′,5,5′-tetrachloro-4,4′-bipyridine, all the six possible non-symmetrical isomers bearing Cl, Br, I or CN in 2- and 2′-positions were efficiently prepared. The isomer bearing bromine and iodine was used as the key component for the improved synthesis of biologically active 2-iodinated 4,4′-bipyridines and for the preparation of 4,4′-bipyridines with two different aryl groups in 2- and 2′-positions. This work opens new directions toward the synthesis of non-symmetrical ligands for the design of supramolecular materials, such as coordination polymers and metal-organic frameworks (MOFs).

Supplementary Materials: The following are available online at https://www.mdpi.com/article/10.3390/compounds1020006/s1, Figure S1: 3,3′,5,5′-Tetrachloro-1,4-dihydro-4,4′-bipyridine (**12**), Figure S2: 3,3′,5,5′-Tetrachloro-[4,4′-bipyridine] 1-oxide (**13**), Figure S3: 3,3′,5,5′-Tetrachloro-[4,4′-bipyridine] 1,1′-dioxide (**14**), Figure S4: 2-Bromo-3,3′,5,5′-tetrachloro-4,4′-bipyridine (**15**), Figure S5: 3,3′,5,5′-Tetrachloro-[4,4′-bipyridine]-2-carbonitrile (**16**), Figure S6: 2-Bromo-3,3′,5,5′-tetrachloro-[4,4′-bipyridine] 1-oxide (**17**), Figure S7: 3,3′,5,5′-Tetrachloro-2′-cyano-[4,4′-bipyridine] 1-oxide (**18**), Figure S8: 2-Bromo-2′,3,3′,5,5′-pentachloro-4,4′-bipyridine (**19**), Figure S9: 2′,3,3′,5,5′-Pentachloro-[4,4′-bipyridine]-2-carbonitrile (**20**), Figure S10: 3,3′,5,5′-Tetrachloro-2′-iodo-[4,4′-bipyridine] 1-oxide (**21**), Figure S11: 2,3,3′,5,5′-Pentachloro-2′-iodo-4,4′-bipyridine (**22**), Figure S12: 2′-Bromo-3,3′,5,5′-tetrachloro-[4,4′-bipyridine]-2-carbonitrile (**23**), Figure S13: 2-Bromo-3,3′,5,5′-tetrachloro-2′-iodo-4,4′-bipyridine (**24**), Figure S14: 3,3′,5,5′-Tetrachloro-2′-iodo-[4,4′-bipyridine]-2-carbonitrile (**25**), Figure S15: 2-Bromo-3,3′,5,5′-tetrachloro-2′-phenyl-4,4′-bipyridine (**26**), Figure S16: 2-Bromo-2′-(4-((tert-butyldimethylsilyl)oxy)phenyl)-3,3′,5,5′-tetrachloro-4,4′-bipyridine (**27**), Figure S17: 2″-(4-((Tert-butyldimethylsilyl)oxy)phenyl)-3′,3″,5′,5″-tetrachloro-4,2′:4′,4″-terpyridine (**31**), Figure S18: 4-(3′,3″,5′,5″-Tetrachloro-[4,2′:4′,4″-terpyridin]-2″-yl)phenol (32), Figure S19: Packing in the solid-state structure of **24**. Interaction between x, y, z, and 2 − x, 1 − y, 1 − z molecules. Representative distance is given in Å. Only the major component disorder is shown, Figure S20: Packing in the solid-state structure of **24**. Interaction between x, y, z, and x, −y + 3/2, z − 1/2 molecules. Representative distance is given in Å. Only the major component disorder is shown, Figure S21: Packing in the solid-state structure of **24**. Interaction between x, y, z, and 1 + x, −y + 3/2, z + 1/2 molecules. Representative distance is given in Å. Only the major component disorder is shown, Figure S22:

Superimposition of the molecular environment about a central molecule (displayed as ball and sticks) showing the isostructural relationship between **18** (colored as atom type) and **13** (light gray), Figure S23: Focus on the region centered on –CN group in **18**, showing the largest differences between **18** (colored as atom type) and **13** (light gray) crystal structures, Figure S24: Packing in the solid-state structure of **18**. Interaction between x, y, z, and $2 - x, 1 - y, 1 - z$ molecules. Representative distances are given in Å, Figure S25: Packing in the solid-state structure of **13**. Interaction between x, y, z, and $2 - x, 1 - y, 1 - z$ molecules. Representative distance is given in Å, Figure S26: Packing in the solid-state structure of **18**. Interaction between x, y, z, and $1 - x, 1 - y, 1 - z$ molecules. Representative distances are given in Å, Figure S27: Packing in the solid-state structure of **13**. Interaction between x, y, z, and $1 + x, -y + 3/2, z + 1/2$ molecules. Representative distances are given in Å, Figure S28: Packing in the solid-state structure of **18**. Interaction between x, y, z, and $x, -y + 3/2, z - 1/2$ molecules. Representative distances are given in Å, Figure S29: Packing in the solid-state structure of **13**. Interaction between x, y, z, and $x, -y + 3/2, z - 1/2$ molecules. Representative distances are given in Å, Figure S30: Packing in the solid-state structure of **13**. Interaction between x, y, z, and $1 - x, y + 1/2, -z + 1/2$ molecules. Representative distances are given in Å, Figure S31: Electrostatic potential mapped on the $\rho = 0.002$ a.u. electron density isosurface of **13**. Coloring from red = -0.05 a.u. to blue = $+0.05$ a.u., Figure S32: Electrostatic potential mapped on the $\rho = 0.002$ a.u. electron density isosurface of **18**. Coloring from red = -0.05 a.u. to blue = $+0.05$ a.u., Figure S33: Integrated Bader atomic charges for isolated 24 molecule, Figure S34: Integrated Bader atomic charges for isolated **13** molecule, Figure S35: Integrated Bader atomic charges for isolated **18** molecule, Figure S36: Tertamer extracted from the experimental structure of **24** and used for the modeling of I/Br substation disorder. HalX/HalX′ are I/Br or Br/I, Table S1: Intermolecular interaction energies (kJ/mol) in the packing of solid state structure of **24**. R is the distance between molecular centroids (mean atomic position) in Å, Table S2: Intermolecular interaction energies (kJ/mol) in the packing of solid state structure of **18**. Largest differences between **18** and **13** are highlighted in bold. R is the distance between molecular centroids (mean atomic position) in Å, Table S3. Intermolecular interaction energies (kJ/mol) in the packing of solid state structure of **13** Largest differences between **18** and **13** are highlighted in bold. R is the distance between molecular centroids (mean atomic position) in Å, Table S4. Electrostatic potential maxima ($V_{S,max}$; kcal/mol) on the molecular surface $\rho = 0.002$ a.u. of **24**, Table S5: Type-II hal$\cdots$hal bonds four membered ring tetramer relative energies. Hal1-2-3-4 numbering corresponds to Figure S18.

Author Contributions: Conceptualization, V.M. and P.P.; methodology, V.M.; X-ray analysis, E.A. and E.W.; data curation, V.M.; writing—original draft preparation, V.M.; writing—review and editing, all authors. All authors have read and agreed to the published version of the manuscript.

Funding: This research received no external funding.

Data Availability Statement: All the data are given in the article and in the associated supporting information.

Acknowledgments: We thank University of Strasbourg and CNRS for financial support. "The Plateforme de Mesures de Diffraction et Diffusion des rayons X of Université de Lorraine" is thanked for providing access to crystallographic facilities. The EXPLOR mesocentre is thanked for computing facilities (Project 2019CPMXX0984).

Conflicts of Interest: The authors declare no conflict of interest.

References

1. Biradha, K.; Sarkar, M.; Rajput, L. Crystal engineering of coordination polymers using 4,4′-bipyridine as a bond between transition metal atoms. *Chem. Commun.* **2006**, *40*, 4169–4179. [CrossRef] [PubMed]
2. Striepe, L.; Baumgartner, T. Viologens and their application as functional materials. *Chem. Eur. J.* **2017**, *23*, 16924–16940. [CrossRef] [PubMed]
3. Papadakis, R. Mono-and di-quaternized 4,4′-bipyridine derivatives as key building blocks for medium- and environment-responsive compounds and materials. *Molecules* **2019**, *25*, 1. [CrossRef] [PubMed]
4. Chelucci, G.; Thummel, R.P. Chiral 2,2′-bipyridines, 1,10-phenanthrolines, and 2,2′:6′,2″-terpyridines: Syntheses and applications in asymmetric homogeneous catalysis. *Chem. Rev.* **2002**, *102*, 3129–3170. [CrossRef]
5. Fletcher, N.C. Chiral 2,2′-bipyridines: Ligands for asymmetric induction. *J. Chem. Soc. Perkin Trans. 1* **2002**, *16*, 1831–1842. [CrossRef]

6. Bednářová, E.; Malatinec, S.; Kotora, M. Applications of Bolm's ligand in enantioselective synthesis. *Molecules* **2020**, *25*, 958. [CrossRef]

7. Sbircea, L.; Sharma, N.D.; Clegg, W.; Harrington, R.W.; Horton, P.N.; Hursthouse, M.B.; Apperley, D.C.; Boyd, D.R.; James, S.L. Chemoenzymatic synthesis of chiral 4,4′-bipyridyls and their metal–organic frameworks. *Chem. Commun.* **2008**, *43*, 5538–5540. [CrossRef]

8. Jouaiti, A.; Hosseini, M.W.; Kyritsakas, N. Non-centrosymmetric packing of 1-d coordination networks based on chirality. *Chem. Commun.* **2002**, *17*, 1898–1899. [CrossRef]

9. Mamane, V.; Aubert, E.; Peluso, P.; Cossu, S. Synthesis, resolution, and absolute configuration of chiral 4,4′-bipyridines. *J. Org. Chem.* **2012**, *77*, 2579–2583. [CrossRef]

10. Rang, A.; Engeser, M.; Maier, N.M.; Nieger, M.; Lindner, W.; Schalley, C.A. Synthesis of axially chiral 4,4′-bipyridines and their remarkably selective self-assembly into chiral metallo-supramolecular squares. *Chem. Eur. J.* **2008**, *14*, 3855–3859. [CrossRef]

11. Aubert, E.; Abboud, M.; Doudouh, A.; Durand, P.; Peluso, P.; Ligresti, A.; Vigolo, B.; Cossu, S.; Pale, P.; Mamane, V. Silver(I) coordination polymers with 3,3′,5,5′-tetrasubstituted 4,4′-bipyridine ligands: Towards new porous chiral materials. *RSC Adv.* **2017**, *7*, 7358–7367. [CrossRef]

12. Peluso, P.; Mamane, V.; Aubert, E.; Dessì, A.; Dallocchio, R.; Dore, A.; Pale, P.; Cossu, S. Insights into halogen bond-driven enantioseparations. *J. Chromatogr. A* **2016**, *1467*, 228–238. [CrossRef]

13. Peluso, P.; Mamane, V.; Dallocchio, R.; Dessì, A.; Villano, R.; Sanna, D.; Aubert, E.; Pale, P.; Cossu, S. Polysaccharide-based chiral stationary phases as halogen bond acceptors: A novel strategy for detection of stereoselective σ-hole bonds in solution. *J. Sep. Sci.* **2018**, *41*, 1247–1256. [CrossRef]

14. Peluso, P.; Gatti, C.; Dessì, A.; Dallocchio, R.; Weiss, R.; Aubert, E.; Pale, P.; Cossu, S.; Mamane, V. Enantioseparation of fluorinated 3-arylthio-4,4′-bipyridines: Insights into chalcogen and π-hole bonds in high-performance liquid chromatography. *J. Chromatogr. A* **2018**, *1567*, 119–129. [CrossRef]

15. Peluso, P.; Dessì, A.; Dallocchio, R.; Sechi, B.; Gatti, C.; Chankvetadze, B.; Mamane, V.; Weiss, R.; Pale, P.; Aubert, E.; et al. Enantioseparation of 5,5′-dibromo-2,2′-dichloro-3-selanyl-4,4′-bipyridines on polysaccharide-based chiral stationary phases: Exploring chalcogen bonds in liquid-phase chromatography. *Molecules* **2021**, *26*, 221. [CrossRef]

16. Weiss, R.; Aubert, E.; Peluso, P.; Cossu, S.; Pale, P.; Mamane, V. Chiral chalcogen bond donors based on the 4,4′-bipyridine scaffold. *Molecules* **2019**, *24*, 4484. [CrossRef]

17. Dessì, A.; Peluso, P.; Dallocchio, R.; Weiss, R.; Andreotti, G.; Allocca, M.; Aubert, E.; Pale, P.; Mamane, V.; Cossu, S. Rational design, synthesis, characterization and evaluation of iodinated 4,4′-bipyridines as new transthyretin fibrillogenesis inhibitors. *Molecules* **2020**, *25*, 2213. [CrossRef]

18. Mamane, V.; Peluso, P.; Aubert, E.; Cossu, S.; Pale, P. Chiral hexahalogenated 4,4′-bipyridines. *J. Org. Chem.* **2016**, *81*, 4576–4587. [CrossRef]

19. Foulger, N.J.; Wakefield, B.J. Polyhalogenoaromatic compounds. XXIX. Hexachloro-5,5′-dilithio-4,4′-bipyridine as an intermediate for organometallic and organic syntheses. *J. Organomet. Chem.* **1974**, *69*, 161–167. [CrossRef]

20. Mamane, V.; Aubert, E.; Peluso, P.; Cossu, S. Lithiation of prochiral 2,2′-dichloro-5,5′-dibromo-4,4′-bipyridine as a tool for the synthesis of chiral polyhalogenated 4,4′-bipyridines. *J. Org. Chem.* **2013**, *78*, 7683–7689. [CrossRef]

21. Kutasevich, A.V.; Perevalov, V.P.; Mityanov, V.S. Recent progress in non-catalytic C–H functionalization of heterocyclic *N*-oxides. *Eur. J. Org. Chem.* **2021**, *2021*, 357–373. [CrossRef]

22. Abboud, M.; Mamane, V.; Aubert, E.; Lecomte, C.; Fort, Y. Synthesis of polyhalogenated 4,4′-bipyridines via a simple dimerization procedure. *J. Org. Chem.* **2010**, *75*, 3224–3231. [CrossRef] [PubMed]

23. Chen, Y.; Huang, J.; Hwang, T.L.; Chen, M.J.; Tedrow, J.S.; Farrell, R.P.; Bio, M.M.; Cui, S. Highly regioselective halogenation of pyridine *N*-oxide: Practical access to 2-halo-substituted pyridines. *Org. Lett.* **2015**, *17*, 2948–2951. [CrossRef] [PubMed]

24. Carson, M.W.; Giese, M.W.; Coghlan, M.J. An intra/intermolecular Suzuki sequence to benzopyridyloxepines containing geometrically pure exocyclic tetrasubstituted alkenes. *Org. Lett.* **2008**, *10*, 2701–2704. [CrossRef] [PubMed]

25. Klapars, A.; Buchwald, S.L. Copper-catalyzed halogen exchange in aryl halides: An aromatic Finkelstein reaction. *J. Am. Chem. Soc.* **2002**, *124*, 14844–14845. [CrossRef] [PubMed]

26. Peluso, P.; Sechi, B.; Lai, G.; Dessì, A.; Dallocchio, R.; Cossu, S.; Aubert, E.; Weiss, R.; Pale, P.; Mamane, V.; et al. Comparative enantioseparation of chiral 4,4′-bipyridine derivatives on coated and immobilized amylose-based chiral stationary phases. *J. Chromatogr. A* **2020**, *1625*, 461303. [CrossRef] [PubMed]

27. Dallocchio, R.; Sechi, B.; Dessì, A.; Chankvetadze, B.; Cossu, S.; Mamane, V.; Weiss, R.; Pale, P.; Peluso, P. Enantioseparations of polyhalogenated 4,4′-bipyridines on polysaccharide-based chiral stationary phases and molecular dynamics simulations of selector–selectand interactions. *Electrophoresis* **2021**. [CrossRef] [PubMed]

28. Perdomo Rivera, R.; Ehlers, P.; Ohlendorf, L.; Torres Rodríguez, E.; Villinger, A.; Langer, P. Chemoselective synthesis of arylpyridines through Suzuki–Miyaura cross-coupling reactions. *Eur. J. Org. Chem.* **2018**, *8*, 990–1003. [CrossRef]

29. Turner, M.J.; McKinnon, J.J.; Wolff, S.K.; Grimwood, D.J.; Spackman, P.R.; Jayatilaka, D.; Spackman, M.A. CrystalExplorer17 (2017). University of Western Australia. Available online: https://hirshfeldsurface.net (accessed on 26 February 2021).

30. Desiraju, G.R.; Parthasarathy, R. The nature of halogen···halogen interactions: Are short halogen contacts due to specific attractive forces or due to close packing of nonspherical atoms? *J. Am. Chem. Soc.* **1989**, *111*, 8725–8726. [CrossRef]

31. Metrangolo, P.; Resnati, G. Type II halogen···halogen contacts are halogen bonds. *IUCrJ* **2014**, *1*, 5–7. [CrossRef]

32. Zhu, X.; Kreutter, K.D.; Hu, H.; Player, M.R.; Gaul, M.D. A novel reagent combination for the oxidation of highly electron deficient pyridines to *N*-oxides: Trifluoromethanesulfonic anhydride/sodium percarbonate. *Tetrahedron Lett.* **2008**, *49*, 832–834. [CrossRef]
33. Caron, S.; Do, N.D.; Sieser, J.E. A practical, efficient, and rapid method for the oxidation of electron deficient pyridines using trifluoroacetic anhydride and hydrogen peroxide–urea complex. *Tetrahedron Lett.* **2000**, *41*, 2299–2302. [CrossRef]

 *compounds*

Review

Waste-Glycerol as a Precursor for Carbon Materials: An Overview

Mary Batista [1,*], **Silvia Carvalho** [1,2,*], **Renato Carvalho** [3], **Moisés L. Pinto** [2] and **João Pires** [1]

1   Centro de Química Estrutural, Institute of Molecular Sciences, Departamento de Química e Bioquímica, Faculdade de Ciências, Universidade de Lisboa, Campo Grande, 1749-016 Lisboa, Portugal
2   CERENA, Departamento de Engenharia Química, Instituto Superior Técnico, Universidade de Lisboa, Campus Alameda, 1049-001 Lisboa, Portugal
3   IBEROL, Sociedade Ibérica de Biocombustíveis e Oleaginosas, 2600-531 Alhandra, Portugal
*   Correspondence: mkbatista@fc.ul.pt (M.B.); silvia.carvalho@tecnico.ulisboa.pt (S.C.)

Abstract: Biodiesel is produced by the transesterification of animal fats and vegetable oils, producing a large amount of glycerol as a by-product. The crude glycerol cannot be used in the food or pharmaceutical industries. It is crucial to transform glycerol into value-added products with applications in different areas to biodiesel be economically viable. One of the possible applications is its use as a precursor for the synthesis of carbon materials. The glycerol-based carbon materials have distinct properties due to the presence of sulfonic acid groups on the material surface, making them efficient catalysts. Additionally, the glycerol-based activated carbon materials show promising results concerning the adsorption of gases and liquid pollutants and recently as capacitors. Despite their potential, currently, little research has been carried out on the synthesis and application of those materials. This review summarized the preparation and application of carbon materials from glycerol, intending to show the potential of these materials.

Keywords: biodiesel production; crude glycerin; carbon materials

Citation: Batista, M.; Carvalho, S.; Carvalho, R.; Pinto, M.L.; Pires, J. Waste-Glycerol as a Precursor for Carbon Materials: An Overview. *Compounds* **2022**, *2*, 222–236. https://doi.org/10.3390/compounds2030018

Academic Editor: Juan C. Mejuto

Received: 23 June 2022
Accepted: 9 September 2022
Published: 16 September 2022

Publisher's Note: MDPI stays neutral with regard to jurisdictional claims in published maps and institutional affiliations.

1. Introduction

Biodiesel is obtained predominantly by transesterification (chemically or enzymatic) of vegetable oils and animal fats and is known for its energy security awareness. It is composed of free fatty acid alkyl esters and has low toxicity and high biodegradability [1]. Biofuels are a clean energy source, whose combustion emits ≈ 35% fewer greenhouse gases compared with diesel fuel [1,2]. However due to technological limitations, biodiesel's cost is still higher than fossil diesel [1].

The transesterification of a variety of materials that contain fatty acids that include various vegetable and animal fats, vegetable oils, and edible oil processing residues such as soybean [3], sunflower [4], palm [5], rapeseed [6], canola [7], Jatropha, and cottonseed [8–10] generates a large amount of glycerol waste. For instance, for each 10 kg of biodiesel produced, 1 kg of glycerol is generated [11]. The increase of biodiesel production has, as a consequence, an increase of glycerol production. Solutions for this glycerol must be found [12–15] to ensure the economic feasibility of biodiesel. Yet this crude glycerol cannot be used in most of the traditional applications of glycerol such as food [16–19] and pharmaceutical industries [20–23], personal care products [24–26], anti-freezers [27,28], e-cigarette liquids [29–31], explosives [32] and many other processes as an intermediate compound [33–35] due to its poor quality. Additionally, the use of crude glycerol has been gaining ground as a component of heavier fuels and in the processes of obtaining acrylic acid [36]. A much less explored possible use of crude glycerol concerns the development of new solid materials generating high-value products. Coal is traditionally produced from waste, thus adding value to the residue [37,38]. However, the use of glycerol as a precursor in the preparation of carbon materials (carbon and activated carbon, Figure 1) is relatively new and unexplored in comparison with other precursors such as rapeseed [37], potato peel [38,39], sugarcane bagasse [40,41], waste coffee residues [42,43], waste rice husk [44,45], and waste corn [46,47], etc. The glycerol-based carbon materials are obtained in one step by in situ partial carbonization and sulfonation of glycerol with sulfuric acid. In turn, glycerol-based activated carbon materials are prepared in two steps: (i) partial carbonization and sulfonation of glycerol in the presence of sulfuric acid; (ii) chemical or thermal activation of the glycerol-based carbon material. The main advantages of the coal obtained from glycerol are the sulfonic acid groups on the material surface, which give them specific properties for diverse environmental applications such as catalyst, capacitor, and adsorbent materials. This review collected the existing data about the production and use of carbons from glycerol. Despite their potential, most of the research is not contemporary. We intended to show the potential and arouse the interest for this type of material, thus allowing the development of more effective glycerol-based carbons. The synthetic method of the glycerol-based carbons investigated is the same with few variations in the experimental conditions. More in-depth research into the synthesis may lead to carbons with improved properties for the desired applications. The use of these carbons in catalysis is the most explored application. Still, activated carbons have much potential as adsorbents of pollutants and capacitors, yet they are unexplored, and new research is needed, as may be seen in this review. In this work, we intended to call the attention of scientists in the area by showing the potentialities of this by-product in developing advanced carbon materials and promote research in an area with great potential.

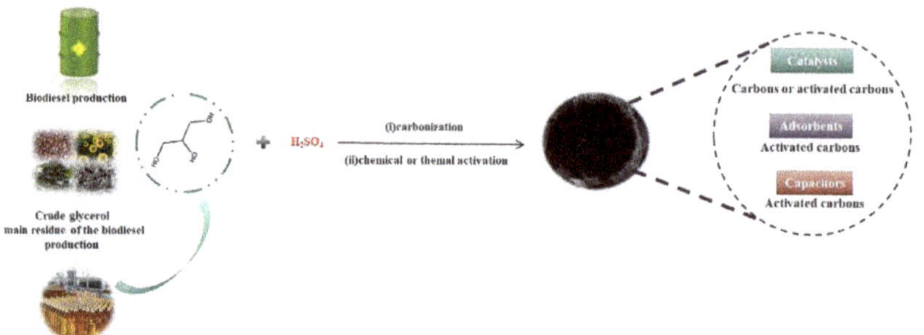

Figure 1. Synthesis of glycerol-based carbons and their main applications.

2. Synthesis of Glycerol-Based Carbon Materials

The synthesis of carbons from glycerol is a relatively new area concerning the use of glycerol. It mainly consists of the partial carbonization and sulfonation of glycerol with sulfuric acid. The main differences observed in the literature concerning the synthesis of carbons are variations in experimental conditions such as glycerol: sulfuric acid mass ratio, reaction temperature, and time. The paper of Devi et al. published in 2009 [48] was the first to synthesize carbon from glycerol. It used the one-pot reaction shown in Figure 1. The partial carbonization and sulfonation of glycerol were carried out with concentrated sulfuric acid (1:4 w/w) using soft experimental conditions. First, the glycerol and sulfuric acid mixture was heated to 180 °C. The mixture was kept at this temperature until foaming ceased. Then, the product was cooled to room temperature and washed with hot water under agitation until reaching neutral pH value. Two years later, the same authors synthesized other carbon using the same procedure but used pitch glycerol as a carbon source and at a different temperature (250 °C) [49]. The reaction yields were 50% and 40%, respectively, and no explanation for why a higher temperature had to be used for this last procedure was presented. The carbons were fully characterized by a great diversity of techniques (elemental analysis, X-ray photoelectron spectroscopy (XPS), X-ray Powder Diffraction (XRD), scanning electron microscopy (SEM), Fourier Transform InfraRed (FTIR), Magic-angle spinning (MAS) NMR ^{13}C, Raman, potentiometric titrations, N_2 isotherms, and thermogravimetry/differential thermal analysis (TG/DTA). The obtained carbons had a non-porous nature (<1 m^2·g^{-1}), a high density of sulfonic acid groups (–SO$_3$H), and their catalytic capacity in the esterification of palmitic acid, tetrahydropyranylation, and dehydropyranylation was evaluated (See Section 3.1 for more details).

Mantovanic et al. [50] and Gonçalves et al. [51] prepared carbons by hydrothermal carbonization using a mixture of glycerol waste and sulfuric acid (different mass ratio) at 150 °C or 180 °C and using several reaction times (0.25–24 h). Interestingly, the authors successfully increased the number of sulfonic groups using sulfuric acid in a post-synthesis treatment [51]. The carbons also had a non-porous nature and a high numbers of acidic surface groups and were tested as catalysts in acetalization and etherification reactions (see Section 3.1 for more details).

The synthetic procedure described gives rise to non-porous carbons whose main application is in catalysis. For other applications, such as adsorption, the development of a porous structure is crucial. Typically, the synthesis of activated carbons from glycerol requires two steps, carbonization, followed by an activation step [52–54]. Some examples of those activated carbons obtained via chemical activation (KOH, ZnCl$_2$, and H$_3$PO$_4$) [52,53] and thermal activation [55,56] may be found in the literature. Different activation agent ratios and temperatures have been tested to vary the porosity of the obtained materials. For instance, the group of Ribeiro et al. [56], after obtaining the carbon using the already described procedure, carried out further calcination (120 °C, 400 °C, 600 °C—60 min in each temperature, plus 800 °C—240 min) under nitrogen flow. The obtained material showed high thermal stability, a basic character due to the decomposition of the sulphonic acid groups, and non-porous nature. This material was then thermally activated under an air atmosphere at different temperatures (150 °C, 200 °C, 300 °C, and 350 °C) for 1 h and generated porosity which increased with the temperature as shown in Figure 2. Additionally, the increase in temperature in the surface oxygen groups (lactones, phenols, and quinones) increases its acid character. Although rare, this work used activated carbons in the catalytic wet peroxide oxidation (CWPO) of 2-nitrophenol (See Section 3.1 for more details).

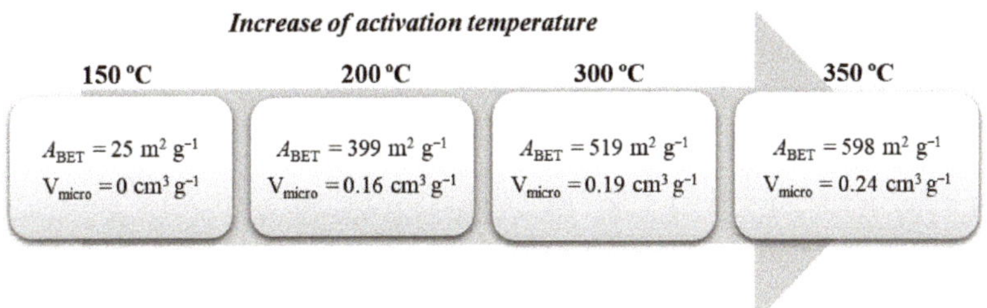

Figure 2. Influence of the activation temperature in the generation of porosity and development of microporosity.

Another way of obtaining porous carbons may be using a pore-forming agent. The work of Lee et al. [57] used a pure glycerol and crude waste glycerol as a carbon precursor for mesoporous carbon. It also explored glycerol as a pore-forming agent for mesoporous silica. The mesoporous carbon was obtained by the carbonization of glycerol–silica nanoparticles at high temperatures (600 °C) under a nitrogen atmosphere. NaOH solution was used for the removal of the silica–nanoparticle framework. By simply changing the silica particle size in glycerol–silica nanocomposites or changing the silica particle size, it was possible to tailor the pore size and volume, surface area, and pore wall thickness of mesoporous carbon. Due to the presence of other components that may also act as a pore-forming agent in the crude waste glycerol, its use in the synthesis leads to a multimodal pore size distribution of micropores smaller than 2 nm, small mesopores centered at 3.8 nm, and large mesopores above 10 nm.

To the best of our knowledge, the work published in 2016 by Álvarez-Torrellas et al. was the first to study the application of glycerol-based activated carbons as adsorbent materials [55]. The synthesis consisted in the partial carbonization of a glycerol–sulfuric acid mixture, followed by thermal activation. A glycerol and sulfuric acid mixture was heated to 180 °C for 20 min. The resulting material was calcinated in a tube furnace under a nitrogen flow (100 $cm^3 \cdot min^{-1}$) at different temperatures (120 °C, 400 °C and 600 °C) during 60 min and 800 °C during 240 min. Then the calcined material (GBCM) was thermally activated at different temperatures (200 °C, 300 °C and 350 °C, for 60 min) in a tube furnace under oxidative atmosphere (flow of 100 $cm^3 \cdot min^{-1}$). The textural properties were studied by N_2 adsorption-desorption isotherms. The presence of the oxygenated groups was investigated by zeta potential and FTIR data. All obtained acid activated carbons (GBCM$_{200}$, GBSM$_{300}$ and GBCM$_{350}$—where the subscript represents the activation temperature) presented high surface area and microporous structure developed. Their adsorption capacities were evaluated through flumequine and tetracycline (See Section 3.2 for more details).

Cui et al. [54] investigated glycerol as a liquid precursor for the preparation of activated carbon. The authors concluded that glycerol pyrolysis in the absence of acid generates no carbon material. This was justified by the evaporation of glycerol (boiling point of 290 °C) before it was carbonized. The description of different acids' roles and the absence of acid in carbon formation was reported. For this effect, glycerol was mixed with an acid (H_2SO_4, H_3PO_4, HCl, or CH_3COOH) at volume ratios (10:1, 10:2 and 10:3 v/v). The solutions were added to a quartz boat and heated on the tube furnace in N_2 atmosphere to 400 °C, 500 °C, 600 °C, 700 °C, or 800 °C for 1 h. The glycerol pyrolysis in the presence of HCl or CH_3COOH did not produce carbon material. However, with H_2SO_4 or H_3PO_4 addition, glycerol pyrolysis generated carbon materials. According to the authors, the glycerol is dehydrated and polymerized when exposed to the presence of acids (H_2SO_4 or H_3PO_4) at moderate temperatures (<200 °C). Both acids induce dehydration of alcohol groups via protonation of the alcoholic oxygen (Figure 3).

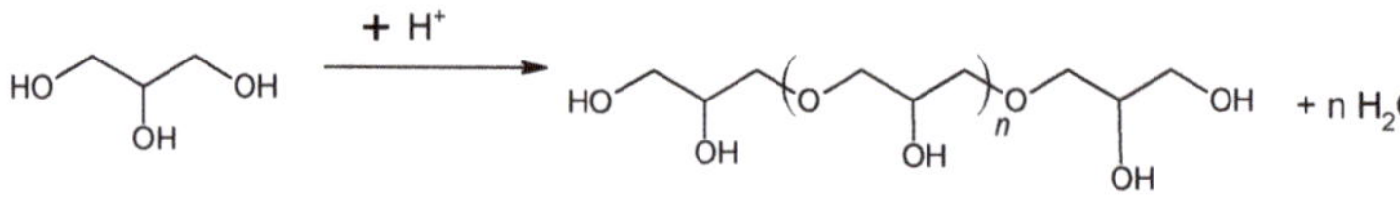

Figure 3. General representation of the polymerization reaction.

In the case of HCl, it has a low boiling point (48 °C), and in the case of CH_3COOH, because it is a weak acid, it cannot initiate the glycerol dehydration. In this context, carbon materials with various functional groups and porosities were prepared via sulfuric or phosphoric acid-mediated polymerization and carbonization followed by steam or CO_2 activation. The porosity in the activated carbons reached surface areas up to 2470 $m^2 \cdot g^{-1}$ and pore volumes up to 1.44 $cm^3 \cdot g^{-1}$. The samples prepared with H_3PO_4 were consistently more mesoporous than samples prepared with H_2SO_4. The adsorption capacity of those materials was evaluated for the removal of gas phase volatile organic compounds (VOCs) and aqueous phase chromium Cr(VI) (See Section 3.2 for more details).

Naverkar et al. [58] prepared glycerol-based carbon by partial carbonization of glycerol using concentrated sulfuric acid (molar ratio 1:4) followed by thermal treatment. The sulfuric acid was added dropwise to glycerol (10 g) and stirred for 20 min at 180 °C. The carbonized material was further treated at 120 °C and 350 °C to obtain the samples GBC-120 and GBC-350. The carbon materials were characterized by XRD, FTIR, thermal analysis (TG/DTG/DTA), pH_{PZC} measurements, SEM, and N_2 adsorption-desorption at low temperature. The samples GBC-120 and GBC-350 presented BET surface areas of 21 $m^2 \cdot g^{-1}$ and 464 $m^2 \cdot g^{-1}$, respectively. They were studied for the adsorption of methylene blue (See Section 3.2 for more details).

Gonçalves et al. [53] also prepared glycerol-based activated carbon via two steps (polymerization + chemical activation). Firstly, the glycerol polymer was prepared by glycerol polymerization under reflux in the presence of sulfuric acid. The glycerol polymer was chemically activated with $ZnCl_2$ or H_3PO_4. The authors also investigated several activated carbon synthesis conditions such as the type of activating agent ($ZnCl_2$ or H_3PO_4), the impregnation ratio ($ZnCl_2$ (X_{Zn} = 0.4 and 0.8) and H_3PO_4 (X_P = 0.3 or 0.6) activating agent mass/polymer mass), activation time, and temperature. They were evaluated as supercapacitor electrode and for the adsorption of organic contaminants (See Sections 3.2 and 3.3 for more details).

Glycerol-based magnetic carbon composites were synthesized by Medeiros et al. [59]. The carbon composites were prepared by mixing glycerol waste and iron(III) salt (heating to 380 °C, 600 °C, or 800 °C) for 3 h in a vertical reflux reactor. The textural properties of GFe3-380 (5 $m^2 \cdot g^{-1}$), GFe3-600 (140 $m^2 \cdot g^{-1}$), and GFe3-800 (136 $m^2 \cdot g^{-1}$) composites were evaluated by N_2 adsorption–desorption isotherms. The carbon composites presented the following surface area 5 $m^2 \cdot g^{-1}$ (GFe3-380), 140 $m^2 \cdot g^{-1}$ (GFe3-600) and 136 $m^2 \cdot g^{-1}$ (GFe3-800). According to the pore size distribution for composites, the samples GFe3-600 and GFe3-800 contain both micropores and mesopores (essentially). The composites (GFe3-600 and GFe3-800) were tested as adsorbents of dyes (methylene blue and indigo carmine) (See Section 3.2 for more detail).

More recently, Batista et al. [52] prepared a series of glycerin-activated carbons from crude glycerin (82% glycerol) for application in the gas separation by adsorption processes. Glycerin-activated carbons were prepared via a two-step procedure involving carbonization followed by chemical activation with KOH. A mixture of industrial crude glycerin (82% glycerol) and concentrated sulfuric acid was prepared using a volume ratio of 1:0.5 (glycerol:H_2SO_4). The acid carbonization process was carried out in a Teflon lined Hydrothermal Autoclave at 180 °C for 6 h in an oven. The carbonized (glycerin-char) was washed with distilled water until the washing was neutral and dried (100 °C); The obtained solid (glycerin-char, crushed to fine powder of dimension < 0.297 mm) was mixed with an activating agent (KOH) in distilled water, been stirring for 2 h (at ambient temperature) and when dried (100 °C). It was used two activation temperatures (700 °C and 800 °C) and weight ratios (1:1, 2:1 and 3:1, KOH:glycerin-char). The mixture (activating agent:glycerin-char) was activated in a horizontal furnace Thermolyne 21100 (under N_2 flow 5 $cm^3 \cdot s^{-1}$ and 10 $°C \cdot min^{-1} \cdot h^{-1}$). The glycerin-activated carbons were washed with distilled water until the washing was neutral and dried at 100 °C. The prepared samples (G@700/1, G@700/2, G@700/3 and G@800/1, G@800/2 and G@800/3—the 700/800 corresponds to the activation temperature and the 1, 2 and 3 to the KOH:glycerin-char ratio) presented high surface areas (1166–2150 $m^2 \cdot g^{-1}$) and pore volumes between 0.63 and 1.03 $cm^3 \cdot g^{-1}$. These glycerin-activated carbons were evaluated as adsorbents for the adsorption separation of ethane and ethylene (See Section 3.2 for more details).

In another work, Batista et al. [60] modified a glycerin-activated carbon and zeolite type A surfaces with chitosan. The purpose of this work was different from the other works presented here. It was to evaluate the potential of those materials as H_2S donors for therapeutic application. The activated carbon (Gta@600) was prepared by a combination of acid carbonization with H_2SO_4 followed by thermal activation (in a nitrogen flow rate = 5 $cm^3 \cdot min^{-1}$ at 600 °C for 1 h). The modification of the material surface was obtained by adding chitosan dissolved in acetic acid solution (1 wt%). to a suspension of Gta@600. The chitosan-based carbon (Gta@600Chi) was characterized (FTIR, SEM, XDR, Elemental analysis and N_2 adsorption–desorption isotherms). The adsorption capacity of H_2S by Gta@600 and Gta@600Chi was performed to evaluate their use as H_2S donors (See Section 3.2 for more information).

3. Principal Uses of Carbons from Glycerol

3.1. Catalysis

The carbons synthesized by R.B.N Prasad et al. [48,49,61–63], using the procedure previously described, were tested as a solid-acid catalyst for a diversity of one-pot reactions (described in the following paragraphs) showing very good performance with their activity maintained during several cycles. Yet, no investigation was carried out about leaching into the reaction medium and only one catalyst concentration (10 wt%) was tested.

The first carbon obtained by partial carbonization and sulfonation of glycerol [48] was tested as a catalyst in the esterification of palmitic acid with methanol at 65 °C. It revealed a high activity (99% conversion in 4 h). The carbon obtained from pitch glycerol using the same experimental procedure was tested as catalyst for tetrahydropyranylation (THP ether synthesis) and dehydropyranylation using a wide variety of alcohols and phenols (17 total) [49]. The tetrahydropyranylation reactions were performed in dichloromethane at room temperature. The THP ethers were obtained in 80–98% yield, the yield being dependent on the alcohol structure, the lowest yields being observed for phenols due to the lower nucleophilic character of the phenolic oxygen. On the other hand, for the dehydropyranylation reactions, the structure of the substrate had no influence, and reactions yields of 95–99% were obtained using methanol as a solvent. Other reactions were also investigated. For instance, highly substituted imidazole derivatives were obtained from 1,2-diketones, aldehydes, NH_4OAc, and amines [61]. A study on the reaction temperature and solvents showed the use of acetonitrile at 50–55 °C gives the highest yields (70–84%). The catalyst was also effective for the synthesis of diverse dihydropyrimidinones in refluxing acetonitrile (yields 80–92%) [63]. The introduction of a halogen onto the aromatic ring yields or replaces urea with thiourea originated the lowest yields. Additionally, the same authors explored the use of the catalysts in the obtention of substituted benzamides (71–78% yield) using different aldehydes and amines as substrates [62]. Many parameters were optimized, and the best condition was attained using acetonitrile at 60–65 °C and Cs_2CO_3 as a base. Finally, using slightly different conditions, two other reactions were investigated using the same type of catalysts [64]. First, the acetylation of alcohols, phenols, and aromatic amines using acetic anhydride at 65 °C (no-solvent and 15% catalyst) was investigated. The acetylation reactions of the alcohols and phenols had a 75–96% yield, while for aromatic amines had a 95–97% yield. The yields and reaction time of the acetylation of alcohol and phenol were related to their structure. The primary and secondary alcohols had higher yields and rapid reaction times, while the phenols showed slow reactions. Interesting, selectivity for the acetylation of the amine group was observed, a fact that may be explained by more nucleophilicity of amines than phenols. Secondly, using different aldehydes and 2,2-bis (hydroxymethyl) propane-1,3-diol different pentaerythritol diacetals were obtained [22]. Among the investigated reaction conditions (catalyst content and reaction temperature), the best results were obtained using 5 wt% catalyst in toluene at 80 °C. Aliphatic aldehydes showed no reactivity. The catalyst showed selectivity towards aromatic aldehydes. Among the aromatic aldehydes, the presence of the electron-donor group showed less reactivity, while the presence of electron-withdrawing groups enhanced the reaction rate. The catalyst was also effective in the deprotection of the diacetals in methanol in reflux within 30 min. Figure 4 summarizes the studies carried out up to this point.

The group of Gonçalves et al. [50,51] used a glycerin-based carbon obtained by hydrothermal carbonization of a mixture of glycerol waste and sulfuric acid as a catalyst in the glycerol acetalization reaction. The acid character attributed to the $-SO_3H$ and $-COOH$ groups at the material surface and the excess of sulfuric acid used during the synthetic procedure affected the final surface chemistry of the material by increasing the surface acidity of these catalysts. The catalyst activity of the carbons through the glycerol acetalization with acetone was dependent on the carbon used as a catalyst. For instance, the 3:1 showed no activity, probably due to the low concentration and absence of $-COOH$ and $-SO_3H$ groups, respectively. The importance of the sulfonic groups on the surface was once more confirmed and was highly related to the high catalytic activity of the carbon. For instance, the 3:1 and 2:1 carbon (the ones with higher sulfonic groups) had an 82% conversion of glycerol with almost complete selectivity for solketal. Further studies, using the 2:1 carbon, were conducted to evaluate the effect of different variables in the catalytic activity. They revealed that an increase in the glycerol: acetone molar ratio provided a relevant increase in the glycerol conversion; additionally, an increase in the glycerol conversion was observed.

Increasing the amount of catalyst by more than 3% brought no advantage. Regards the temperature reaction, it was observed that at room temperature, the reaction was slightly slower than at 40 °C and 65 °C. However, the process reaches equilibrium at the same level of conversion independent of temperature. The leaching tests showed no appreciable leaching of any active groups present over the surface of the solids.

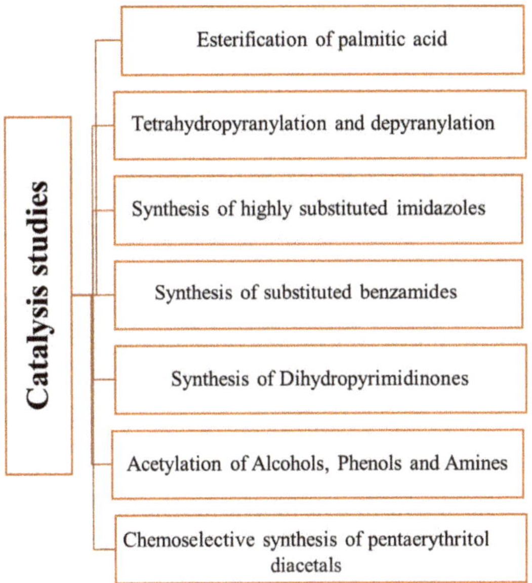

Figure 4. Reactions studied [22,33,48,49,61–64] using carbons obtained from glycerol as catalysts.

The same group used carbons obtained by hydrothermal carbonization of a mixture of glycerol waste and sulfuric acid and post-synthesis modified as catalysts for the glycerol etherification reaction [65]. The catalytic activity was evaluated through glycerol etherification with tert-butyl alcohol using a 5 wt% catalyst at 120 °C. The 10:1 carbon showed negligible catalytic activity while the 1:3 carbon had a high catalytic activity (better than Amberlyst resin), a fact that may be attributed to the lower concentration of acidic sites on the 10:1 carbon surface, which is essential for etherification reactions. Interestingly, the post-synthesis modification of the 10:1 carbon led to a substantial improvement in the catalytic (similar to the 1:3 carbon). This improvement was attributed to the introduction of sulfonic acid groups but also to other surface functional groups, such as carboxylic acids. Figure 5 shows the reactions studied by Gonçalves et al. [50,51,65].

The different reactions in which carbons from glycerol may be used as catalysts are shown in Table 1.

The previously described work concerns the use of carbons without activation for the use as catalysts. As already referred, the use of activated carbon is not so usual, and only one work exists that effectively used activated carbons as a catalyst [56]. The catalytic activity of carbons was investigated in the catalytic wet peroxide oxidation (CWPO) of 2-nitrophenol. Adsorption studies revealed that some adsorption on the material surface occurred, yet catalytic activity of the carbons was observed, especially to carbon activated at 300 °C. The characteristics of these materials (developed porosity allied to high basicity and lower oxygen content) seemed to explain their catalytic activity. On the other hand, the higher removal of 2-nitrophenol by the material activated at 350 °C may be explained by a high contribution of the adsorption process. Further studies with the carbon activated at 300 °C revealed its catalytic efficiency was increased when the CWPO process was conducted under intensified conditions (T = 50 °C, pH = 3, stoichiometric amount of H_2O_2

and a pollutant/catalyst mass ratio = 2). However, the recyclability of the catalyst was studied. The catalyst lost activity after the first cycle due to the adsorption process and the deactivation of the carbon active sites responsible for hydrogen peroxide decomposition, yet its activity may be restored by a simple oxidative thermal regeneration.

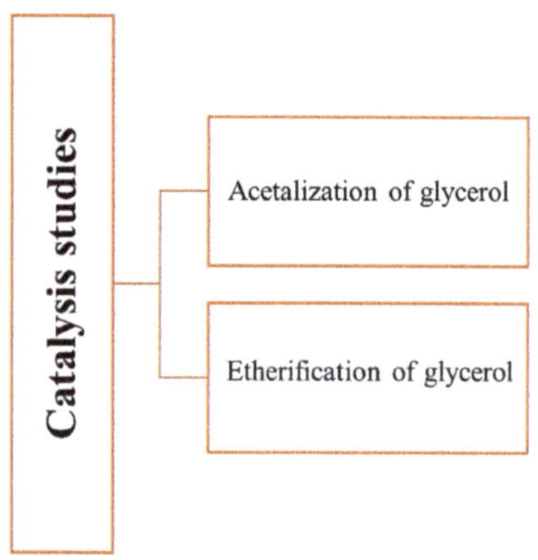

Figure 5. Reactions studied [50,51,65] using carbons obtained from glycerol as catalysts.

Table 1. Review of the most relevant data for the indicated reactions using glycerol-based non-porous carbon as catalysts.

Product	Catalyst Loading (%)	Solvent	Temperature (°C)	Reaction Time (h)	Yield (%)	Recyclability of Catalyst (n° of Cycles Studied)	Reference
Biodiesel	10	Methanol	65	4	99 *	8	[48]
THP ether	10	Dichloromethane	R.T.	2	80–98	8	[49]
Alcohols	10	Methanol	R.T.	0.5	95–99	8	[49]
Substituted imidazole derivatives	10	Acetonitrile	50–55	7	70–84	3	[61]
Substituted 3,4-dihydropyridine-2-(1H)-ones	10	Acetonitrile	reflux	4–4.5	80–92	3	[63]
Substituted benzamides	10	Acetonitrile	60–65	-	71–78	3	[61]
Acetylated alcohol and phenols	15	No-solvent	65	0.5–2	75–96	5	[64]
Acetylated amines	15	No-solvent	65	0.5	92–97	-	[64]
Pentaerythritol diacetals	5	Toluene	80	1.5–8.5	94–98	5	[22]
Glycerol acetal +	3	-	40/65	1	82 *	5	[51]
Glycerol etherification	5	-	120	6	52 (MTBG), 22 (DTBG + TTBG)	8	[65]

R.T.—Room Temperature; * Refers to the reagent conversion, not yield; +—best result; MTBG—mono-tert-butyl glycerol; DTBG—di-tert-butyl glycerol, TTBG—tri-tert-butyl glycerol.

3.2. Adsorption

As mentioned, glycerol can be converted into a material that has promising properties for application as adsorbent materials. Their adsorbent capacity was examined to different adsorbates such as medicines (flumequine, tetracycline and paracetamol) [54,65], aqueous phase chromium Cr(VI), dyes (methylene blue and indigo carmine), VOCs (toluene and hexane), and ethene, ethylene. The adsorption studies (Figure 6) in aqueous solutions are different from the gas adsorption, which normally requires special equipment based on gravimetric or volumetric methods. The schematic representation of a volumetric apparatus is shown in Figure 7a, for the ethene/ethylene separation, and in Figure 7b, for H_2S adsorption.

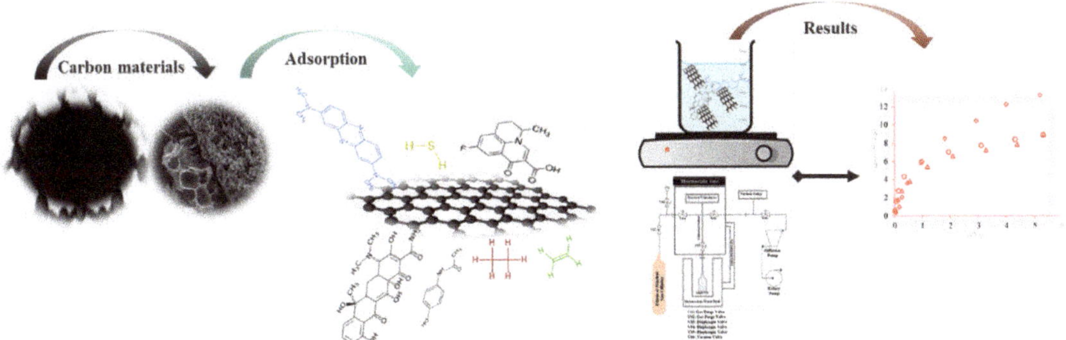

Figure 6. Representation of the adsorption studies.

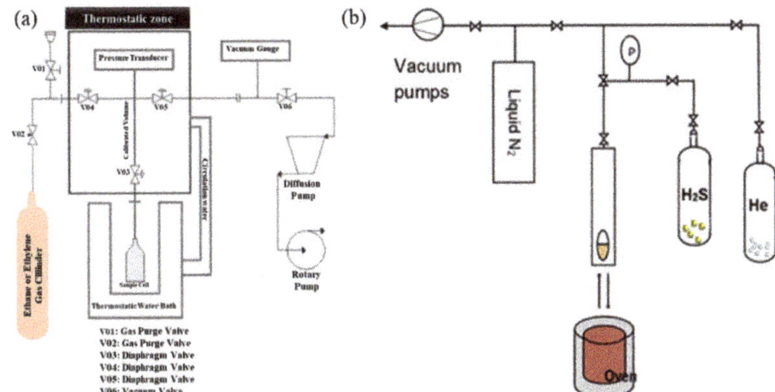

Figure 7. Volumetric system used for ethane/ethylene separation (**a**) and H_2S adsorption (**b**) studies. Reproduced with permission from [66], Wyley, 2022.

As already mentioned, to the best of our knowledge, the work published in 2016 by Álvarez-Torrellas et al. was the first to study the application of glycerol-based activated carbons as adsorbent materials [55]. The authors focused on the preparation of 3 activated carbon ($GBCM_{200}$, $GBSM_{300}$ and $GBCM_{350}$) and their application as adsorbent materials for the removal of the antibiotic compounds (flumequine and tetracycline) from aqueous solution. The adsorption of flumequine was found to be dependent on the textural properties of the glycerol-based activated carbon materials. The maximum adsorption capacity ($41.5 \ mg \cdot g^{-1}$) was verified onto sample $GBCM_{350}$. The sequence of flumequine adsorption capacity in the glycerol-activated carbon series was the following: $GBCM_{350} > GBCM_{300} > GBCM_{200}$ with adsorption capacities of $41.5 \ mg \cdot g^{-1}$, $33.7 \ mg \cdot g^{-1}$ and $0.9 \ mg \cdot g^{-1}$, respectively. For tetracycline this sequence was $GBCM_{350} > GBCM_{200} > GBCM_{300}$ ($58.1 \ mg \cdot g^{-1}$, $53.9 \ mg \cdot g^{-1}$ and $51.3 \ mg \cdot g^{-1}$, respectively). The activated carbons showed a higher adsorption capacity for tetracycline and its adsorption was almost the same for all three activated carbons, showing that the adsorption of this antibiotic was not dependent on the structural differences obtained at the different activation temperatures used. Additionally, no relation between the antibiotic structure and activated carbon properties was found.

Cui et al. [54] investigated the obtained activated carbon from liquid glycerol as adsorbent for the removal of gas phase volatile organic compounds (VOCs) and aqueous phase chromium Cr(VI). The adsorption capacities reported for toluene, hexane, and Cr(VI) were $1.5 \ g \cdot g^{-1}$, $1.1 \ g \cdot g^{-1}$, and $56 \ mg \cdot g^{-1}$, respectively. The adsorption of the compound in aqueous solutions was much lower than the VOCs, probably due to the competing adsorption of water, making the comparison among them difficult.

Naverkar et al. [58] examined the adsorption of methylene blue by glycerol-based carbons (GBC-120 and GBC-350). The samples GBC-120 and GBC-350 presented BET surface areas of 21 $m^2 \cdot g^{-1}$ and 464 $m^2 \cdot g^{-1}$, respectively. The sample (GBC-120) exhibited maximum methylene blue adsorption of 1050 $mg \cdot g^{-1}$. According to the authors, the higher equilibrium adsorption of 1050 mg^{-1} on GBC-120 was attributed to the presence of a large amount of $-SO_3H$ groups compared with GBC-350, where several surface functionalities were lost upon thermal treatment.

Gonçalves et al. [53] also investigated the adsorption of organic contaminants from water: a dye (methylene blue) and a drug (paracetamol) on glycerol-based carbons. The activated carbons from glycerol were also tested as capacitor materials (described in Section 3.3). More recently, glycerol-based magnetic carbon composites were synthesized by Medeiros et al. [59]. The composites (GFe3-600 and GFe3-800) were tested as adsorbents of dyes (methylene blue and indigo carmine). The sample GFe3-800 showed a higher adsorption capacity than GFe3-600 for methylene blue, adsorbed up to 82% and 62% in 60 min, respectively.

In 2021, Batista et al. [52] prepared a series of glycerin-activated carbons from crude glycerin (82% glycerol) for gas separation by adsorption processes. The glycerin-activated carbons were evaluated as adsorbents for the adsorption of ethane and ethylene. All the adsorbents were shown to be ethane selective. The materials exhibited a higher adsorption capacity of ethane (8.92–14.81 $mmol \cdot g^{-1}$) than ethylene (8.27–12.63 $mmol \cdot g^{-1}$). The glycerin-activated carbons (except for the sample G@700/3) after two regeneration cycles presented ~100% of the adsorption capacity. In addition, in another work, M. Batista et al. [59] used the glycerin-based activated carbon (Gta@600) and its chitosan-based carbon (Gta@600Chi) as H_2S adsorbents. The chitosan-based carbon (Gta@600Chi) presented a H_2S insignificant release due to its chemical adsorption. However, the Gta@600 adsorbed a significant amount of H_2S and it could be investigated for other applications such as natural gas purification.

The results already available in the literature clearly show activated carbons obtained from glycerol can adsorb compounds with different structures and properties (Table 2), and therefore indicate the importance to extend the research to the adsorption of other class of chemicals. Most of the studies may have applications in environmental problems. However, as was shown they may also be used for separation processes. Nevertheless, and despite their potential, more studies should be conducted, namely regeneration studies. At this moment no commercial products exist, and their development is dependent on more research to be possible obtain high effective products at low production cost. In Table 2 is presented the adsorption data on glycerol-based carbons.

3.3. Capacitors

The ability that activated carbons obtained from glycerol may have as capacitors has been investigated very recently and a brief overview of the work done so far is presented. The first reference to this possibility was in 2019 by Gonçalves et al. and is a small part of the work concerning the adsorption capacity of activated carbon described before [53]. The authors selected three activated carbons: two with larger surface areas activated differently and one with the higher micropores/mesopores ratio. Electrodes were prepared by pressing a mixture of activated, multiwalled carbon nanotubes. The more suitable activated carbon was obtained from using $ZnCl_2$ as the activating agent. It presented the higher micropores/mesopores ratio and not the largest surface. This characteristic was attributed to the more suitable pore distribution in this carbon and the higher micropores/mesopores ratio. More recently, Narvekar et al. [67] synthesized carbon from glycerol which was chemically activated with KOH at 800 °C under N_2 atmosphere for 2 h. Cyclic voltammetry studies showed the activated carbon had a much higher capacitance than the commercial carbons (Vulcan XC-72 or CNT), a fact that was attributed to the carbonyl and sulphonyl surface functionalities and large surface area with favorable pore size distribution wherein the pores are accessible to form an extended electrical double layer. More recently, Juchen et al. [68] synthesized KOH activated carbon from crude glycerol. The electrodes were prepared by mixing 90 wt% of chemically activated carbon and 10 wt% of Polyvinylidene fluoride (PVDF) in n-methyl-pyrrolidone (NMP) solvent and used for the desalination of brackish water. Figure 8 shows the results of cyclic voltammetry experiences showing the electrode capacitance, resistivity, and mass transfer effects in the desalination process. The electrodes remained stable over 50 desalination/regeneration cycles applying potentials lower than 1.2 V.

Table 2. Review of the textural properties and adsorption capacity of glycerol-activated carbon materials.

Sample	A_{BET} $(m^2 \cdot g^{-1})$	Vmicro $(cm^3 \cdot g^{-1})$	Adsorption Capacity		Reference
			Flumequine	Tetracycline	
GBCM$_{200}$	352	0.17	0.9 mmol·g^{-1}	53.9 mmol·g^{-1}	[55]
GBCM$_{300}$	391	0.19	33.7 mmol·g^{-1}	51.3 mmol·g^{-1}	
GBCM$_{350}$	436	0.22	41.5 mmol·g^{-1}	58.1 mmol·g^{-1}	
			Toluene	Hexane	Cr(VI)	
S3-steam	2470	0.80	–	–	30 mg·g^{-1}	[54]
S3-CO$_2$	1050	0.38	–	–	15 mg·g^{-1}	
P1-steam	1420	0.41	–	–	39 mg·g^{-1}	
P1-CO$_2$	1590	0.50	1.5 g·g^{-1}	1.1 g·g^{-1}	56 mg·g^{-1}	
			Methylene blue		
GBC-120	21	0.06	1050 mg·g^{-1}		[58]
GBC-350	464	0.10	139 mg·g^{-1}		
			Methylene blue	Paracetamol	
AC$_{Zn}$-847	500	–	109 mol·g^{-1}	39 mol·g^{-1}	
AC$_{Zn}$-447	680	–	151 mol·g^{-1}	88 mol·g^{-1}	
AC$_{Zn}$-425	800	–	200 mol·g^{-1}	81 mol·g^{-1}	[53]
AC$_P$-646	420	–	263 mol·g^{-1}	28 mol·g^{-1}	
AC$_P$-644	460	–	370 mol·g^{-1}	23 mol·g^{-1}	
AC$_P$-346	390	–	256 mol·g^{-1}	28 mol·g^{-1}	
			Methylene blue	Indigo carmine	
GFe3-800	136	–	80%	71%	[59]
GFe3-600	140	–	62%	30%	
			Ethane	Ethylene	
G@700/3	1564	0.69	8.98 mol·g^{-1}	8.62 mol·g^{-1}	
G@700/2	1441	0.64	13.24 mol·g^{-1}	12.63 mol·g^{-1}	
G@700/1	1166	0.63	8.92 mol·g^{-1}	8.27 mol·g^{-1}	[52]
G@800/3	2150	1.03	13.46 mol·g^{-1}	10.88 mol·g^{-1}	
G@800/2	1895	0.95	14.81 mol·g^{-1}	12.19 mol·g^{-1}	
G@800/1	1720	0.76	12.64 mol·g^{-1}	11.67 mol·g^{-1}	
			H$_2$S		
Gta@600	466	–	0.02 mol·g^{-1}		[60]
Gta@600Chi	<5	–	0.012 mol·g^{-1}		

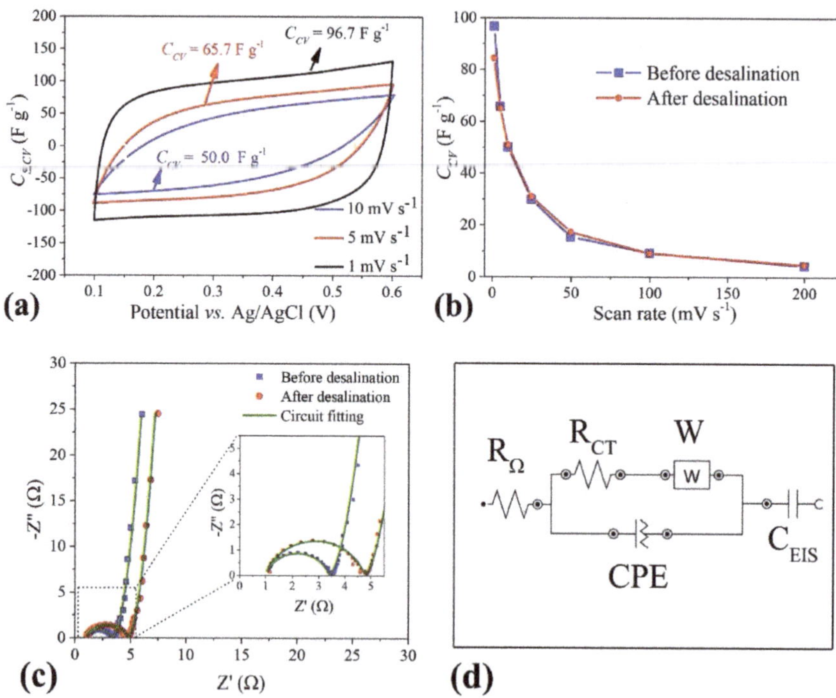

Figure 8. (**a**) Specific capacitance from cyclic voltammograms recorded at different scan rates, before desalination; (**b**) total specific capacitance, as a function of scan rate, before and after desalination applying 1.2 V; (**c**) Nyquist plots before and after desalination applying 1.2 V; (**d**) modified Randle equivalent circuit. Working and counter electrodes: PGAC. Electrolyte: 1 mol·L^{-1} NaCl. Reproduced with permission from [68], Elsevier, 2022.

4. Summary and Outlook

The amount of glycerol produced as a by-product in the biodiesel industry has been increasing. In addition, the use of waste fats (waste and residues), for sustainability reasons, by the biodiesel industry originated glycerol, which may contain unwanted compounds (contaminants). This causes this glycerol not to be used in certain applications such as food or cosmetics, because they do not have the kosher certification as demanded by the food, pharmaceutical, and cosmetic industries. This fact reinforces the need to quickly discover other applications for this glycerol and its use for the synthesis of carbons may be a solution, as may be seen by the work developed so far and their wide applications. The carbons from glycerol have been successfully used in a wide range of applications such as catalyst for a wide range of reactions such as acetylation, etherification, synthesis of substituted imidazoles, and benzamides, among others. The activated carbons have been used as adsorbent of gases (H_2S, VOCs, ethene and ethylene) and liquid (dyes and pharmaceuticals) pollutants, and capacitor materials. Nevertheless, this research is still in its initial stages in comparison with other carbons, and optimization of the synthetic procedures by changing the activated agent, temperatures, and pressure may give rise to more effective materials for a given application. Other possibilities could be surface functional group variation on the activated carbon surface, which may be achieved using different treatment parameters, or by post-synthesis modification, a possibility that has not been investigated so far. A systematic study of surface modification may help in obtaining better materials for the intended application.

Concerning practical applications, the adsorption process is the most promising for the glycerol-based active carbons. Adsorption technology is known for its simplicity, reliability, and low energy and maintenance costs, and it is already being used in many situations. The viability of this process is very dependent on the adsorbent. The use of glycerol-based activated carbons as adsorbent will depend on the possibility of producing this material using an energetic and environmentally sustainable processes.

Author Contributions: Conceptualization, M.B., S.C., J.P., M.L.P. and R.C.; methodology, M.B. and S.C.; validation, M.B. and S.C.; formal analysis, M.B. and S.C.; investigation, M.B. and S.C.; writing—original draft preparation, M.B. and S.C.; writing—review and editing, M.B., S.C., J.P., M.L.P. and R.C.; funding acquisition, J.P. and M.L.P.; M.B. and S.C. contributes equally for the manuscript. All authors have read and agreed to the published version of the manuscript.

Funding: This research was financed by Fundação para a Ciência e a Tecnologia (FCT) developed in the scope of the projects UIDB/00100/2020 (CQE), UIDB/04028/2020 & UIDP/04028/2020 (CERENA).

Data Availability Statement: No applicable.

Acknowledgments: This work has received funding from Fundação para a Ciência e a Tecnologia (FCT) in the scope of the projects UIDB/00100/2020 (CQE), UIDB/04028/2020 (CQE), IMS—LA/P/0056/2020 & UIDP/04028/2020 (CERENA). MB and SC acknowledge for FCT-Investigator contract-DL57 and PTDC/MEDQUI/28271/2017, respectively.

Conflicts of Interest: The authors declare no conflict of interest.

References

1. Mathew, G.M.; Raina, D.; Narisetty, V.; Kumar, V.; Saran, S.; Pugazhendi, A.; Sindhu, R.; Pandey, A.; Binod, P. Recent advances in biodiesel production: Challenges and solutions. *Sci. Total Environ.* **2021**, *794*, 148751. [CrossRef] [PubMed]
2. Niekurzak, M. Determining the Unit Values of the Allocation of Greenhouse Gas Emissions for the Production of Biofuels in the Life Cycle. *Energies* **2021**, *14*, 8394. [CrossRef]
3. Noureddini, H.; Zhu, D. Kinetics of transesterification of soybean oil. *J. Am. Oil Chem. Soc.* **1997**, *74*, 1457–1463. [CrossRef]
4. Porte, A.F.; Schneider, R.D.C.D.S.; Kaercher, J.A.; Klamt, R.A.; Schmatz, W.L.; da Silva, W.L.T.; Filho, W.A.S. Sunflower biodiesel production and application in family farms in Brazil. *Fuel* **2010**, *89*, 3718–3724. [CrossRef]
5. Darnoko, D.; Cheryan, M. Kinetics of palm oil transesterification in a batch reactor. *J. Am. Oil Chem. Soc.* **2000**, *77*, 1263–1267. [CrossRef]
6. Kusdiana, D.; Saka, S. Kinetics of transesterification in rapeseed oil to biodiesel fuel as treated in supercritical methanol. *Fuel* **2001**, *80*, 693–698. [CrossRef]
7. Jang, M.G.; Kim, D.K.; Park, S.C.; Lee, J.S.; Kim, S.W. Biodiesel production from crude canola oil by two-step enzymatic processes. *Renew. Energy* **2012**, *42*, 99–104. [CrossRef]
8. Folaranmi, J. Production of Biodiesel (B100) from Jatropha Oil Using Sodium Hydroxide as Catalyst. *J. Pet. Eng.* **2013**, *2013*, 430–438. [CrossRef]
9. Sharma, L.; Grover, N.K.; Bhardwaj, M.; Kaushal, I. Comparison of Engine Performance of Mixed Jatropha and Cottonseed Derived Biodiesel Blends with Conventional Diesel. *Int. J. Emerg. Technol.* **2012**, *3*, 29–32.
10. Köse, Ö.; Tüter, M.; Aksoy, H.A. Immobilized Candida antarctica lipase-catalyzed alcoholysis of cotton seed oil in a solvent-free medium. *Bioresour. Technol.* **2002**, *83*, 125–129. [CrossRef]
11. Katryniok, B.; Paul, S.; Bellière-Baca, V.; Rey, P.; Dumeignil, F. Glycerol dehydration to acrolein in the context of new uses of glycerol. *Green Chem.* **2010**, *12*, 2079–2098. [CrossRef]
12. Dosuna-Rodríguez, I.; Gaigneaux, E. Glycerol acetylation catalysed by ion exchange resins. *Catal. Today* **2012**, *195*, 14–21. [CrossRef]
13. Choi, W.J. Glycerol-Based Biorefinery for Fuels and Chemicals. *Recent Pat. Biotechnol.* **2008**, *2*, 173–180. [CrossRef] [PubMed]
14. Melero, J.; Vicente, G.; Morales, G.; Paniagua, M.; Moreno, J.; Roldán, R.; Ezquerro, A.; Pérez, C. Acid-catalyzed etherification of bio-glycerol and isobutylene over sulfonic mesostructured silicas. *Appl. Catal. A Gen.* **2008**, *346*, 44–51. [CrossRef]
15. Kong, P.S.; Aroua, M.K.; Daud, W.M.A.W. Conversion of crude and pure glycerol into derivatives: A feasibility evaluation. *Renew. Sustain. Energy Rev.* **2016**, *63*, 533–555. [CrossRef]
16. Xia, L.-Z.; Yang, M.; He, M.; Jiang, M.-Z.; Qin, C.; Wei, Z.-J.; Gao, H.-T. Food emulsifier glycerin monostearate aggravates phthalates' testicular toxicity by disrupting tight junctions' barrier function in rats. *Food Qual. Saf.* **2021**, *5*, 1–9. [CrossRef]
17. Wang, Z.W.; Saini, M.; Lin, L.-J.; Chiang, C.-J.; Chao, Y.-P. Systematic Engineering of Escherichia coli for d-Lactate Production from Crude Glycerol. *J. Agric. Food Chem.* **2015**, *63*, 9583–9589. [CrossRef]
18. Feng, X.; Ding, Y.; Xian, M.; Xu, X.; Zhang, R.; Zhao, G. Production of optically pure d -lactate from glycerol by engineered Klebsiella pneumoniae strain. *Bioresour. Technol.* **2014**, *172*, 269–275. [CrossRef]

19. Khan, A.; Bhide, A.; Gadre, R. Mannitol production from glycerol by resting cells of Candida magnoliae. *Bioresour. Technol.* **2009**, *100*, 4911–4913. [CrossRef]

20. Pagliaro, M.; Rossi, M. Glycerol: Properties and production. In *The Future of Glycerol*, 2nd ed.; Green Chemistry Series; RSC: Cambridge, UK, 2010; pp. 1–187.

21. Ayoub, M.; Abdullah, A.Z. Critical review on the current scenario and significance of crude glycerol resulting from biodiesel industry towards more sustainable renewable energy industry. *Renew. Sustain. Energy Rev.* **2012**, *16*, 2671–2686. [CrossRef]

22. Ummadisetti, C.; Rachapudi, D.N.P.; Dethala, L.A.P.D. Glycerol-based SO3H-Carbon Catalyst: A green recyclable catalyst for the chemoselective synthesis of pentaerythritol diacetals. *Eur. J. Chem.* **2014**, *5*, 536–540. [CrossRef]

23. Hejna, A.; Kosmela, P.; Formela, K.; Piszczyk, Ł.; Haponiuk, J.T. Potential applications of crude glycerol in polymer technology–Current state and perspectives. *Renew. Sustain. Energy Rev.* **2016**, *66*, 449–475. [CrossRef]

24. Hara, M.; Verkman, A.S. Glycerol replacement corrects defective skin hydration, elasticity, and barrier function in aquaporin-3-deficient mice. *Proc. Natl. Acad. Sci. USA* **2003**, *100*, 7360–7365. [CrossRef] [PubMed]

25. Milani, M.; Sparavigna, A. The 24-hour skin hydration and barrier function effects of a hyaluronic 1%, glycerin 5%, and *Centella asiatica* stem cells extract moisturizing fluid: An intra-subject, randomized, assessor-blinded study. *Clin. Cosmet. Investig. Dermatol.* **2017**, *10*, 311–315. [CrossRef] [PubMed]

26. Becker, L.C.; Bergfeld, W.F.; Belsito, D.V.; Hill, R.A.; Klaassen, C.D.; Liebler, D.C.; Marks, J.J.G.; Shank, R.C.; Slaga, T.J.; Snyder, P.W.; et al. Safety Assessment of Glycerin as Used in Cosmetics. *Int. J. Toxicol.* **2019**, *38*, 6S–22S. [CrossRef]

27. Silva, P.H.; Gonçalves, V.L.; Mota, C.J. Glycerol acetals as anti-freezing additives for biodiesel. *Bioresour. Technol.* **2010**, *101*, 6225–6229. [CrossRef]

28. Trifoi, A.R.; Agachi, P.; Pap, T. Glycerol acetals and ketals as possible diesel additives. A review of their synthesis protocols. *Renew. Sustain. Energy Rev.* **2016**, *62*, 804–814. [CrossRef]

29. Li, L.; Lee, E.S.; Nguyen, C.; Zhu, Y. Effects of propylene glycol, vegetable glycerin, and nicotine on emissions and dynamics of electronic cigarette aerosols. *Aerosol Sci. Technol.* **2020**, *54*, 1270–1281. [CrossRef] [PubMed]

30. Ooi, B.G.; Dutta, D.; Kazipeta, K.; Chong, N.S. Influence of the E-Cigarette Emission Profile by the Ratio of Glycerol to Propylene Glycol in E-Liquid Composition. *ACS Omega* **2019**, *4*, 13338–13348. [CrossRef]

31. Woodall, M.; Jacob, J.; Kalsi, K.K.; Schroeder, V.; Davis, E.; Kenyon, B.; Khan, I.; Garnett, J.P.; Tarran, R.; Baines, D.L. E-cigarette constituents propylene glycol and vegetable glycerin decrease glucose uptake and its metabolism in airway epithelial cells in vitro. *Am. J. Physiol.-Lung Cell. Mol. Physiol.* **2020**, *319*, L957–L967. [CrossRef]

32. Rarata, G.; Smętek, J. Explosives Based on Hydrogen Peroxide—A Historical Review and Novel Applications. *High-Energ. Mater.* **2016**, *8*, 56–62.

33. Hong, X.; McGiveron, O.; Kolah, A.K.; Orjuela, A.; Peereboom, L.; Lira, C.T.; Miller, D.J. Reaction kinetics of glycerol acetal formation via transacetalization with 1,1-diethoxyethane. *Chem. Eng. J.* **2013**, *222*, 374–381. [CrossRef]

34. Nanda, M.R.; Yuan, Z.; Qin, W.; Ghaziaskar, H.S.; Poirier, M.-A.; Xu, C.C. A new continuous-flow process for catalytic conversion of glycerol to oxygenated fuel additive: Catalyst screening. *Appl. Energy* **2014**, *123*, 75–81. [CrossRef]

35. Nanda, M.R.; Yuan, Z.; Qin, W.; Ghaziaskar, H.S.; Poirier, M.-A.; Xu, C.C. Thermodynamic and kinetic studies of a catalytic process to convert glycerol into solketal as an oxygenated fuel additive. *Fuel* **2014**, *117*, 470–477. [CrossRef]

36. Ahmad, M.Y.; Basir, N.I.; Abdullah, A.Z. A review on one-pot synthesis of acrylic acid from glycerol on bi-functional catalysts. *J. Ind. Eng. Chem.* **2021**, *93*, 216–227. [CrossRef]

37. Batista, M.K.S.; Mestre, A.S.; Matos, I.; Fonseca, I.M.; Carvalho, A.P. Biodiesel production waste as promising biomass precursor of reusable activated carbons for caffeine removal. *RSC Adv.* **2016**, *6*, 45419–45427. [CrossRef]

38. Bernardo, M.; Rodrigues, S.; Lapa, N.; Matos, I.; Lemos, F.; Batista, M.K.S.; Carvalho, A.P.; Fonseca, I. High efficacy on diclofenac removal by activated carbon produced from potato peel waste. *Int. J. Environ. Sci. Technol.* **2016**, *13*, 1989–2000. [CrossRef]

39. Osman, A.I.; Blewitt, J.; Abu-Dahrieh, J.K.; Farrell, C.; Al-Muhtaseb, A.H.; Harrison, J.; Rooney, D.W. Production and characterisation of activated carbon and carbon nanotubes from potato peel waste and their application in heavy metal removal. *Environ. Sci. Pollut. Res.* **2019**, *26*, 37228–37241. [CrossRef]

40. Guo, Y.; Tan, C.; Sun, J.; Li, W.; Zhang, J.; Zhao, C. Porous activated carbons derived from waste sugarcane bagasse for CO2 adsorption. *Chem. Eng. J.* **2020**, *381*, 122736. [CrossRef]

41. Ruiz, M.; Rolz, C. Activated Carbons from Sugar Cane Bagasse. *Ind. Eng. Chem. Prod. Res. Dev.* **1971**, *10*, 429–432. [CrossRef]

42. Kemp, K.; Baek, S.-B.; Lee, W.-G.; Meyyappan, M.; Kim, K.S. Activated carbon derived from waste coffee grounds for stable methane storage. *Nanotechnology* **2015**, *26*, 385602. [CrossRef] [PubMed]

43. Pagalan, E., Jr.; Sebron, M.; Gomez, S.; Salva, S.J.; Ampusta, R.; Macarayo, A.J.; Joyno, C.; Ido, A.; Arazo, R. Activated carbon from spent coffee grounds as an adsorbent for treatment of water contaminated by aniline yellow dye. *Ind. Crops Prod.* **2020**, *145*, 111953. [CrossRef]

44. Riyanto; Astuti, R.; Mukti, B.I. Simple preparation of rice husk activated carbon (RHAC) and applications for laundry and methylene blue wastewater treatment. *AIP Conf. Proc.* **2017**, *1911*, 20033. [CrossRef]

45. Sharath, D.; Ezana, J.; Shamil, Z. Production of activated carbon from solid waste rice peel (husk) using chemical activation. *J. Ind. Pollut. Control* **2017**, *33*, 1132–1139.

46. Tsai, W.; Chang, C.; Lee, S. A low cost adsorbent from agricultural waste corn cob by zinc chloride activation. *Bioresour. Technol.* **1998**, *64*, 211–217. [CrossRef]

47. Medhat, A.; El-Maghrabi, H.H.; Abdelghany, A.; Abdel Menem, N.M.; Raynaud, P.; Moustafa, Y.M.; Elsayed, M.A.; Nada, A.A. Efficiently activated carbons from corn cob for methylene blue adsorption. *Appl. Surf. Sci. Adv.* **2021**, *3*, 100037. [CrossRef]

48. Devi, B.L.A.P.; Gangadhar, K.N.; Prasad, P.S.S.; Jagannadh, B.; Prasad, R.B.N. A Glycerol-based Carbon Catalyst for the Preparation of Biodiesel. *ChemSusChem* **2009**, *2*, 617–620. [CrossRef]

49. Prabhavathi Devi, B.L.A.; Gangadhar, K.N.; Siva Kumar, K.L.N.; Shiva Shanker, K.; Prasad, R.B.N.; Sai Prasad, P.S. Synthesis of sulfonic acid functionalized carbon catalyst from glycerol pitch and its application for tetrahydropyranyl protection/deprotection of alcohols and phenols. *J. Mol. Catal. A Chem.* **2011**, *345*, 96–100. [CrossRef]

50. Mantovani, M.; Aguiar, E.M.; Carvalho, W.A.; Mandelli, D.; Gonçalves, M. Utilization of biodiesel waste for acid carbon preparation with high catalyst activity in the glycerol etherification reaction. *Quim. Nova* **2015**, *38*, 526–532. [CrossRef]

51. Gonçalves, M.; Rodrigues, R.; Galhardo, T.S.; Carvalho, W.A. Highly selective acetalization of glycerol with acetone to solketal over acidic carbon-based catalysts from biodiesel waste. *Fuel* **2016**, *181*, 46–54. [CrossRef]

52. Batista, M.; Pinto, M.L.; Carvalho, R.; Pires, J. Glycerin-based adsorbents for the separation of ethane and ethylene. *Colloids Surf. A Physicochem. Eng. Asp.* **2022**, *634*, 127975. [CrossRef]

53. Gonçalves, M.; Castro, C.S.; Boas, I.K.V.; Soler, F.C.; Pinto, E.D.C.; Lavall, R.L.; Carvalho, W.A. Glycerin waste as sustainable precursor for activated carbon production: Adsorption properties and application in supercapacitors. *J. Environ. Chem. Eng.* **2019**, *7*, 103059. [CrossRef]

54. Cui, Y.; Atkinson, J.D. Tailored activated carbon from glycerol: Role of acid dehydrator on physiochemical characteristics and adsorption performance. *J. Mater. Chem. A* **2017**, *5*, 16812–16821. [CrossRef]

55. Álvarez-Torrellas, S.; Ribeiro, R.; Gomes, H.; Ovejero, G.; García, J. Removal of antibiotic compounds by adsorption using glycerol-based carbon materials. *Chem. Eng. J.* **2016**, *296*, 277–288. [CrossRef]

56. Ribeiro, R.S.; Silva, A.M.; Pinho, M.T.; Figueiredo, J.L.; Faria, J.L.; Gomes, H.T. Development of glycerol-based metal-free carbon materials for environmental catalytic applications. *Catal. Today* **2015**, *240*, 61–66. [CrossRef]

57. Lee, D.-W.; Jin, M.-H.; Park, J.C.; Lee, C.-B.; Oh, D.-K.; Lee, S.-W.; Park, J.-W.; Park, J.-S. Waste-Glycerol-Directed Synthesis of Mesoporous Silica and Carbon with Superior Performance in Room-Temperature Hydrogen Production from Formic Acid. *Sci. Rep.* **2015**, *5*, 15931. [CrossRef]

58. Narvekar, A.A.; Fernandes, J.; Tilve, S. Adsorption behavior of methylene blue on glycerol based carbon materials. *J. Environ. Chem. Eng.* **2018**, *6*, 1714–1725. [CrossRef]

59. Medeiros, M.A.; Ardisson, J.D.; Lago, R.M. Preparation of magnetic mesoporous composites from glycerol and iron(III) salt. *J. Chem. Technol. Biotechnol.* **2020**, *95*, 1038–1045. [CrossRef]

60. Batista, M.; Pinto, M.L.; Antunes, F.; Pires, J.; Carvalho, S. Chitosan Biocomposites for the Adsorption and Release of H$_2$S. *Materials* **2021**, *14*, 6701. [CrossRef]

61. Ramesh, K.; Murthy, S.N.; Karnakar, K.; Nageswar, Y.V.D.; Vijayalakhshmi, K.; Prabhavathi Devi, B.L.A.; Prasad, R.B.N. A novel bioglycerol-based recyclable carbon catalyst for an efficient one-pot synthesis of highly substituted imidazoles. *Tetrahedron Lett.* **2012**, *53*, 1126–1129. [CrossRef]

62. Ramesh, K.; Murthy, S.N.; Karnakar, K.; Reddy, K.H.V.; Nageswar, Y.V.D.; Vijay, M.; Devi, B.P.; Prasad, R.B.N. A mild and expeditious synthesis of amides from aldehydes using bio glycerol-based carbon as a recyclable catalyst. *Tetrahedron Lett.* **2012**, *53*, 2636–2638. [CrossRef]

63. Konkala, K.; Sabbavarapu, N.M.; Katla, R.; Durga, N.Y.V.; Kumar Reddy, T.V.; Prabhavathi, P.D.; Rachapudi, B.N.P. Revisit to the Biginelli reaction: A novel and recyclable bioglycerol-based sulfonic acid functionalized carbon catalyst for one-pot synthesis of substituted 3,4-dihydropyrimidin-2-(1H)-ones. *Tetrahedron Lett.* **2012**, *53*, 1968–1973. [CrossRef]

64. Gangadhar, K.N.; Vijay, M.; Prasad, R.B.N.; Devi, B.L.A.P. Glycerol-Based Carbon-SO3H Catalyzed Benign Synthetic Protocol for the Acetylation of Alcohols, Phenols and Amines under Solvent-Free Conditions. *Green Sustain. Chem.* **2013**, *03*, 122–128. [CrossRef]

65. Gonçalves, M.; Mantovani, M.; Carvalho, W.A.; Rodrigues, R.; Mandelli, D.; Albero, J.S. Biodiesel wastes: An abundant and promising source for the preparation of acidic catalysts for utilization in etherification reaction. *Chem. Eng. J.* **2014**, *256*, 468–474. [CrossRef]

66. Pinto, R.V.; Carvalho, S.; Antunes, F.; Pires, J.; Pinto, M.L. Emerging Nitric Oxide and Hydrogen Sulfide Releasing Carriers for Skin Wound Healing Therapy. *ChemMedChem* **2022**, *17*, e202100429. [CrossRef] [PubMed]

67. Narvekar, A.A.; Fernandes, J.; Naik, S.; Tilve, S. Development of glycerol based carbon having enhanced surface area and capacitance obtained by KOH induced thermochemical activation. *Mater. Chem. Phys.* **2021**, *261*, 124238. [CrossRef]

68. Juchen, P.T.; Barcelos, K.M.; Oliveira, K.S.; Ruotolo, L.A. Using crude residual glycerol as precursor of sustainable activated carbon electrodes for capacitive deionization desalination. *Chem. Eng. J.* **2022**, *429*, 132209. [CrossRef]

 *compounds*

Article

DNA and BSA Interaction Studies and Antileukemic Evaluation of Polyaromatic Thiosemicarbazones and Their Copper Complexes

Giorgio Pelosi [1,2,3], **Silvana Pinelli** [4] **and Franco Bisceglie** [1,2,3,*]

1   Department of Chemistry, Life Sciences and Environmental Sustainability, University of Parma, 43124 Parma, Italy; giorgio.pelosi@unipr.it
2   C.E.R.T. Centro di Eccellenza per la Ricerca Tossicologica, University of Parma, 43124 Parma, Italy
3   C.O.M.T. Centre for Molecular and Translational Oncology, University of Parma, 43124 Parma, Italy
4   Department of Medicine and Surgery, University of Parma, 43126 Parma, Italy; silvana.pinelli@unipr.it
*   Correspondence: franco.bisceglie@unipr.it

Abstract: Some ten million cancer deaths occurred in 2020, highlighting the fact that the search for new anticancer drugs remains extremely topical. In the search for new coordination compounds with relevant biological properties, the choice of a metal ion is important for the design of the complex. In this regard, copper plays a peculiar role, thanks to its distinct properties. Thiosemicarbazones are, analogously, a unique class of ligands because they are easily modifiable, and therefore, extremely versatile in terms of modulating molecular properties. In this work, we synthesized and characterized, by means of X-ray diffraction, four new naphthaldehyde and anthraldehyde thiosemicarbazone derivatives and their copper complexes to be used in interaction studies with biological systems. The objective was to evaluate the antileukemic activity of these compounds. Reactions of these ligands with Cu(II) salts produced unexpected oxidation products and the isolation of Cu(I) metal complexes. One ligand and its related Cu(I) complex, which is stable in physiological conditions, were subjected to in vitro biological tests (UV-Vis and CD titration). An important interaction with DNA and an affinity toward BSA were observed in FT-IR experiments. Preliminary in vitro biological tests against a histiocytic lymphoma cell line revealed an interestingly low IC_{50} value, i.e., 5.46 μM, for the Cu(I) complex.

Keywords: thiosemicarbazone; metal complexes; DNA interactions; biological activity

Citation: Pelosi, G.; Pinelli, S.; Bisceglie, F. DNA and BSA Interaction Studies and Antileukemic Evaluation of Polyaromatic Thiosemicarbazones and Their Copper Complexes. *Compounds* **2022**, *2*, 144–162. https://doi.org/10.3390/compounds2020011

Academic Editor: Juan Mejuto

Received: 11 April 2022
Accepted: 17 May 2022
Published: 23 May 2022

1. Introduction

It was recently reported that worldwide, an estimated 19.3 million new cancer cases and almost 10.0 million cancer deaths occurred in 2020 [1]. It is therefore apparent that the search for new anticancer drugs is still extremely topical. The discovery of *cis*-{[PtCl$_2$(NH$_3$)$_2$]}, cisplatin, a platinum-based drug, was a major breakthrough in cancer treatment strategies [2]. However, the toxicity and drug resistance associated with this compound steered drug discovery research toward the rational development of metal-containing agents with more specific activity and less toxicity, and a mode of action different from *cis*-{[PtCl$_2$(NH$_3$)$_2$]} and its derivatives. Metallodrugs have been used for centuries, but only now are methods and techniques becoming available to characterize such drugs more precisely, to identify their target sites, and to elucidate their often unique mechanisms of action [3]. It is also noteworthy that a better understanding of the roles played by metal compounds at a mechanistic level will help in the implementation of new metal-based therapies by providing an alternative, targeted, and rational approach to supplement non-targeted screening of novel chemical entities for biological activity [4]. In the search for new coordination compounds with promising biological properties, the choice of the metal ion is crucial [3,4]. Amongst metal ions, copper plays an important role thanks to its distinct properties [5,6]. Copper is a bioessential element in biology with truly unique chemical characteristics in its two biologically relevant oxidation states, i.e., +1 and +2. Its most notable features are its almost exclusive function in the metabolism of O_2 or N/O compounds (NO_2^-, N_2O) and its frequent association with the oxidation/generation of organic and inorganic radicals such as tyrosyl, semiquinones, superoxide, or nitrosyl [7]. Many ligands can be chosen to bind copper and create valuable coordination complexes. Thiosemicarbazones are particularly interesting ligands because they present at least a couple of N, S donor atoms that can modulate the hard and soft character. Moreover, they can be suitably modified to increase the denticity of the ligand or the number of donor atoms, or to adjust parameters such as solubility and the partition coefficient [8–11], thereby modulating the biological activity of the compound in question. Thiosemicarbazones are a class of compounds which are known to exert different biological properties, e.g., catalysis [12] antibacterial [13,14], antifungal [15–21], antiparasitic [22], antiviral [23,24] and anticancer [25–32]. The antitumor activity provided by thiosemicarbazones is usually enhanced upon complexation. Many mechanisms of action have been proposed, including ribonucleotide reductase and topoisomerase II inhibitors, ROS generators (which, it is assumed, interact with DNA) and others which are attributed, for example, to their strong iron chelating ability [8,29]. With these hypotheses, we decided to synthesize new thiosemicarbazones and their copper complexes and perform interaction studies with in vitro biological systems in order to preliminarily evaluate their antitumor activity. As mentioned, DNA is a major target for both anticancer therapy and metal based drugs, and it is also known that polycyclic aromatic hydrocarbons tend to intercalate into DNA nitrogenous bases stackings [33]. Based on these hypotheses, analogues of naphthaldehyde and anthraldehyde thiosemicarbazone derivatives were synthesized and characterized, because naphthaldehyde [34–41] and anthraldehyde [37,42–48] have already shown interesting and promising chemical and biological properties. Structural modifications on the thiosemicarbazide terminal nitrogen, which seems to play a relevant role in its biological activity, have been investigated. Unexpectedly, reaction of the ligands with Cu(II) salts produced ligand oxidation products and the isolation of Cu(I) metal complexes. The nature of these compounds, formed upon reduction of copper, was assessed by means of X-ray crystallography. The ligand and its Cu(I) complex were subjected to biological tests (UV-Vis and CD titration) and showed important interaction with DNA which was not ascribable to intercalation. The same compounds also showed affinity toward BSA, as established by FT-IR experiments. Preliminary in vitro biological tests against a histiocytic lymphoma cell line resulted in a very low IC_{50} value, i.e., 5.46 µM, for the Cu(I) complex, highlighting the interesting behavior of this compound.

2. Materials and Methods

^{1}H NMR were recorded on a Bruker Anova spectrometer at 300 MHz, with chemical shift reported in δ units (ppm). NMR spectra were referenced relative to residual NMR solvent peaks. Coupling constants (J) are reported in hertz (Hz). The solvent used in the acquisitions of spectra was DMSO-d_6.

The FT-IR measurements were recorded on Perkin Elmer's Spectrum Two in the 4000–400 cm^{-1} range, equipped with the ATR accessory. The shapes and signal intensities are reported as w (weak), m (medium), s (strong), sh (sharp), b (broad).

Elemental analyses were performed using Flashsmart CHNS Elemental Analyzer (Thermofisher Scientific, Waltham, MA, USA).

Mass analyses were carried out using a Waters Acquity Ultraperformance ESI-MS spectrometer with Single Quadrupole Detector (Mode used: Flow Injection; Source temperature (°C) 150; Desolvation Temperature (°C) 300; Cone Gas Flow (L/Hr) 100; Desolvation Gas Flow (L/Hr) 480; Solvent Flow (mL/min) 0.2; Capillary voltage (kV) 3, Cone voltage (V) 20/50/80). The compounds were dissolved in MeOH.

Melting points were determined using a SPM3 apparatus (Stuart Scientific, Nicosia, Cyprus).

Circular dichroism spectra were recorded with a Jasco J-715 spectropolarimeter.

UV-Vis spectra were collected using Thermofisher Scientific's Evolution 260 Bio Spectrophotometer in a quartz cuvette.

The crystallographic data of compounds L_1, L_2, L_3, L_5 and $[Cu^I{}_2(SO_4)(L_2)_5]$ (**2**) were collected with a SMART APEX2 diffractometer using Mo-Kα radiation and a graphite crystal monochromator [λ(Mo-Kα) 0.71073 Å]. Intensities data for compounds L_4 and $[Cu^I(L_1)_2](HSO_4)$ (**1**) were collected on a Siemens AED diffractometer using Cu-Kα radiation [λ(Cu-Kα) 1.54178 Å]. For the data collected on the SMART APEX2 diffractometer, the SAINT [49] software was used for integrating reflection intensities and scaling, and SADABS [50] for absorption correction. The structures were solved by direct methods using SHELXS [51]] and refined by full-matrix least-squares on all F2 using SHELXL97 [52] implemented in the OLEX package [53]. The structure drawings were obtained with the ORTEPIII [54] and Mercury [55] programs.

2.1. Synthetic Procedures

2.1.1. General Information

The following compounds were used: 4-methyl-3-thiosemicarbazide, 97% (Aldrich, St. Louis, MO, USA), 4,4-dimethyl-3-thiosemicarbazide, 98% (TCI Europe N.V., Zwijndrecht, Belgium), 2-naphthaldehyde (Aldrich), 10-chloro-9-anthraldehyde, 97% (Aldrich), $Cu(SO_4)_2 \cdot 5H_2O$ (Aldrich), disodium salt of calf thymus DNA (CT DNA) (Serva), bovine serum albumin (BSA) (Aldrich).

2.1.2. Synthesis of the Ligands

L_1, 2-naphthaldehyde 4,4-dimethyl-3-thiosemicarbazone was synthesized as follows.

First, 2-naphthaldehyde (0.0826 g, 0.529 mmol) was placed in a round bottomed flask with 25 mL of absolute ethanol. Subsequently, a slightly larger amount of 4,4-dimethyl-3-thiosemicarbazide (0.0815 g, 0.684 mmol) was added to the reaction flask. A representation of the synthesis is presented in Scheme 1. The mixture was gently heated until dissolution of both reagents, and then the flask was placed in an ice bath to limit the formation of by-products for one day under magnetic stirring. The solution took on a more intense yellow color over time, until a yellow suspension formed which was filtered on Buchner and then analyzed. The analyses highlighted the purity of the product, which was then recrystallized from acetonitrile, yielding straw yellow, needle-like crystals which were subjected to X-ray diffractometric analysis (Figure 1). Crystal data details are reported in the Supplementary Materials.

R^1	R^2	Product
CH₃–	CH₃–	L_1
CH₃–	H–	L_2

Scheme 1. Representation of the syntheses of ligands L_1 and L_2.

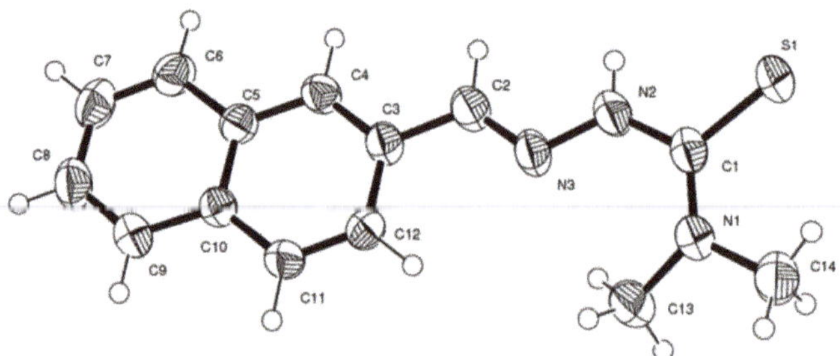

Figure 1. X-ray structure of L$_1$ with ellipsoid probability at 50%.

Yield: 52%.

^{1}H NMR (300 MHz, ppm, DMSO d$_6$): 11.05 (s, 1H, N–NH–C=S), 8.36 (s, 1H, CH=N), 7.97 (m, 5H, aromatic), 7.55 (q, 2H, aromatic), 3.33 (s, 6H, N–(CH$_3$)$_2$).

FT–IR: 3139 cm^{-1}, m, sh, ν N–H; 3008 cm^{-1}, w, broad, ν sp^2 C–H; 2980 cm^{-1}, w, broad, ν CH$_3$; 1550 cm^{-1}, s, ν C=C; 1520 cm^{-1}, s, ν C=N; 1282 cm^{-1}, s, ν C–N; 1121 cm^{-1}, m, ν C=S.

L$_2$, 2-naphthaldehyde 4-methyl-3-thiosemicarbazone was synthesized as follows.

First, 2-naphthaldehyde (0.0947 g, 0.606 mmol) was placed in a round bottomed flask with 25 mL of ethanol and a slight excess amount of 4-methyl-3-thiosemicarbazide (0.0769 g, 0.727 mmol) was added. A representation of the synthesis is reported in Scheme 1. The mixture was then gently heated to dissolve both reagents, and then the reaction flask was placed in an ice bath under magnetic stirring for three days, monitoring the reaction by means of TLC. The product was then extracted from the solution by evaporation of the solvent under vacuum and analyzed. Due to the presence of reagent impurities, the product was then subjected to a purification silica column (ethyl acetate/cyclohexane 1/5 as mobile phase). The central fraction was then dried and recrystallized from acetonitrile, yielding white crystals in the shape of rice grains, which were suitable for diffractometric analysis (Figure 2). Crystal data details are reported in the Supplementary Materials.

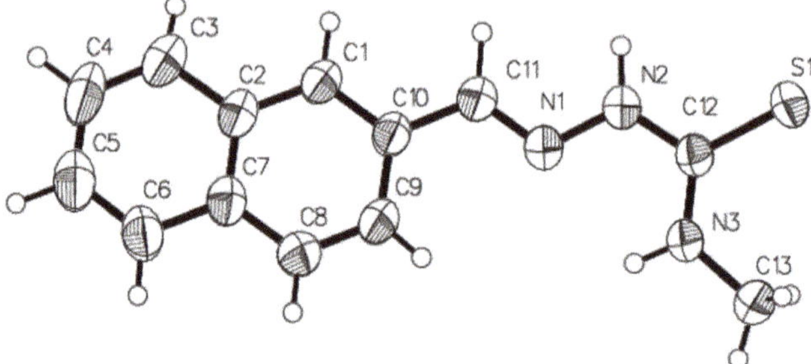

Figure 2. X-ray structure of L$_2$ with ellipsoid probability at 50%.

Yield: 61%.

^{1}H NMR (300 MHz, ppm, DMSO d$_6$): 11.58 (s, 1H, N–NH–C=S); 8.60 (d, J = 4.75 Hz, 1H, S=C–NH–CH$_3$); 8.19 (m, 2H, aromatic); 8.10 (s, 1H, CH=N); 7.95 (m, 3H, aromatic); 7.55 (q, 2H, aromatic); 3.05 (d, J = 4.57 Hz, 3H, NH–CH$_3$).

FT–IR: 3294 cm^{-1}, w, ν N–H; 3167 cm^{-1}, w, b, ν sp^2 C–H; 2980 cm^{-1}, w, b, ν CH$_3$; 1550 cm^{-1}, s, ν C=C; 1517 cm^{-1}, s, ν C=N; 1282 cm^{-1}, s, ν C–N; 1093 cm^{-1}, m, ν C=S.

L$_3$, 10-chloro-9-anthraldehyde 4,4-dimethyl-3-thiosemicarbazone was synthesized as follows.

First, 10-chloro-9-anthraldehyde (0.105 g, 0.436 mmol) was mixed with an equimolar amount of 4,4-dimethyl-3-thiosemicarbazide, in 25 mL isopropanol in a round bottomed flask. A representation of the synthesis is presented in Scheme 2. A few drops of glacial acetic acid were added and the suspension was left with magnetic stirring at a reflux temperature for 2 h. Next, the mixture was allowed to reach room temperature and left under stirring for three days during which the formation of a precipitate was observed. The solid product, in the form of a red powder, was then filtered by gravity and finally extracted three times using acetone (3 × 25 mL). From an acetone saturated solution, red-orange crystals were obtained which were suitable for X-ray diffraction (Figure 3). Crystal data details are reported in the Supplementary Materials.

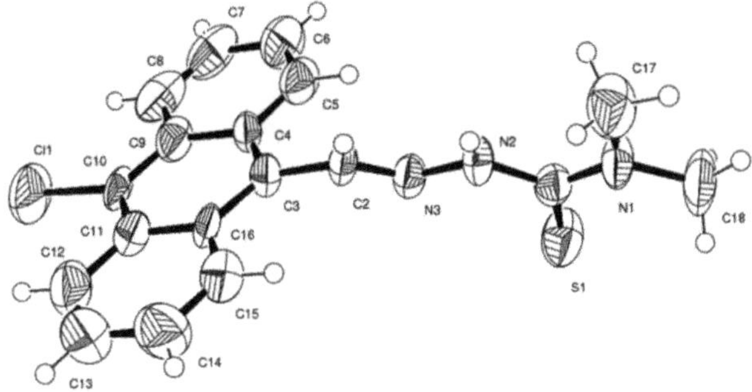

R¹	R²	Product
CH₃–	CH₃–	L₃
CH₃–	H–	L₄

Scheme 2. Representation of the syntheses of ligands L₃ and L₄.

Figure 3. X-ray structure of L₃ with ellipsoid probability at 50%.

Yield: 74%.

^{1}H NMR (300 MHz, ppm, DMSO d$_6$): 11.245 (s, 1H, N–NH–C=S), 9.47 (s, 1H, CH=N), 8.92–8.50 (m, 4H, aromatic), 7.82–7.69 (m, 4H, aromatic), 3.38 (s, 6H, N(CH₃)₂).

FT–IR: 3156 cm^{-1}, w, b, ν N–H; 2947 cm^{-1}, w, b, ν CH₃; 1545 cm^{-1}, s, broad ν C=C/ν C=N; 1258 cm^{-1}, m, ν C–N; 1061 cm^{-1}, m, ν C=S; 750 cm^{-1}, s, sh, ν C–Cl.

L₄, 10-chloro-9-anthraldehyde 4-methyl-3-thiosemicarbazone was synthesized as follows.

First, 10-chloro-9-anthraldehyde (0.2676 g, 1.11 mmol) together with 4-methyl-3-thiosemicarbazide (0.1609 g, 1.53 mmol) were placed in a 25 mL round bottomed flask. A representation of the synthesis is presented in Scheme 2. Then, 10 mL of ethyl acetate was added, a solvent in which 10-chloro-9-anthraldehyde is very soluble, while 4-methyl-3-thiosemicarbazide is relatively insoluble. The suspension was left for three days at room temperature under vigorous stirring until the formation of a pale green precipitate was observed, which was then filtered using a Buchner. To obtain crystals suitable for an XRD analysis, a small quantity of product was dissolved in a CCl_4-$CHCl_3$-CH_2Cl_2 mixture (1-1-1 *v/v*) in a test tube. Pale green square microcrystals were obtained; these were used as crystallization germs for supersaturated solutions until a crystal of suitable size was obtained (Figure 4). Crystal data details are reported in the Supplementary Materials.

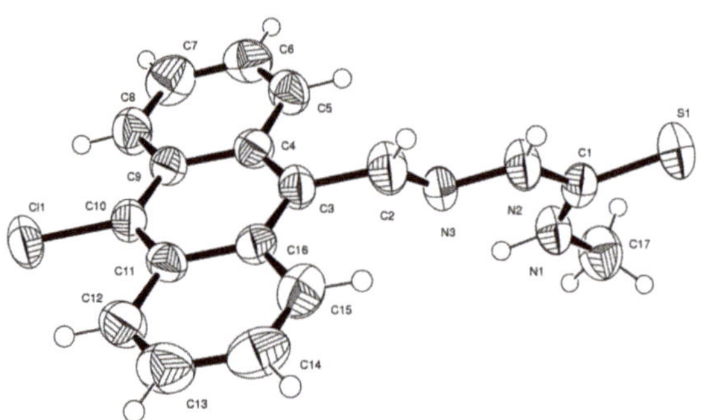

Figure 4. X-ray structure of L_4 with ellipsoid probability at 50%.

Yield: 76%.

1H NMR (300 MHz, DMSO d_6): ppm 11.78 (s, 1H, N–NH–C=S), 9.21 (s, 1H, CH=N), 8.55–8.49 (m, 4H, aromatic), 8.40 (q, J = 4.50 Hz, 1H, NH–CH$_3$), 7.82–7.70 (m, 4H, aromatic), 3.00 (d, J = 4.50 Hz, 3H, NH–CH$_3$).

FT–IR: 3342/3169 cm^{-1}, m, sh, ν N–H; 2960 cm^{-1}, w, broad, ν CH$_3$; 1550 cm^{-1}, s, ν C=C; 1517 cm^{-1}, s, ν C=N; 1238 cm^{-1}, s, ν C–N; 1079 cm^{-1}, m, ν C=S, 745 cm^{-1}, s, sh, ν C–Cl.

2.1.3. Synthesis of the Complexes

[Cu^I(L$_1$)$_2$](HSO$_4$) (1)

Ligand L$_1$ (0.0318 g 0.124 mmol) was placed in a round bottomed flask together with CuSO$_4$·5H$_2$O (0.0164 g, 0.0657 mmol) in order to obtain a 2:1 = ligand:metal stoichiometry. Then, 12 mL of 1/1 acetonitrile/methanol mixture was added. After the addition of the solvent, the solution immediately turned light brown and traces of a brown precipitate remained on the bottom. The solution was then filtered on Buchner and the mother liquors were left to evaporate slowly. Red crystals in the shape of prisms mixed with a powdery brown solid were obtained. The crystals were subjected to XRD analysis, which characterized them as a copper (I) complex, having a 2:1 = ligand:metal stoichiometry (Figure 5). Crystal data details are reported in the Supplementary Materials.

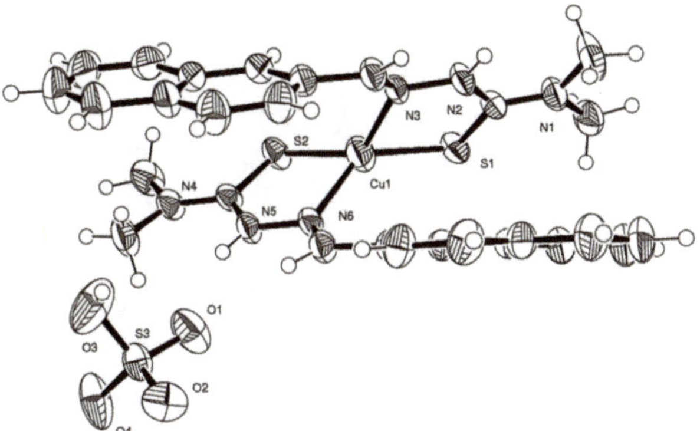

Figure 5. X-ray structure of $[Cu^I(L_1)_2](HSO_4)$ **(1)** with ellipsoid probability at 50%.

The brown powder was then dissolved in $CHCl_3$ and the product was allowed to crystallize, giving rise to crystals which were suitable for X-ray diffraction. This compound was identified as L_5 (Figure 6). Crystal data details are reported in the Supplementary Materials.

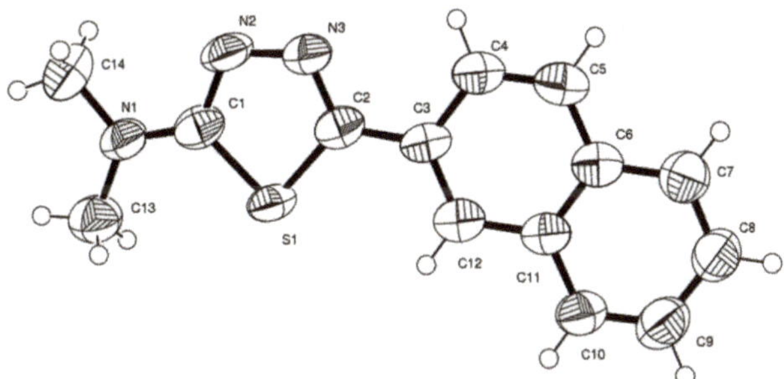

Figure 6. X-ray structure of L5 with ellipsoid probability at 50%.

Yield: 49%.

FT-IR: 3142 cm^{-1}, m, sh, ν N–H; 3005 cm^{-1}, w, broad, ν sp^2 C–H; 2984 cm^{-1}, w, broad, ν CH$_3$; 1551 cm^{-1}, s, ν C=C; 1505 cm^{-1}, m, ν C=N; 1274 cm^{-1}, m, ν C–N; 1084 cm^{-1}, m, ν C=S.

Melting point: 250 °C

$[Cu^I{}_2(SO_4)(L_2)_5]$ **(2)**

L_2 (0.0522 g, 0.215 mmol) dissolved in 10 mL of acetonitrile and $CuSO_4 \cdot 5H_2O$ (0.0287 g, 0.115 mmol), dissolved in 10 mL of methanol, were combined in a 25 mL flask. The two solutions, respectively initially colorless and light blue, once joined gave rise to a clear light green colored solution. The resulting mixture was left under magnetic stirring at room temperature for 30 min and then placed in a crystallizer to evaporate. The slow evaporation of the solvent took place over about 10 days, leaving a heterogeneous mixture, in which light yellow crystals and brown/green powdery agglomerates (L_6) were present. The crystals were identified as $[Cu^I{}_2(SO_4)(L_2)_5]$ (Figure 7).

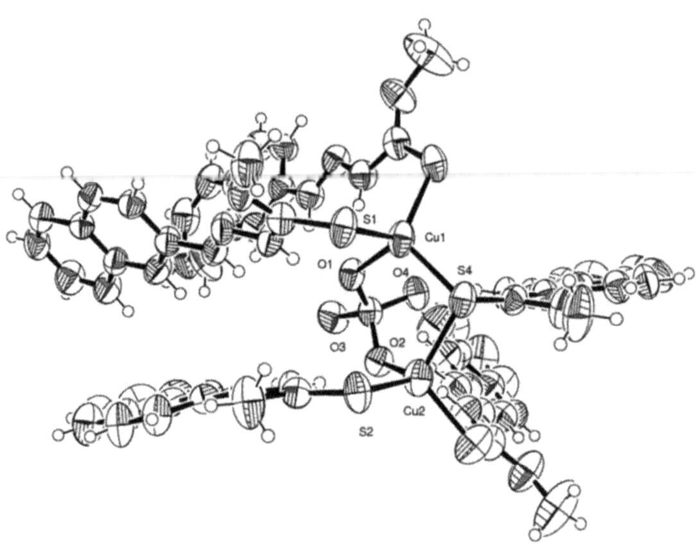

Figure 7. X-ray structure of [Cu$^{\mathrm{I}}_2$(SO$_4$)(L$_2$)$_5$] (**2**) with ellipsoid probability at 50%.

FT–IR: 3290 cm^{-1}, w, ν N-H; 3165 cm^{-1}, w, b, ν sp^2 C–H; 2983 cm^{-1}, w, b, ν CH$_3$; 1550 cm^{-1}, s, ν C=C; 1500 cm^{-1}, s, ν C=N; 1278 cm^{-1}, s, ν C-N; 1081 cm^{-1}, m, ν C=S.
Melting point: 188 °C.
L$_5$
FT–IR: 3008 cm^{-1}, w, broad, ν sp^2 C–H; 2985 cm^{-1}, w, broad, ν CH$_3$; 1552 cm^{-1}, s, ν C=C; 1520 cm^{-1}, s, ν C=N; 1275 cm^{-1}, s, ν C–N; 1125 cm^{-1}.

2.2. Measurements

L$_1$
Melting point: 172 °C.
Elemental analysis. Calculated C(65.34%) H(5.87%) N(16.33%) S(12.46%). Found: C(65.85%) H(5.54%) N(16.75%) S(12.33%).
ESI-MS: 258.15 (MH$^+$).
Soluble in: acetonitrile, toluene, DMSO, partially in ethanol and methanol.
L$_2$
Melting point: 221 °C.
Elemental analysis. Calculated: C(64.17%) H(5.39%) N(17.27%) S(13.18%). Found C(64.38) H(5.20), N(17.25), S(13.44).
ESI-MS: 244.18 (MH$^+$).
Soluble in: acetonitrile, DMSO, partly in ethanol and methanol.
L$_3$
Melting point: 204 °C.
Elemental analysis. Calculated: C(63.24%), H(4.72%), N(12.29%), S(9.38%). Found: C(63.55%), H(4.64%), N(12.40%), S(9.15%).
ESI-MS: 342.78 (MH$^+$).
Soluble in: THF, 1,4-dioxane, DMSO, partially in acetone and chloroform.
L$_4$
Melting point: 238 °C. (A decomposition into a glassy dark red substance occurs around 220 °C which then melts normally at the indicated temperature).
Elemental analysis. Calculated: C(62.28%), H(4.30%), N(12.82%), S(9.78%). Found C(62.43%), H(4.28%), N(12.58%), S(9.66%).
ESI-MS: 328.48 (MH$^+$).

Soluble in: THF, 1,4-dioxane, partially in acetone/methanol (1/1 *v/v*), chloroform, very weakly in pure alcohols.

L_5

Melting point: 180 °C

Elemental analysis. Calculated C(65.85%) H(5.13%) N(16.46%) S(12.56%). Found: C(65.63%) H(5.24%) N(16.37%) S(12.34%).

ESI-MS: 256.40 (MH$^+$).

$[Cu^I(L_1)_2](HSO_4)$ **(1)**

Elemental analysis. Calculated C(49.80%) H(4.63%) N(12.44%) S(14.24%). Found: C(50.08%) H(4.95%) N(12.35%) S(14.48%).

ESI-MS: 578.10 (M$^+$-HSO$_4{}^-$).

Circular dichroism (CD) spectra were recorded at 25 °C with buffer compensation. Each spectrum is the average of three independent measurements. Cuvettes with a 1-cm path-length quartz were used. CT-DNA was used as received and stored at 4 °C. Solutions of DNA in 10 mM of PBS (pH = 7.4) 137 mM NaCl, 2.7 mM KCl gave a ratio of UV absorbance at 260 and 280 nm, A260/A280, of 1.9, indicating that the DNA was sufficiently free of protein. The concentration of stock solutions of DNA, expressed in moles of nucleotide phosphate [NP] was determined by UV absorbance at 260 nm. The extinction coefficient, ε_{260}, was taken as 6600 M^{-1} cm^{-1} [56]. Stock solutions were stored at 4 °C and used after no more than four days. Tested compounds were dissolved in DMSO. The final concentration of DMSO in the buffered solution never exceeded 5%. Ligand L_1 and its copper complex $[Cu^I(L_1)_2](HSO_4)$ **(1)** were subjected to analyses. The effect of the ligand and its complex on the conformation of the DNA secondary structure, explored with CD, was studied, keeping the CT-DNA concentration constant at 4.5×10^{-5} M for the ligand and at 5×10^{-5} M for its copper complex. The concentration of the studied molecules in the 10 mL buffer solution of PBS (pH = 7.4) varied following the ratio r = [ligand or complex]/[DNA] = 0, 0.1, 0.2, 0.4, 0.6. The spectrum of CT-DNA and those with the added substances were monitored in the 220–320 nm range. Each spectrum is the average of 3.

Interactions of compounds with CT-DNA were also investigated by means of UV/Vis titrations in order to find the binding constants. Measured volumes of the ligand and, separately, its complex at known concentrations were added to the PBS solution containing different amounts of DNA.

The titrations took place at room temperature with an incubation time of approximately 4 h at 37 °C. Changes in absorbance were monitored at absorption maxima of 313 nm for the ligand and 393 nm for the complex.

Binding constants for the interaction of the studied compounds with nucleic acid were determined as described in [33] by means of UV–vis titrations. The intrinsic binding constant Kb for the interaction of the compounds under study with CT-DNA was calculated by absorption spectra titration data using the following equation:

$$1/\Delta\varepsilon_{ap} = 1/(\Delta\varepsilon KbD) + 1/\Delta\varepsilon \tag{1}$$

where $\Delta\varepsilon_{ap} = |\varepsilon_A - \varepsilon_f|$, $\Delta\varepsilon = |\varepsilon_B - \varepsilon_f|$, D = [DNA], and ε_A, ε_B, and ε_f are the apparent, bound, and free extinction coefficients of the compound, respectively. The constant Kb is given by the ratio of the slope to intercept when it is reported in plot form [DNA]/$(\varepsilon_A - \varepsilon_f)$ versus [DNA], and it is expressed as M^{-1}. The previous equation, originally used to calculate the binding constants for hydrophobic derivatives, is now broadly used to investigate a wide variety of metal complexes containing phenanthroline and its derivatives, and has subsequently been adopted to evaluate binding constant values from metal complexes with different ligands [34,57–62]. Fixed amounts of the ligands and complexes were dissolved in DMSO because their high solubility in this solvent allowed us to prepare concentrated solutions, and therefore, to utilize reduced volumes in titrations. It was also verified that the DMSO percentage added to the DNA solution did not interfere with the nucleic acid; in fact, the 260 nm absorption band was not subject to modifications in intensity and position. Calculated amounts of stock solutions were taken to final concentration values of 10 mM of PBS, and increasing amounts of DNA over a range of ratios r = [DNA]/[complex] = 0, 0.5, 1, 1.5, 1.9. The final concentration of the ligand was kept constant at 2.57×10^{-5} M, while that of the complex was kept constant at 3.10×10^{-5} M. The changes in absorbance of an intraligand (IL) band upon each addition of DNA were monitored at the maximum wavelengths 313 for L_1, and 393 nm for $[Cu^I(L_1)_2](HSO_4)$ (**1**).

Drug effects on cell viability were analyzed using a 3-(4,5-dimethylthiazol-2-yl)-2,5-diphenyltetrazolium bromide (MTT) colorimetric assay, based upon the ability of metabolically active cells to reduce MTT into formazan by the action of mitochondrial dehydrogenases. U937 (150,000 mL^{-1}) cells were seeded into 96-well plates overnight, and then exposed to compounds at indicated concentrations. At the end of the treatment, MTT was added (final concentration 0.5 mg $mL \cdot mL^{-1}$) for 3 h at 37 °C and formazan crystals were dissolved in 100 mL for each well of acidic isopropanol (0.08 N HCl). After mixing, absorbance was evaluated using a Multiskan Ascent microwell plate reader equipped with a 550 nm filter (Thermo Labsystems, Helsinki, Finland). At least three independent experiments were performed with eight replicate wells per sample. The percentage of cell viability was calculated using the following equation: % viability = (Mean OD_{sample})/(Mean OD_{blank}) × 100. The half maximal inhibitory concentration (IC_{50}) was determined as the concentration resulting in 50% cell growth reduction compared with untreated control cells (Figure S1).

3. Results and Discussion

3.1. Synthetic Comments

During the synthesis of L_2 an alternative strategy was attempted that produced an already pure product obtained in a single step but with lower yields (55%), by mixing a stoichiometric quantity of reagents in ethanol with a few drops of acetic acid and heating the mixture at reflux temperature for 24 h. The product precipitates and can be filtered.

A second procedure was attempted also for the synthesis of L_4, that gave better results in terms of speed and yield (90%). Equimolar amounts of reagents in absolute ethanol at reflux temperature were placed in a round bottomed flask, together with a few drops of glacial acetic acid as catalyst. The aldehyde is insoluble as well as the product, which was obtained in the form of a pale green powder that was filtered on Buchner within 4 h.

The brown/green powder agglomerates, named L_6, obtained while synthesizing $[Cu^I_2(SO_4)(L_2)_5]$ (**2**), were subject to IR, AE and MS analyses but it was not possible to clearly identify their nature. Different attempts were made to redissolve the heterogeneous powdery system in an acetonitrile/methanol mixture. The most of the product was solubilized, but a dark brown precipitate remained on the bottom of the beaker; the two phases were then separated by gravity filtration. The soluble part was placed in a crystallizer to evaporate while an IR was made of the insoluble part, which showed that it was mostly composed of $CuSO_4$ and traces of the ligand. The soluble part was dried in a few days, and the IR spectrum was very similar to both the spectra of the crystals and of the agglomerates. An elemental analysis was performed on it but the result was not perfectly congruent with none of the hypotheses proposed. Neither MS could help.

The major surprises raised from the syntheses of the copper complexes with L_1 and L_2 ligands that produced unexpected but reproducible results. It is noteworthy that thanks to X-ray diffractometry it has been possible to understand the nature of the isolated pure products. By reacting the naphthyl derivatives with the Cu(II) salt, in fact, only Cu(I) complexes were isolated. Unfortunately, it was not possible to isolate the byproduct L_6 in a crystal form apt for X-ray diffractometry. Also its characterization by means of spectroscopic, spectrometric and elemental analysis did not allow to get a stoichiometry of the compound that contain the oxidation products associated to the reduction of the complex. Likely, it contains a mixture of oxidized ligand together with a part of a Cu(II) complex, but unfortunately it was not possible to separate the different species present in the mixture. Also in the case of the syntheses of the complexes with the anthraldehyde derivatives, it could not be possible to isolate the products probably because of the formation of a mixture of different redox products promoted by the electron withdrawing behaviour of the chloroanthraldehyde that could not be properly separated.

The red $[Cu^I(L_1)_2](HSO_4)$ (**1**) crystals exposed to air, do not undergo to a re-oxidation process and therefore are kinetically stable. The same crystals dissolved in a methanol/acetonitrile = 1/1 solution, seem stable for a period of *circa* 3 days, but with the slow evaporation of the solvent and the concentration of the mixture, it can be observed the formation of the red $[Cu^I(L_1)_2](HSO_4)$ (**1**) crystals together with the green/brown powder associated to L_5. This leads to the conclusion that the copper (I) complex in crystalline form is kinetically stable in air. In solution, in the presence of oxygen for a medium/long period of time, $[Cu^I(L_1)_2](HSO_4)$ (**1**) partially degrades. The exact degradation mechanism is still unknown.

3.2. Description of the Structures

For what L_1 and L_2 is concerned, from the X-ray analysis it can be seen that the molecule exhibits an E conformation around the C1-N2 bond (Figure 1) as found in other uncoordinated thiosemicarbazones.

The mean planes of the thiosemicarbazone group and that of the naphthyl group form a dihedral angle of 15.9°. Also L_2 shows the same behaviour (Figure 2).

On the contrary, the L_3 ligand shows a Z conformation around the C1-N2 (Figure 3), therefore the sulfur and the iminic nitrogen atoms are already in the proper position for chelation. The mean planes of the thiosemicarbazone group and that of the anthracene aromatic structure form a dihedral angle of about 70°.

For what L_4 is concerned, if the dihedral angle is still of about 70°, the conformation around the C1–N2 bond is E instead (Figure 4).

The X-ray analysis of the compound obtained by reaction of L_1 with Cu(II) ions revealed the copper complex in its reduced oxidation state. The structure of $[Cu^I(L_1)_2](HSO_4)$ (**1**) (Figure 5) is formed by a Cu(I) ion which coordinates two molecules of non-deprotonated S,N bidentate ligand to form a distorted tetrahedron. The positive charge of the Cu(I) ion is neutralized by the negative charge of a hydrogensulfate ion. The coordination distances are Cu1-S1 = 2.252(1) Å, Cu1-S2 = 2.246(1) Å, Cu1-N3 = 2.153(2) Å, Cu1-N6 = 2.150(3) Å. The ligand molecule is strongly distorted due to the coordination. In fact the dihedral angle between the mean plane of thiosemicarbazide and that of naphthalene is 50.9°. The packing is determined by hydrogen bonds between the hydrazine nitrogen atoms and two oxygen atoms of the hydrogen sulfate ion (N5-H.....O1e N2-H.....O4 (x − 1, y, z) with the value of 2.885(5) Å) and by a short hydrogen bond O3-H . . . ..O3 (−x + 1, −y + 1, −z + 1) between two hydrogen sulfate ions of only 2.615(5) Å.

The oxidation product isolated L_5 (Figure 6) has a thiadiazole aromatic heterocyclic ring coplanar with the naphthyl moiety. The bond distances are within the mean values observed in similar structures.

Comparing the dimethylated derivative with the monomethylated one, the structure of the corresponding copper complex $[Cu^I{}_2(SO_4)(L_2)_5]$ (**2**) (Figure 7 and Figure S2), dramatically changes. In fact, the crystalline structure is formed by two independent copper atoms in the oxidation state +1 and by five ligand molecules that behave as monodentates through the sulfur atom. One of the five ligands acts as a bridging ligand through the sulfur atom between two Cu(I) ions. All the ligand molecules are neutral and the positive charge of the complex is neutralized by the sulfate group arranged as a bridge between the two copper atoms. Each Cu(I) therefore has a tetrahedral coordination determined by three sulfur atoms of the ligands (one of which is a bridge) and by an oxygen atom of the sulfate group. The coordination distances are respectively Cu1-S: 2.308(2), 2.305(2), 2.276(2) Å; Cu1-O1: 2.281(2) Å and Cu2-S: 2.349(2), 2.271(2), 2.265(2) Å; Cu2-O2: 2.293(2) Å. The five molecules of the ligand, unlike what has been observed for the $[Cu^I(L_1)_2](HSO_4)$ (**1**) complex, are markedly planar, in fact the dihedral angles between the thiosemicarbazidic group and the naphthalene plane in the five ligands are 3.51°, 8.76°, 11.63°, 11.69° and 15.66°. The crystalline structure contains three water molecules of crystallization that contribute to the packing.

Beside the reduction of copper, an oxidation of the ligands occurred that in the case of the L_1 derivative could also be detected and clarified by means of X-ray diffraction (L_5, Figure 6). The oxidation of the thiosemicarbazone leading to a 1,3,4-thiadiazole derivative promoted by metal salts is an already known process [63–65]. Usually, this kind of oxidation is promoted by Fe(III) as metal ion, while Cu(II) has been observed to produce only the 1,2,4-triazolines species [65]. It has also been observed that very different results—both in terms of reaction regiochemistry and yields, and in terms of reaction kinetics—can be achieved by the use of different salts, depending on the hardness/softness of the metal cation, as well as on its intrinsic oxidizing strength. The formation of the 1,3,4-thiadiazole ring seems to be induced by an electrophilic attack of the metal cation as a Lewis acid on the imine nitrogen atom, followed by the ring closure step, and finally by the metal reduction and-deprotonation step [66]. Also the role of the substituent plays a fundamental role in the oxidation process [67].

3.3. DNA and BSA Interaction Studies and Cell Viability Assay

Among the synthesized compounds, we decided to carry out biological testing only on the L_1 ligand and its $[Cu^I(L_1)_2](HSO_4)$ (**1**) complex, since the main interest of our research was to verify whether complexation with copper could significantly improve the effect of the ligand on the biological substrate. The $[Cu^I_2(SO_4)(L_2)_5]$ (**2**) complex had to be excluded because it proved to be poorly soluble and stable in the biological medium and the anthraldehyde derivatives were also not taken into consideration due to the impossibility to isolate their well characterized copper complexes.

As a first biological approach, we started by studying if these compounds could give direct interactions with DNA. To establish in detail whether the interaction of the molecules under study leads to a significant conformational change of the DNA double helix, Circular Dichroism (CD) spectra were recorded as the compound/CT-DNA ratio increased. The CD spectrum observed for calf thymus DNA consists of a positive band at 280 nm (UV: λ max, 260 nm) due to the nitrogenous base stacking and a negative band at 250 nm, due to helicity. This behaviour is characteristic of right-handed DNA in B form. It is known that at wavelengths over 230 nm, the CD spectrum of DNA in its B form consists of a positive band at longer wavelengths and a negative band at shorter wavelengths of almost equal magnitude with the point of intersection at maximum absorption [31]. As shown in Figure 8 (left) regarding the L_1 ligand, the CD spectrum of DNA shows a moderate monotone decrease in the band to ca. 250 nm with a blueshift of a few nm.

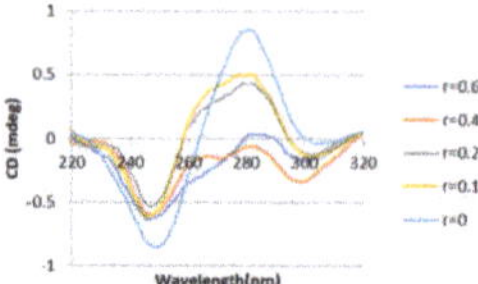
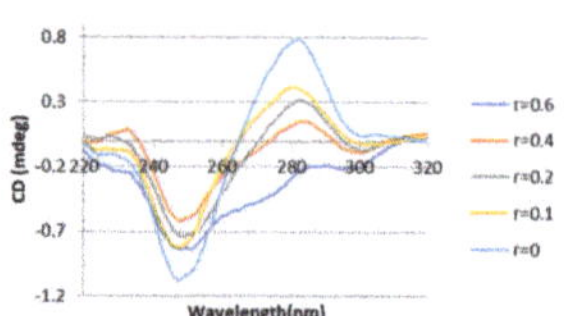

Figure 8. Circular dichroism titration of CT-DNA with L_1 (**left**) and $[Cu^I(L_1)_2](HSO_4)$ (**1**) (**right**) at different *r* values (*r* = [compound]/[DNA]).

The most important feature is observed for the positive band at approx. 280 nm. Usually, a slight interaction with the DNA groove and electrostatic interactions with small molecules show small perturbations or even no interaction on the bands attributed to base stacking or helicity [68]. In our case, both the ligand and its complex reveal an interaction with the nucleic acid. For both compounds, the observed behaviour at circular dichroism is consistent with a possible conformational change of DNA from B to C [69]. This aspect is more marked in the case of the ligand in which the positive band becomes practically absent as the concentration increases. For both compounds, the absence of isodichroic points also suggests the simultaneous presence of more than one mode of interaction with respect to DNA.

This last aspect could also justify the different spectral behaviour of the negative band, especially when the DNA interacts with the complex (Figure 8, right). In fact, in the CD spectrum of DNA in form C the negative band is similar in position and size to that of form B. In our case, the slightest deviation in the case of the interaction with the ligand and the more marked one in the case of the complex could lead us to think to an interaction of the condensed aromatic system with the nitrogenous bases of DNA. In conclusion, the results of the circular dichroism studies are indicative of a conformational change of the DNA double helix following the interaction of the DNA macromolecule with the tested compounds, and an intercalating interaction can be excluded.

From the data obtained from Figure 9; Figure 10 the constants Kb have been obtained, which are 1.3×10^4 M^{-1} and 2.1×10^2 M^{-1} for the ligand and the complex respectively. These values are significantly lower than those found for classical intercalators (for example ethidium bromide has a Kb of 1.4×10^6 in 25 mM Tris-HCl/40 mM NaCl buffer, pH = 7.8 [70]), and indicate that the compounds studied have a low affinity for DNA.

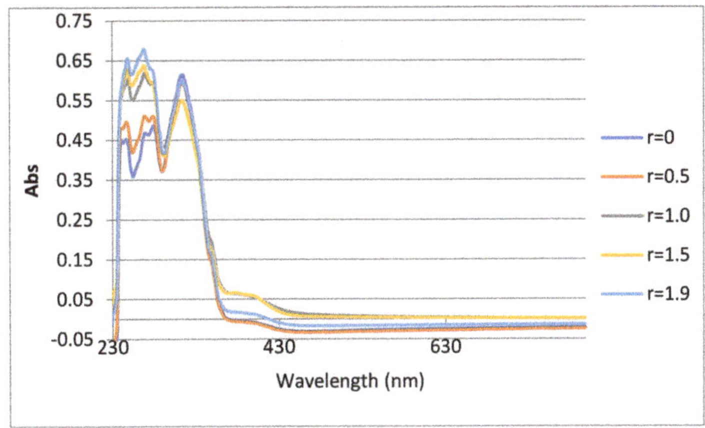

Figure 9. UV–vis spectra of CT-DNA treated with L$_1$ at different [DNA]/[ligand] ratios.

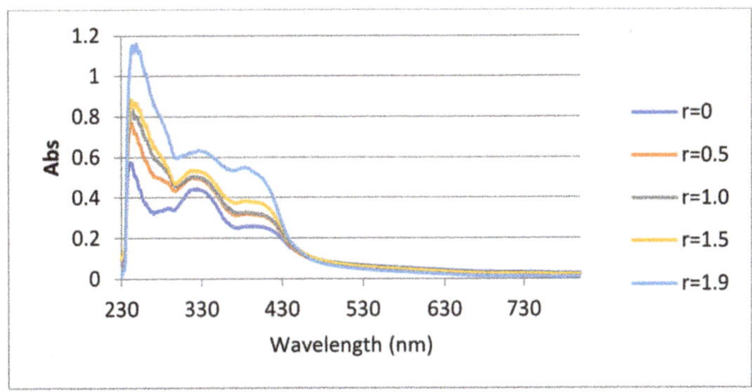

Figure 10. UV–vis spectra of CT-DNA treated with [Cu^I(L$_1$)$_2$](HSO$_4$) (1) at different [DNA]/[complex] ratios.

It is very interesting to note that the metal complex binds to DNA with less affinity than the free ligand. This data can be justified by the ionic character of the complex which could exert a greater electrostatic affinity towards the negatively charged nucleic acid. Furthermore, the overall structure of the compound strongly deviates from the planarity, as verified by X-ray diffraction, accentuating the distortions already present in the ligand (clearly visible in the packing that both molecules assume in the crystalline phase), making the aromatic condensed moiety less accessible to a stacking with the nitrogenous bases.

Hypochromism and red shift (bathochromism) in the absorption spectra of DNA bound to different compounds is generally attributed to intercalation, involving a strong stacking between the aromatic chromophores and the DNA base pairs. In our case this phenomenon is observed only for the ligand and in any case up to values of $r = 1.5$. This behaviour shows that a partial intercalation cannot be excluded for the ligand. In any case, this phenomenon is strongly dependent on the concentration. With the increase of the dilution of the ligand with respect to DNA, weaker interactions prevail which justify the not excessively high Kb values and also confirm the possibility that the ligand has to act in different ways against DNA, as also already verified by CD experiments. In the case of the complex, a marked hyperchromism is observed. This phenomenon, not yet well understood, suggests an external mode of action towards DNA, determined more by electrostatic interactions [71] and by the possibility to form a great network of hydrogen bonds. Since an hypochromic increase is normally associated with an increase in hydrophobic character, again this is in agreement with the ionic nature of the complex [72]. In conclusion, with our UV experiments we observe a strong interaction with the nucleic acid, as already verified in the CD experiments. However, the Kb values and the spectral behaviour exclude an intercalative action for the complex, behaviour that can only be partially provided by the ligand.

To better understand if the ligand or its complex could have a higher affinity for cellular proteins than for DNA, FT-IR spectra of bovine albumin (BSA) titrated with the ligand and the complex were recorded and reported in Figure 11.

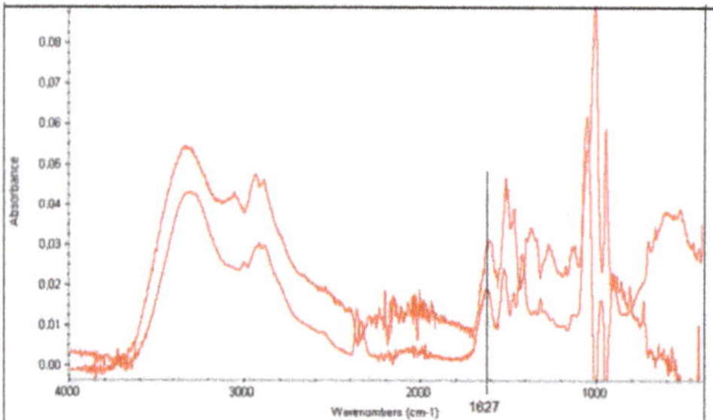
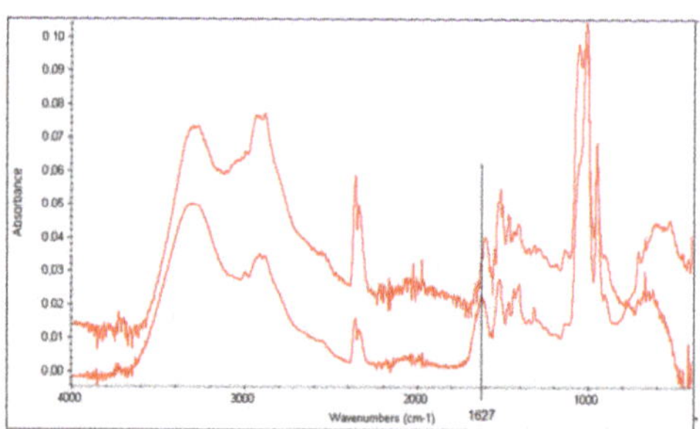

Figure 11. IR spectra of BSA alone (lower spectrum) and after incubation with L_1 (**left**) and $[Cu^I(L_1)_2](HSO_4)$ (**1**) (**right**) (higher spectrum).

The studies were performed following the procedure reported in the literature [73]. The infrared spectra of BSA, BSA with the compound (5 μM solution) in a 1:1 molar ratio and the compound alone were recorded after evaporating the solvent. The BSA-compound solutions were incubated for 24 h prior to measurement. The absorbance of the buffer was subtracted from the spectra of the solutions. Then, difference spectra were calculated using the instrument's software package.

The infrared spectra of proteins [73] show the amide I bond (attributed to the C=O stretching of the functional group of the peptide moiety) between 1600 and 1700 cm^{-1}. The exact position is determined by the backbone conformation. Since the band is associated with the secondary structure of the protein, perturbations of the bond on interaction with the tested compounds can provide a qualitative assessment of the binding, assuming that the binding induces a conformation change. As can be seen from Figure 11, the considered peak undergoes a red shift of about 20 cm^{-1} (1606 cm^{-1} for L_1, 1605 cm^{-1}

for $[Cu^I(L_1)_2](HSO_4)$ (**1**)). Since this peak is associated with the secondary structure of the protein, a significant perturbation of this type probably indicates that both the ligand and the complex have an affinity for BSA to induce a modification in its conformation. Although there could be other explanations, it is probably the thiosemicarbazone part of the compounds that interacts, since there is no large shift difference in the spectra of BSA treated with the ligand or with the complex.

The last biological test that has been performed is the study on tumour cell viability. The U937 cell line, originally established from a histiocytic lymphoma, has been widely used as a powerful in vitro model for hematological studies [74]. In our experiments, we found no activity exerted by the ligand alone on the cell line, while its complex $[Cu^I(L_1)_2](HSO_4)$ (**1**) showed an important inhibition on cell viability, reaching remarkably low IC_{50} value of 5.46 μM, considering that on the same cell line, the most famous metal based antitumor drug, *cis*-{$[PtCl_2(NH_3)_2]$}, owns an IC_{50} value of 3 μM [75].

4. Conclusions

A set of four thiosemicarbazone ligands derived from naphthaldehyde and anthraldehyde have been synthesized and characterized also X-ray diffraction. The reactivity of the ligands toward Cu(II) ions revealed unexpected behavior. Cu^{2+} was able to oxidize a part of the naphthaldehyde thiosemicarbazones to a 1,3,4-thiadiazole derivative; this was unusual, given that this reactivity is usually associated with Fe^{3+} ions and not with other redox metal ions. The remaining naphthaldehyde thiosemicarbazone molecules present in solution yielded a Cu(I) complex. The anthraldehyde thiosemicarbazones were even more reactive towards Cu(II) ions and were oxidized to multiple species (not clearly identifiable) that probably contained not only 1,3,4-thiadiazole derivatives, but also 1,2,4-triazolines derivatives. Between the two Cu(I) complexes isolated, only $[Cu^I(L_1)_2](HSO_4)$ (**1**) was stable under the tested conditions, and moreover, did not undergo reoxidation, at least for the duration of the biological tests. Therefore, it was selected, together with its parent ligand, for further biological investigations. Both the ligand and its complex revealed an interaction with DNA. For both compounds, the observed circular dichroism behavior was consistent with a possible conformational change of DNA from B to C, and with the simultaneous presence of more than one mode of interaction. The UV experiments confirmed multiple ways for both compounds to interact with DNA, excluding complex intercalation as the prevalent interaction, and suggesting an external mode of action toward DNA, as determined by electrostatic interactions and by the formation of a great network of hydrogen bonds. Both the ligand and the complex had an affinity for BSA; this was probably due to the thiosemicarbazone moiety of the compounds that interacted with the protein, since there was no significant difference between the ligand and the complex behavior. The most interesting outcome of this research is that complex $[Cu^I(L_1)_2](HSO_4)$ (**1**) showed an important inhibition effect on U937 leukemic cell line viability, reaching a very low IC_{50} value of 5.46 μM, a property that is absent in the ligand alone.

Supplementary Materials: The following supporting information can be downloaded at: https://www.mdpi.com/article/10.3390/compounds2020011/s1, Crystal data and Cell viability curve after complex treatment. Figure S1: U937 cell viability curve after treatment with $[CuI(L_1)_2](HSO_4)$ (**1**) at different concentrations; Figure S2: Representation of complex $[CuI_2(SO_4)(L_2)_5]$ (**2**).

Author Contributions: Conceptualization, G.P. and F.B.; Data curation, F.B.; Formal analysis, G.P., S.P. and F.B.; Methodology, F.B.; Supervision, G.P. and F.B.; Writing—original draft, F.B.; Writing—review & editing, G.P. All authors have read and agreed to the published version of the manuscript.

Funding: This research was funded by UNIPR Fil.

Institutional Review Board Statement: Not applicable.

Informed Consent Statement: Not applicable.

Data Availability Statement: CCDC 2164721-2164726 contain the supplementary crystallographic data. These data can be obtained free of charge via http://www.ccdc.cam.ac.uk/conts/retrieving.html (accessed on 6 April 2022), or from the Cambridge Crystallographic Data Centre, 12 Union Road, Cambridge CB2 1EZ, UK; Fax: (+44)-1223-336-033, or E-mail: deposit@ccdc.cam.ac.uk.

Conflicts of Interest: The authors declare no conflict of interest.

References

1. Sung, H.; Ferlay, J.; Siegel, R.L.; Laversanne, M.; Soerjomataram, I.; Jemal, A.; Bray, F. Global Cancer Statistics 2020: GLOBOCAN Estimates of Incidence and Mortality Worldwide for 36 Cancers in 185 Countries. *CA Cancer J. Clin.* **2021**, *71*, 209–249. [CrossRef] [PubMed]
2. Kelland, L. The Resurgence of Platinum-Based Cancer Chemotherapy. *Nat. Rev. Cancer* **2007**, *7*, 573–584. [CrossRef] [PubMed]
3. Anthony, E.J.; Bolitho, E.M.; Bridgewater, H.E.; Carter, O.W.L.; Donnelly, J.M.; Imberti, C.; Lant, E.C.; Lermyte, F.; Needham, R.J.; Palau, M.; et al. Metallodrugs Are Unique: Opportunities and Challenges of Discovery and Development. *Chem. Sci.* **2020**, *11*, 12888–12917. [CrossRef] [PubMed]
4. Boros, E.; Dyson, P.J.; Gasser, G. Classification of Metal-Based Drugs According to Their Mechanisms of Action. *Chem* **2020**, *6*, 41–60. [CrossRef]
5. More, M.S.; Joshi, P.G.; Mishra, Y.K.; Khanna, P.K. Metal Complexes Driven from Schiff Bases and Semicarbazones for Biomedical and Allied Applications: A Review. *Mater. Today Chem.* **2019**, *14*, 100195. [CrossRef]
6. Denoyer, D.; Clatworthy, S.A.S.; Cater, M.A. Copper Complexes in Cancer Therapy. In *Metallo-Drugs: Development and Action of Anticancer Agents*; Sigel, A., Sigel, H., Freisinger, E., Sigel, R.K.O., Eds.; De Gruyter: Berlin, Germany; Boston, MA, USA, 2018; Volume 18, pp. 469–506.
7. Kaim, W.; Rall, J. Copper—A"Modern" Bioelement. *Angew. Chem. Int. Edit.* **1996**, *35*, 43–60. [CrossRef]
8. Pelosi, G. Thiosemicarbazone Metal Complexes: From Structure to Activity. *Open Crystallogr. J.* **2010**, *3*, 16–28. [CrossRef]
9. Kostas, I.D.; Steele, B.R. Thiosemicarbazone Complexes of Transition Metals as Catalysts for Cross-Coupling Reactions. *Catalysts* **2020**, *10*, 1107. [CrossRef]
10. Bisceglie, F.; Pinelli, S.; Alinovi, R.; Tarasconi, P.; Buschini, A.; Mussi, F.; Mutti, A.; Pelosi, G. Copper(II) Thiosemicarbazonate Molecular Modifications Modulate Apoptotic and Oxidative Effects on U937 Cell Line. *J. Inorg. Biochem.* **2012**, *116*, 195–203. [CrossRef]
11. Bisceglie, F.; Tavone, M.; Mussi, F.; Azzoni, S.; Montalbano, S.; Franzoni, S.; Tarasconi, P.; Buschini, A.; Pelosi, G. Effects of Polar Substituents on the Biological Activity of Thiosemicarbazone Metal Complexes. *J. Inorg. Biochem.* **2018**, *179*. [CrossRef]
12. Priyarega, S.; Haribabu, J.; Karvembu, R. Development of Thiosemicarbazone-Based Transition Metal Complexes as Homogeneous Catalysts for Various Organic Transformations. *Inorg. Chim. Acta* **2022**, *532*, 120742. [CrossRef]
13. Egorova, A.; Jackson, M.; Gavrilyuk, V.; Makarov, V. Pipeline of Anti-Mycobacterium Abscessus Small Molecules: Repurposable Drugs and Promising Novel Chemical Entities. *Med. Res. Rev.* **2021**, *41*, 2350–2387. [CrossRef]
14. Bisceglie, F.; Bacci, C.; Vismarra, A.; Barilli, E.; Pioli, M.; Orsoni, N.; Pelosi, G. Antibacterial Activity of Metal Complexes Based on Cinnamaldehyde Thiosemicarbazone Analogues. *J. Inorg. Biochem.* **2020**, *203*, 110888. [CrossRef]
15. Bajaj, K.; Buchanan, R.M.; Grapperhaus, C.A. Antifungal Activity of Thiosemicarbazones, Bis(Thiosemicarbazones), and Their Metal Complexes. *J. Inorg. Biochem.* **2021**, *225*, 111620. [CrossRef]
16. Lin, Y.; Betts, H.; Keller, S.; Cariou, K.; Gasser, G. Recent Developments of Metal-Based Compounds against Fungal Pathogens. *Chem. Soc. Rev.* **2021**, *50*, 10346–10402. [CrossRef]
17. Zani, C.; Bisceglie, F.; Restivo, F.M.; Feretti, D.; Pioli, M.; Degola, F.; Montalbano, S.; Galati, S.; Pelosi, G.; Viola, G.V.C.; et al. A Battery of Assays as an Integrated Approach to Evaluate Fungal and Mycotoxin Inhibition Properties and Cytotoxic/Genotoxic Side-Effects for the Prioritization in the Screening of Thiosemicarbazone Derivatives. *Food Chem. Toxicol.* **2017**, *105*, 498–505. [CrossRef]
18. Degola, F.; Bisceglie, F.; Pioli, M.; Palmano, S.; Elviri, L.; Pelosi, G.; Lodi, T.; Restivo, F.M. Structural Modification of Cuminaldehyde Thiosemicarbazone Increases Inhibition Specificity toward Aflatoxin Biosynthesis and Sclerotia Development in Aspergillus Flavus. *Appl. Microbiol. Biotechnol.* **2017**, *101*, 6683–6696. [CrossRef]
19. Degola, F.; Morcia, C.; Bisceglie, F.; Mussi, F.; Tumino, G.; Ghizzoni, R.; Pelosi, G.; Terzi, V.; Buschini, A.; Restivo, F.M.; et al. In Vitro Evaluation of the Activity of Thiosemicarbazone Derivatives against Mycotoxigenic Fungi Affecting Cereals. *Int. J. Food Microbiol.* **2015**, *200*, 104–111. [CrossRef]
20. Bartoli, J.; Montalbano, S.; Spadola, G.; Rogolino, D.; Pelosi, G.; Bisceglie, F.; Restivo, F.M.; Degola, F.; Serra, O.; Buschini, A.; et al. Antiaflatoxigenic Thiosemicarbazones as Crop-Protective Agents: A Cytotoxic and Genotoxic Study. *J. Agric. Food Chem.* **2019**, *67*, 10947–10953. [CrossRef]
21. Dallabona, C.; Pioli, M.; Spadola, G.; Orsoni, N.; Bisceglie, F.; Lodi, T.; Pelosi, G.; Restivo, F.M.; Degola, F. Sabotage at the Powerhouse? Unraveling the Molecular Target of 2-Isopropylbenzaldehyde Thiosemicarbazone, a Specific Inhibitor of Aflatoxin Biosynthesis and Sclerotia Development in Aspergillus Flavus, Using Yeast as a Model System. *Molecules* **2019**, *24*, 2971. [CrossRef]

22. Rogolino, D.; Gatti, A.; Carcelli, M.; Pelosi, G.; Bisceglie, F.; Restivo, F.M.; Degola, F.; Buschini, A.; Montalbano, S.; Feretti, D.; et al. Thiosemicarbazone Scaffold for the Design of Antifungal and Antiaflatoxigenic Agents: Evaluation of Ligands and Related Copper Complexes. *Sci. Rep.* **2017**, *7*, 11214. [CrossRef]

23. Gupta, O.; Pradhan, T.; Bhatia, R.; Monga, V. Recent Advancements in Anti-Leishmanial Research: Synthetic Strategies and Structural Activity Relationships. *Eur. J. Med. Chem.* **2021**, *223*, 113606. [CrossRef]

24. Moharana, A.K.; Dash, R.N.; Subudhi, B.B. Thiosemicarbazides: Updates on Antivirals Strategy. *Mini-Rev. Med. Chem.* **2021**, *20*, 2135–2152. [CrossRef]

25. Pelosi, G.; Bisceglie, F.; Bignami, F.; Ronzi, P.; Schiavone, P.; Re, M.C.; Casoli, C.; Pilotti, E. Antiretroviral Activity of Thiosemicarbazone Metal Complexes. *J. Med. Chem.* **2010**, *53*, 8765–8769. [CrossRef]

26. Shakya, B.; Yadav, P.N. Thiosemicarbazones as Potent Anticancer Agents and Their Modes of Action. *Mini-Rev. Med. Chem.* **2020**, *20*, 638–661. [CrossRef]

27. Singh, N.K.; Kumbhar, A.A.; Pokharel, Y.R.; Yadav, P.N. Anticancer Potency of Copper(II) Complexes of Thiosemicarbazones. *J. Inorg. Biochem.* **2020**, *210*, 111134. [CrossRef]

28. Chekmarev, J.; Azad, M.G.; Richardson, D.R. The Oncogenic Signaling Disruptor, NDRG1: Molecular and Cellular Mechanisms of Activity. *Cells* **2021**, *10*, 2382. [CrossRef]

29. Babak, M.V.; Ahn, D. Modulation of Intracellular Copper Levels as the Mechanism of Action of Anticancer Copper Complexes: Clinical Relevance. *Biomedicines* **2021**, *9*, 852. [CrossRef]

30. Wijesinghe, T.P.; Dharmasivam, M.; Dai, C.C.; Richardson, D.R. Innovative Therapies for Neuroblastoma: The Surprisingly Potent Role of Iron Chelation in up-Regulating Metastasis and Tumor Suppressors and down-Regulating the Key Oncogene, N-Myc. *Pharmacol. Res.* **2021**, *173*, 105889. [CrossRef]

31. Bisceglie, F.; Orsoni, N.; Pioli, M.; Bonati, B.; Tarasconi, P.; Rivetti, C.; Amidani, D.; Montalbano, S.; Buschini, A.; Pelosi, G. Cytotoxic Activity of Copper(Ii), Nickel(Ii) and Platinum(Ii) Thiosemicarbazone Derivatives: Interaction with DNA and the H2A Histone Peptide. *Metallomics* **2019**, *11*, 1729–1742. [CrossRef]

32. Baruffini, E.; Ruotolo, R.; Bisceglie, F.; Montalbano, S.; Ottonello, S.; Pelosi, G.; Buschini, A.; Lodi, T. Mechanistic Insights on the Mode of Action of an Antiproliferative Thiosemicarbazone-Nickel Complex Revealed by an Integrated Chemogenomic Profiling Study. *Sci. Rep.* **2020**, *10*, 10524. [CrossRef] [PubMed]

33. Baldini, M.; Belicchi-Ferrari, M.; Bisceglie, F.; Capacchi, S.; Pelosi, G.; Tarasconi, P. Zinc Complexes with Cyclic Derivatives of α-Ketoglutaric Acid Thiosemicarbazone: Synthesis, X-ray Structures and DNA Interactions. *J. Inorg. Biochem.* **2005**, *99*, 1504–1513. [CrossRef] [PubMed]

34. Wolfe, A.; Shimer, G.H.; Meehan, T. Polycyclic Aromatic Hydrocarbons Physically Intercalate into Duplex Regions of Denatured DNA. *Biochemistry* **1987**, *26*, 6392–6396. [CrossRef] [PubMed]

35. Kalaivani, P.; Prabhakaran, R.; Poornima, P.; Huang, R.; Hornebecq, V.; Dallemer, F.; Vijaya Padma, V.; Natarajan, K. Synthesis and Structural Characterization of New Ruthenium(Ii) Complexes and Investigation of Their Antiproliferative and Metastatic Effect against Human Lung Cancer (A549) Cells. *RSC Adv.* **2013**, *3*, 20363. [CrossRef]

36. Hernández, W.; Paz, J.; Carrasco, F.; Vaisberg, A.; Spodine, E.; Manzur, J.; Hennig, L.; Sieler, J.; Blaurock, S.; Beyer, L. Synthesis and Characterization of New Palladium(II) Thiosemicarbazone Complexes and Their Cytotoxic Activity against Various Human Tumor Cell Lines. Bioinorg. *Chem. Appl.* **2013**, *2013*, 524701. [CrossRef]

37. Prabhakaran, R.; Kalaivani, P.; Huang, R.; Poornima, P.; Vijaya Padma, V.; Dallemer, F.; Natarajan, K. DNA Binding, Antioxidant, Cytotoxicity (MTT, Lactate Dehydrogenase, NO), and Cellular Uptake Studies of Structurally Different Nickel(II) Thiosemicarbazone Complexes: Synthesis, Spectroscopy, Electrochemistry, and X-ray Crystallography. *J. Biol. Inorg. Chem.* **2013**, *18*, 233–247. [CrossRef]

38. Saswati, S.; Chakraborty, A.; Dash, S.P.; Panda, A.K.; Acharyya, R.; Biswas, A.; Mukhopadhyay, S.; Bhutia, S.K.; Crochet, A.; Patil, Y.P.; et al. Synthesis, X-ray Structure and in Vitro Cytotoxicity Studies of Cu(I/II) Complexes of Thiosemicarbazone: Special Emphasis on Their Interactions with DNA. *Dalton Trans.* **2015**, *44*, 6140–6157. [CrossRef]

39. Qi, J.; Gou, Y.; Zhang, Y.; Yang, K.; Chen, S.; Liu, L.; Wu, X.; Wang, T.; Zhang, W.; Yang, F. Developing Anticancer Ferric Prodrugs Based on the N-Donor Residues of Human Serum Albumin Carrier IIA Subdomain. *J. Med. Chem.* **2016**, *59*, 7497–7511. [CrossRef]

40. Subhashree, G.R.; Haribabu, J.; Saranya, S.; Yuvaraj, P.; Anantha Krishnan, D.; Karvembu, R.; Gayathri, D. In Vitro Antioxidant, Antiinflammatory and in Silico Molecular Docking Studies of Thiosemicarbazones. *J. Mol. Struct.* **2017**, *1145*, 160–169. [CrossRef]

41. Bai, Y.-L.; Zhang, Y.-W.; Xiao, J.-Y.; Guo, H.-W.; Liao, X.-W.; Li, W.-J.; Zhang, Y.-C. Oxovanadium Phenanthroimidazole Derivatives: Synthesis, DNA Binding and Antitumor Activities. *Trans. Met. Chem.* **2018**, *43*, 171–183. [CrossRef]

42. Rajendran, N.; Periyasamy, A.; Kamatchi, N.; Solomon, V. Biological Evaluation of Copper(II) Complexes on N (4)−substituted Thiosemicarbazide Derivatives and Diimine Co-Ligands Using DNA Interaction, Antibacterial and in Vitro Cytotoxicity. *J. Coord. Chem.* **2019**, *72*, 1937–1956. [CrossRef]

43. Lewis, N.A.; Liu, F.; Seymour, L.; Magnusen, A.; Erves, T.R.; Arca, J.F.; Beckford, F.A.; Venkatraman, R.; González-Sarrías, A.; Fronczek, F.R.; et al. Synthesis, Characterisation, and Preliminary In Vitro Studies of Vanadium(IV) Complexes with a Schiff Base and Thiosemicarbazones as Mixed Ligands. *Eur. J. Inorg. Chem.* **2012**, *2012*, 664–677. [CrossRef]

44. Khan, A.; Paul, K.; Singh, I.; Jasinski, J.P.; Smolenski, V.A.; Hotchkiss, E.P.; Kelley, P.T.; Shalit, Z.A.; Kaur, M.; Banerjee, S.; et al. Copper(I) and Silver(I) Complexes of Anthraldehyde Thiosemicarbazone: Synthesis, Structure Elucidation, in Vitro Anti-Tuberculosis/Cytotoxic Activity and Interactions with DNA/HSA. *Dalton Trans.* **2020**, *49*, 17350–17367. [CrossRef]

45. Beckford, F.A.; Leblanc, G.; Thessing, J.; Shaloski, M.; Frost, B.J.; Li, L.; Seeram, N.P. Organometallic Ruthenium Complexes with Thiosemicarbazone Ligands: Synthesis, Structure and Cytotoxicity of [(H6-p-Cymene)Ru(NS)Cl]+ (NS = 9-Anthraldehyde Thiosemicarbazones). *Inorg. Chem. Comm.* **2009**, *12*, 1094–1098. [CrossRef]

46. Beckford, F.A. Reaction of the Anticancer Organometallic Ruthenium Compound, [(η^6-*p*-Cymene)Ru(ATSC)Cl]PF$_6$ with Human Serum Albumin. *Int. J. Inorg. Chem.* **2010**, *2010*, 975756. [CrossRef]

47. Beckford, F.A.; Brock, A.; Gonzalez-Sarrías, A.; Seeram, N.P. Cytotoxic Gallium Complexes Containing Thiosemicarbazones Derived from 9-Anthraldehyde: Molecular Docking with Biomolecules. *J. Mol. Struct.* **2016**, *1121*, 156–166. [CrossRef]

48. Beckford, F.A.; Shaloski, M., Jr.; Leblanc, G.; Thessing, J.; Lewis-Alleyne, L.C.; Holder, A.A.; Li, L.; Seeram, N.P. Microwave Synthesis of Mixed Ligand Diimine–Thiosemicarbazone Complexes of Ruthenium(Ii): Biophysical Reactivity and Cytotoxicity. *Dalton Trans.* **2009**, *48*, 10757. [CrossRef]

49. Beebe, S.J.; Celestine, M.J.; Bullock, J.L.; Sandhaus, S.; Arca, J.F.; Cropek, D.M.; Ludvig, T.A.; Foster, S.R.; Clark, J.S.; Beckford, F.A.; et al. Synthesis, Characterization, DNA Binding, Topoisomerase Inhibition, and Apoptosis Induction Studies of a Novel Cobalt(III) Complex with a Thiosemicarbazone Ligand. *J. Inorg. Biochem.* **2020**, *203*, 110907. [CrossRef]

50. *Bruker SAINT*; Bruker AXS Inc.: Madison, WI, USA, 2012.

51. Sheldrick, G.M. *SADABS—Bruker Nonius Area Detector Scaling and Aabsorption Correction—V2016/2*; Bruker AXS Inc.: Madison, WI, USA, 2016.

52. Sheldrick, G.M. SHELXT—Integrated Space-Group and Crystal-Structure Determination. *Acta Crystallogr. A* **2015**, *71*, 3–8. [CrossRef]

53. Sheldrick, G.M. *SHELXL-97—Programs for Crystal Structure Analysis (Release 97-2)*; Institut fur Anorganishe Chemie der Universitat: Goettingen, Germany, 1998.

54. Dolomanov, O.V.; Bourhis, L.J.; Gildea RJ, J.; Howard, A.K.; Puschmann, H. OLEX2: A complete structure solution, refinement and analysis program. *J. Appl. Cryst.* **2009**, *42*, 339–341. [CrossRef]

55. Johnson, C.K.; Burnett, M.N. ORTEPIII. *Report ORNL-6895*; Oak Ridge National Laboratory: Oak Ridges, TN, USA, 1996.

56. Macrae, C.F.; Sovago, I.; Cottrell, S.J.; Galek, P.T.A.; McCabe, P.; Pidcock, E.; Platings, M.G.; Shields, P.; Stevens, J.S.; Towler, M.; et al. *Mercury 4.0*: From visualization to analysis, design and prediction. *J. Appl. Cryst.* **2020**, *53*, 226–235. [CrossRef]

57. Reichmann, M.E.; Rice, S.A.; Thomas, C.A.; Doty, P. A Further Examination of the Molecular Weight and Size of Desoxypentose Nucleic Acid. *J. Am. Chem. Soc.* **1954**, *76*, 3047–3053. [CrossRef]

58. Kalsbeck, W.A.; Thorp, H.H. Determining Binding Constants of Metal Complexes to DNA by Quenching of the Emission of Pt$_2$(pop)$_4{}^{4-}$ (pop= P$_2$O$_5$H$_2{}^{2-}$). *J. Am. Chem. Soc.* **1993**, *115*, 7146–7151. [CrossRef]

59. Baldini, M.; Belicchi-Ferrari, M.; Bisceglie, F.; Pelosi, G.; Pinelli, S.; Tarasconi, P. Cu(II) Complexes with Heterocyclic Substituted Thiosemicarbazones: The Case of 5-Formyluracil. Synthesis, Characterization, X-ray Structures, DNA Interaction Studies, and Biological Activity. *Inorg. Chem.* **2003**, *42*, 2049–2055. [CrossRef]

60. Liu, J.; Zhang, T.; Lu, T.; Qu, L.; Zhou, H.; Zhang, Q.; Ji, L. DNA-Binding and Cleavage Studies of Macrocyclic Copper(II) Complexes. *J. Inorg. Biohem.* **2002**, *91*, 269–276. [CrossRef]

61. Mudasir; Yoshioka, N.; Inoue, H. DNA Binding of Iron(II) Mixed-Ligand Complexes Containing 1,10-Phenanthroline and 4,7-Diphenyl-1,10-Phenanthroline. *J. Inorg. Biohem.* **1999**, *77*, 239–247. [CrossRef]

62. Beckford, F.; Dourth, D.; Shaloski, M.; Didion, J.; Thessing, J.; Woods, J.; Crowell, V.; Gerasimchuk, N.; Gonzalez-Sarrías, A.; Seeram, N.P. Half-Sandwich Ruthenium–Arene Complexes with Thiosemicarbazones: Synthesis and Biological Evaluation of [(H6-p-Cymene)Ru(Piperonal Thiosemicarbazones)Cl]Cl Complexes. *J. Inorg. Biohem.* **2011**, *105*, 1019–1029. [CrossRef]

63. Hassan, A.A.; Shawky, A.M.; Shehatta, H.S. Chemistry and Heterocyclization of Thiosemicarbazones. *J. Heterocycl. Chem.* **2012**, *49*, 21–37. [CrossRef]

64. Shaban, M.A.E.; Mostafa, M.A.; Nasr, A.Z. Oxidative Cyclization of D-Fructose Thiosemicarbazones to 2-Amino-5- (d-Arabino-1,2,3,4-Tetrahydroxybut-1-Yl)-1,3,4-Thiadiazoles through Carboncarbon Bond Cleavage of the Sugar Chain. *Pharmazie* **2003**, *58*, 367–371.

65. Buscemi, S.; Gruttadauria, M. Photocyclization Reaction of Some 2-Methyl-4-Phenyl- Substituted Aldehyde Thiosemicarbazones. Mechanistic Aspects. *Tetrahedron* **2000**, *56*, 999–1004. [CrossRef]

66. Meo, P.L.; Gruttadauria, M.; Noto, R. Oxidative Cyclization of Aldehyde Thiosemicarbazones Induced by Potassium Ferricyanide and by Tris(p-Bromophenyl)Amino Hexachloroantimoniate. A Joint Experimental and Computational Study. *Arkivoc* **2005**, *2005*, 114–129. [CrossRef]

67. Noto, R.; Buccheri, F.; Cusmano, G.; Gruttadauria, M.; Werber, G. Substituent Effect on Oxidative Cyclization of Aldehyde Thiosemicarbazones with Ferric Chloride. *J. Heterocycl. Chem.* **1991**, *28*, 1421–1427. [CrossRef]

68. Ivanov, V.I.; Minchenkova, L.E.; Schyolkina, A.K.; Poletayev, A.I. Different Conformations of Double-Stranded Nucleic Acid in Solution as Revealed by Circular Dichroism. *Biopolymers* **1973**, *12*, 89–110. [CrossRef] [PubMed]

69. Nordén, B.; Tjerneld, F. Structure of Methylene Blue-DNA Complexes Studied by Linear and Circular Dichroism Spectroscopy. *Biopolymers* **1982**, *21*, 1713–1734. [CrossRef]

70. Lepecq, J.-B.; Paoletti, C. A Fluorescent Complex between Ethidium Bromide and Nucleic Acids. *J. Mol. Biol.* **1967**, *27*, 87–106. [CrossRef]

71. Song, Y.; Zhong, D.; Luo, J.; Tan, H.; Chen, S.; Li, P.; Wang, L.; Wang, T. Binding Characteristics and Interactive Region of 2-Phenylpyrazolo[1,5-c]Quinazoline with DNA. *Luminescence* **2014**, *29*, 1141–1147. [CrossRef]

72. Sahoo, D.K.; Jena, S.; Dutta, J.; Chakrabarty, S.; Biswal, H.S. Critical Assessment of the Interaction between DNA and Choline Amino Acid Ionic Liquids: Evidences of Multimodal Binding and Stability Enhancement. *ACS Cent. Sci.* **2018**, *4*, 1642–1651. [CrossRef]
73. Byler, D.M.; Susi, H. Examination of the Secondary Structure of Proteins by Deconvolved FTIR Spectra. *Biopolymers* **1986**, *25*, 469–487. [CrossRef]
74. Minafra, L.; di Cara, G.; Albanese, N.N.; Cancemi, P. Proteomic Differentiation Pattern in the U937 Cell Line. *Leukemia Res.* **2011**, *35*, 226–236. [CrossRef]
75. Akiyama, M.; Horiguchi-Yamada, J.; Saito, S.; Hoshi, Y.; Yamada, O.; Mizoguchi, H.; Yamada, H. Cytostatic Concentrations of Anticancer Agents Do Not Affect Telomerase Activity of Leukaemic Cells in Vitro. *Eur. J. Cancer* **1999**, *35*, 309–315. [CrossRef]

 *compounds*

Article

Defect-Induced Luminescence Quenching of 4H-SiC Single Crystal Grown by PVT Method through a Control of Incorporated Impurity Concentration

Seul-Ki Kim [1,†], **Eun Young Jung** [2,*,†] **and Myung-Hyun Lee** [1,*]

1   Semiconductor Materials Center, Korea Institute of Ceramic Engineering & Technology, Jinju 52851, Korea; skkim@kicet.re.kr

2   School of Electronic and Electrical Engineering, College of IT Engineering, Kyungpook National University, Daegu 41566, Korea

*   Correspondence: eyjung@knu.ac.kr (E.Y.J.); mhlee@kicet.re.kr (M.-H.L.)

†   These authors contributed equally to this work.

Abstract: The structural defect effect of impurities on silicon carbide (SiC) was studied to determine the luminescence properties with temperature-dependent photoluminescence (PL) measurements. Single 4H-SiC crystals were fabricated using three different 3C-SiC starting materials and the physical vapor transport method at a high temperature and 100 Pa in an argon atmosphere. The correlation between the impurity levels and the optical and fluorescent properties was confirmed using Raman spectroscopy, X-ray diffraction, inductively coupled plasma atomic emission spectroscopy (ICP-OES), UV-Vis-NIR spectrophotometry, and PL measurements. The PL intensity was observed in all three single 4H-SiC crystals, with the highest intensities at low temperatures. Two prominent PL emission peaks at 420 and 580 nm were observed at temperatures below 50 K. These emission peaks originated from the impurity concentration due to the incorporation of N, Al, and B in the single 4H-SiC crystals and were supported by ICP-OES. The emission peaks at 420 and 580 nm occurred due to donor–acceptor-pair recombination through the incorporated concentrations of nitrogen, boron, and aluminum in the single 4H-SiC crystals. The results of the present work provide evidence based on the low-temperature PL that the mechanism of PL emission in single 4H-SiC crystals is mainly related to the transitions due to defect concentration.

Keywords: silicon carbide (SiC); 3C-SiC powder; 4H-SiC crystal; impurities; photoluminescence

Citation: Kim, S.-K.; Jung, E.Y.; Lee, M.-H. Defect-Induced Luminescence Quenching of 4H-SiC Single Crystal Grown by PVT Method through a Control of Incorporated Impurity Concentration. *Compounds* **2022**, *2*, 68–79. https://doi.org/10.3390/compounds2010006

Academic Editor: Juan Mejuto

Received: 30 December 2021
Accepted: 21 February 2022
Published: 2 March 2022

Publisher's Note: MDPI stays neutral with regard to jurisdictional claims in published maps and institutional affiliations.

1. Introduction

Silicon carbide (SiC) has attracted the attention of many researchers due to its outstanding electrical, mechanical, and thermal properties. SiC has been used in many industries for power devices and optoelectronic applications [1–8]. There are more than 200 polytypes of Si-C, with cubically (3C-SiC) and hexagonally (4H-SiC or 6H-SiC) modified compounds being the most used. The differences come from the stacking sequence of the hexagonal structure bonded in the Si-C bilayers [1–8]. The existence of polytypes implies that many different stable atomic arrangements and symmetries can be obtained, including hexagonal, cubic, and rhombohedral arrangements [6,7].

Si-C absorbs UV light and is transparent to visible light, making it an ideal material for optically based sensors or photodetectors at UV wavelengths [8]. A thorough analysis of the carrier recombination mechanisms in SiC is needed to understand the underlying physics of the luminescence phenomena for industrial applications. Photoluminescence (PL) is a standard method for characterizing the emission properties of semiconductor materials and can provide information about defect-related carrier transport dynamics [8]. Of the few studies reporting the optical properties of SiC, many have looked at the 3C-, 6H-, and 4H-SiC structures [4–8]. The 4H-SiC structure has been shown to have several defects and PL peaks in the band gap. The PL spectra in 4H-SiC originate from a combination of phonon-instigated electronic transitions caused by defects in SiC [9–11]. This observation of luminescence quenching is not evidence of electronic doping [12,13]. Therefore, it is necessary to determine the additional factors that affect the PL quenching and luminescence properties of 4H-SiC by characterizing it fully.

In this study, we investigated the structural and optical properties of the luminescence quenching of 4H-SiC crystals grown with the physical vapor transport (PVT) method. Control of the structure was achieved by using differently treated starting materials of 3C-SiC and by controlling the boron, aluminum, and nitrogen concentrations. The structural defects of the 4H-SiC crystals were analyzed to correlate them with the impurity concentration and optical properties. The 4H-SiC prepared with the PVT method was characterized using Raman spectroscopy, X-ray diffraction (XRD), inductively coupled plasma optical emission spectrometry (ICP-OES), UV-Vis-NIR spectrophotometry, and PL to compare the changes in impurities and structural properties.

2. Materials and Methods

2.1. Preparation of Starting Material

The 3C-SiC structure was synthesized using chemical vapor deposition involving vaporization, pyrolysis, nucleation, oxidation–reduction, and substitution [14]. The precursor gas consisted of commercial methyltrichlorosilane (MTS), ammonia (NH_3), and carbon dioxide (CO_2). MTS was the silicon precursor, and ethylene (C_2H_4) and propane (C_3H_8) were used as the hydrocarbon precursors. A mixture of H_2 and Ar was used as a carrier gas. The 3C-SiC synthesized was black, indicating the presence of impurities.

The free carbon and silica present in the synthesized 3C-SiC were removed through pulverization (pristine), acid leaching and decarburization (step A), and denitrification treatment (step B), as shown in Figure 1 and Table 1. In the pristine sample, the synthesized 3C-SiC was pulverized. A mixed acid solution dissolved the metal; nitric acid acted as a powerful oxidizer that dissolved to form metal ions (M_3^+), which reacted with hydrochloric acid to produce chorine anions in the solution. Then, the addition of HF allowed the free silica and free silicon in the mixed acid solution to react and generate the SiF_6 gas and $SiCl_6$ in a soluble solution. To decarbonize the powder, oxygen gas was passed through it to react with the free carbon, thus forming CO_2 vapor.

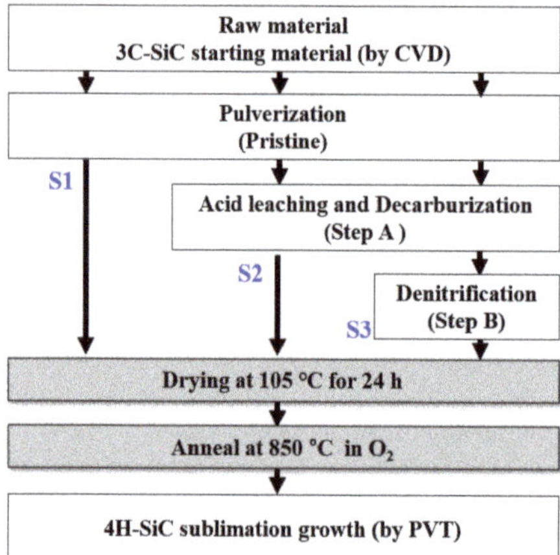

Figure 1. Flowchart for the fabrication of 4H-SiC crystals with the PVT process with three differently treated starting materials of 3C-SiC (S1, S2, and S3).

Table 1. Experimental conditions and identification for the growth of the 4H-SiC crystal samples.

Sample Conditions	Treatment Method	Identification
Pristine	Pulverization	S1
Step A	Acid leaching and decarburization	S2
Step B	Denitrification	S3

Volatile chloride compounds and chlorine gas were produced in step A. The residual acid salt was purified through volatilization, thus removing the excess oxygen present in SiC through a denitrification treatment. This was done by heating at 105 °C in a reduced-pressure evaporation in an argon environment. Other byproduct impurities present in the synthesized SiC were removed through heat treatment at 850 °C for 1 h in an oxygen environment. The dried and purified 3C-SiC powder obtained was green in color and had a purity greater than 90%. As shown in Figure 1 and Table 1, these purification processes were performed using the three purification processes of S1, S2, and S3 [15].

2.2. Growth of 4H-SiC Crystals

The 4H-SiC crystals were fabricated by using the generalized PVT method with three differently treated 3C-SiC powders (S1, S2, and S3) as starting materials. This PVT method has already been explained in detail by other researchers [16,17]. Briefly, the 3C-SiC powder was placed at the bottom of the crucible in a PVT chamber. In this process, the growth process was carried out in an argon environment. The growth of the 4H-SiC crystals took place at a growth temperature in the range of 1900–2100 °C and a pressure of 10^{-2} to 10^{-3} mbar in an argon atmosphere. The grown crystal sample was detached from the crucible and ground to two-inch diameter by slicing and then polishing to prepare the SiC wafers. A detailed flowchart for the fabrication of the 4H-SiC crystals with a commercialized PVT process with three differently treated 3C-SiC starting materials is shown in Figure 1 and Table 1.

2.3. X-ray Diffraction

The crystalline phases of the 3C-SiC powders were characterized using XRD (D/max-2500V/PC, Rigaku, Tokyo, Japan) with Cu Kα radiation at 30 mA and 40 kV. In addition, the crystalline orientation of the 4H-SiC crystal samples was characterized using a multifunction X-ray diffractometer (XRD; PANalytical, Malvern, Worcestershire, UK) and high-resolution two-dimensional (2D) XRD (Bruker, D8 Discover, Billerica, MA, USA) at the Korea Basic Science Institute (KBSI, Daegu, Korea). To obtain the oriented diffraction plane of the main and minor XRD peaks, the XRD result was obtained with a θ-2θ scan using multi-function XRD with Cu Kα radiation at 30 mA and 40 kV.

2.4. Inductively Coupled Plasma Optical Emission Spectrometry

The elemental compositions of the 4H-SiC crystal samples were determined using inductively coupled plasma optical emission spectrometry (ICP-OES; Optima 5300DV, Perkin Elmer, Waltham, MA, USA). The operation conditions were used at a radio-frequency power of 1.6 kW and a plasma argon gas flow rate of 14.0 L/min. Before measurement, the 4H-SiC crystal samples were pre-treated. To remove the organic components or contaminations in the 4H-SiC crystal samples, nitric acid (HNO_3) in an amount of 2–5 mL was put into the sample, and then the resulting products were dried and concentrated. Next, the resulting products were dissolved in the mixture of HCl and HNO_3. Finally, the dissolved products were treated by using the prepared aqueous solution with H_2SO_4 and HF. To fabricate the microwave-assisted acid digestion of the SiC samples, sample preparations were performed by using a microwave digestion system (Milestone Srl - START D, Sorisole, Italy) with PTFE vessels. The quantification of the elemental concentration was performed by using the certified reference material (ECRM 780-1), which was calibrated with a lower standard deviation of 1% for accuracy. For the preparation of the boron analysis, the resulting samples were put into a beaker. Then, the prepared solutions of hydrofluoric acid and nitric acid were put into the PTFE bottle. After that, they were maintained for 24 h to be dissolved in the mixed acid solution.

2.5. Elemental Analyzer

The elemental composition of the 4H-SiC crystal samples was determined using an elemental analyzer (EMGA-920, Horiba, Kyoto, Japan). The 4H-SiC crystal samples were placed in a graphite crucible. The operations were performed under the conditions of a power of 5.5 kW and a high temperature at 2650 °C. The 4H-SiC crystal samples were introduced into a graphite crucible that was placed between two electrodes. Then, to achieve complete combustion and transfer the generated gas into the detector, the 4H-SiC crystal samples were heated at a high temperature. The generated gas was directly extracted into a thermal conductivity detector (TCD).

2.6. Raman Spectrometry

The structural phase identification of all of the samples was performed using a Raman spectrometer (Renishew, Wotton-under-Edge, UK) with a 514 nm laser as the excitation source. The Raman spectra were collected over the wavenumber range of 120–2000 cm^{-1} with a four-stage Peltier cooled CCD detector (UV-Vis-NIR range). The objective lens of the microscope (DM500, LEICA, Wetzlar, Germany) had a magnification of 50×, and the exposure time for accumulation was 5 s. The power intensity of the laser beam was 5.0 ± 0.1 mW.

2.7. UV-Vis-NIR Spectrophotometer

The optical transmittance and absorbance spectra of the 4H-SiC crystal samples in the wavelength range of 200–1000 nm were measured at room temperature (298 K) using an ultraviolet–visible (UV-vis) spectrophotometer (LAMBDA 950, Perkin Elmer, Waltham, MA, USA).

2.8. Photoluminescence Spectrophotometry

Photoluminescence (PL) spectra were collected at both room temperature (298 K) and an extremely low temperature of 50 K using a PL spectrometer (HORIBA, LabRAM HR Evolution, Kyoto, Japan) with a He-Cd laser with a wavelength of 325 nm and power ranging from 0.15 to 15 mW as the excitation source. The power density ranged from approximately 0.023 to 23.6 kW/cm^2. The laser was focused on the sample using a 50× objective lens.

3. Results

Figure 2a,b show a photo image of a two-inch 4H-SiC crystal grown with the PVT process and a visible-light-emitting luminescence of a fluorescent 4H-SiC sample excited by a 325 nm pulsed laser source. The spot diameter of the 325 nm laser source was about 10 mm, and the greenish light was emitted from inside the 4H-SiC crystal sample shown in Figure 2b. The 4H-SiC crystal that was grown was cut in the form of a wafer and was subjected to slicing and then polishing in a direction perpendicular to the c-axis on the Si face before the PL measurement.

The XRD patterns of S1, S2, and S3 (pristine, A, and B, respectively) showed peaks corresponding to 3C-SiC (β-SiC phase), as shown in Figure 3. These peaks at 35.7°, 41.2°, 59.9°, and 71.6° are attributed to the (111), (200), (220), and (311) planes of the β-SiC phase, respectively [18,19].

Figure 4 shows two prominent peaks at 36° and 76°, which are the reflections from the (0004) and (0008) planes, which correspond to 4H-SiC [20,21]. In addition, all peaks are good agreement with ICSD card 98-016-4971. Furthermore, a few peaks of low intensity that were separated by almost equal intervals were also observed. These weak diffraction peaks were due to the (0005), (0006), and (0007) planes [21]. Their appearance was due to the double diffraction effect, as explained by other researchers [21]. Herein, for the highly purified 4H-SiC, the small periodic peaks between the main peaks that were related to the double diffraction effect could indicate the polytype of a SiC crystal with small periodic peaks due to the periodic stacking layers in the c-direction on the crystal [21]. For our XRD results, after the purification process (S2 and S3), small periodic peaks were also observed, as shown in Figure 4. In addition, the S1 sample had no small periodic peaks due to the large amounts of impurities. However, these XRD results could not be evaluated for 4H-SiC crystals with lower impurity doping levels. Thus, in this work, we investigated 4H-SiC crystal samples in terms of the trace impurity doping level by using the Raman and PL techniques.

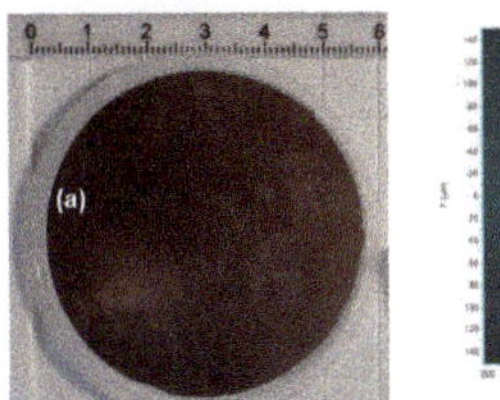
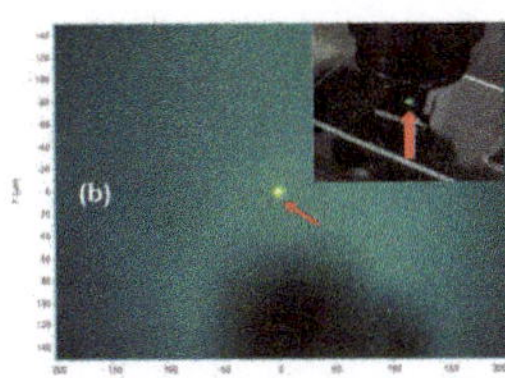

Figure 2. (**a**) Photo image of a two-inch 4H-SiC crystal grown with the PVT process with 3C-SiC powder and (**b**) a visible-light-emitting fluorescent 4H-SiC sample excited by a 325 nm laser source.

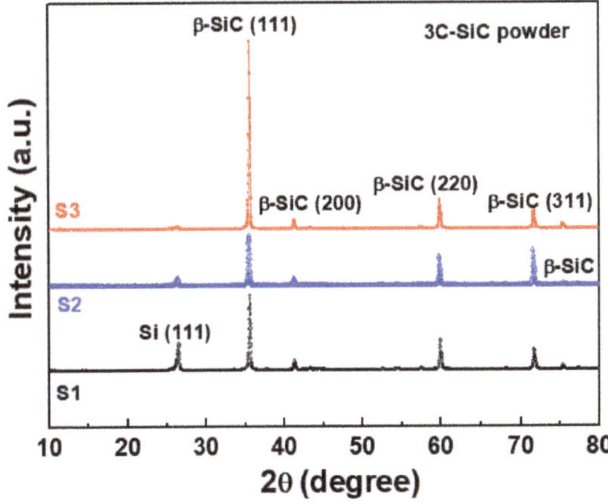

Figure 3. X-ray diffraction (XRD) patterns of three differently treated starting materials of 3C-SiC powder (S1, S2, and S3).

The Raman spectra of the 4H-SiC crystals that were grown are shown in Figure 5. The three characteristic peaks of the 4H-SiC samples were detected at approximately 795, 800, and 970 cm^{-1} [22–24]. In Figure 5a, the two characteristic peaks at 795 and 800 cm^{-1} correspond to the transverse optical (TO) phonon [22,23]. The peak at 970 cm^{-1} is the longitudinal optical (LO) phonon mode of 4H-SiC [22,23].

Figure 5b shows that the LO peaks (from 991 to 981 cm^{-1}) in the Raman spectra of the three 4H-SiC crystal samples were altered with respect to the starting material (S1, S2, and S3). The Raman shifts of the LO peak for two samples (S2 and S3) occurred at significantly lower wavenumbers than those of S1. It may be considered that the reason was that the LO phonon mode also caused a shift in the peaks toward lower frequencies, which could probably be attributed to the decrease in the grain size, internal stress from impurities, and the atomic size effect [22–28]. The peak breadth and reduced intensity of the S1 sample were the result of the increase in the free carrier concentration, as shown in

Figure 5b. The Raman features of the LO (or LOPC) modes are very useful in determining the structural properties of SiC and have been shown to allow the estimation of the nitrogen concentration [29]. As the starting material was purified, S1 to S3, the Raman spectra showed a decreasing nitrogen concentration with the shift toward lower energies, as well as a change in the intensity of the LOPC. The peak intensity of the LOPC increased and the peak position shifted to lower wavenumbers through the elimination of the nitrate element under various purification conditions, such as oxidation and reduction reactions.

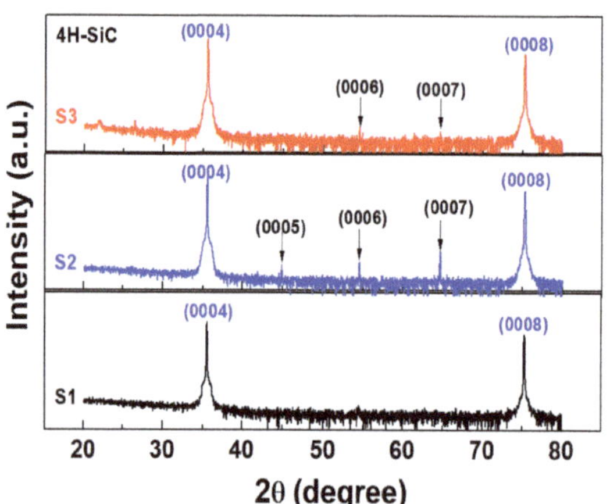

Figure 4. X-ray diffraction (XRD) patterns of the 4H-SiC crystal samples prepared with the PVT process with three different starting materials (S1, S2, and S3).

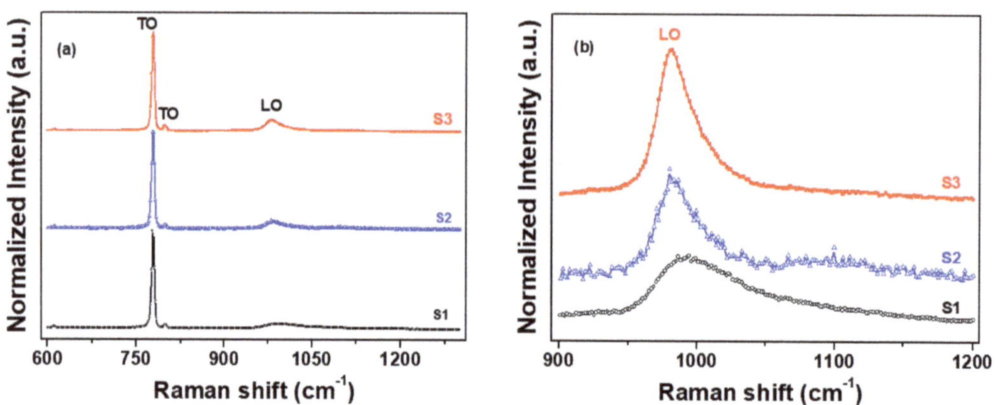

Figure 5. Raman spectra of the three 4H–SiC crystal samples: (**a**) wide scan and (**b**) narrow scan of the LO peak.

Figure 6a shows the PL spectra of the 4H-SiC crystal samples prepared using different starting materials at room temperature (T = 298 K). The measurements were carried out with a 325 nm exciting laser source, and the PL emission spectra are shown in Figure 6a. All of the samples showed a wide non-Gaussian symmetric peak located at 533 nm that corresponds to the N-B donor–acceptor pair (DAP) emission [22,23]. The peak intensity of the N-B DAP emissions for S2 was observed to be high, resulting in an increase in the N-B DAP density at room temperature (T = 298 K) [30–32]. The N-B DAP emissions of S1 were the weakest, although the concentrations of N and B were the highest among the three 4H-SiC crystal samples according to the ICP-OES data in Table 2. The reason could be that the N-B DAP emissions of S1 were affected by the N-Al DAP emissions at 420 nm [30–32]. These luminescence properties are directly correlated with energy-level transitions in semiconductors. The recombination of donor–acceptor pairs for 4H-SiC crystals with an indirect band gap forms a free exciton and a phonon. This type of recombination introduces a complex donor–acceptor recombination mechanism. Thus, the impurity concentration may have a critical influence on the luminescence properties [30–32].

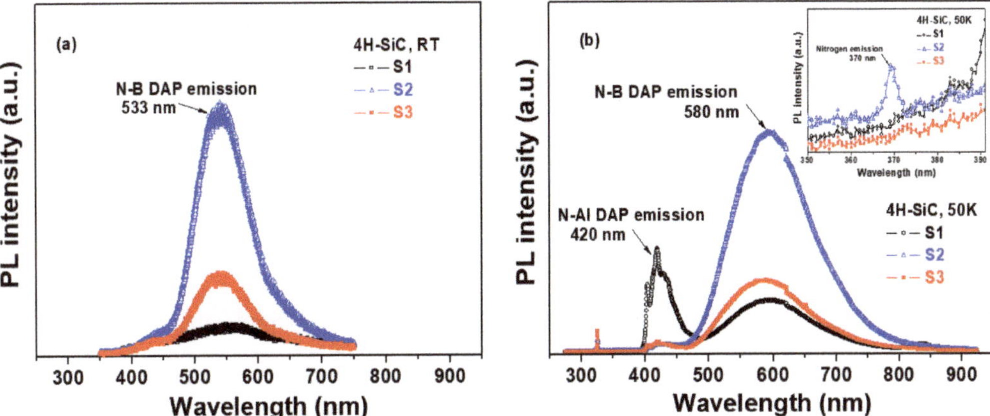

Figure 6. PL spectra of 4H-SiC crystal samples measured at (**a**) room temperature (*T* = 298 K) and (**b**) an extremely low temperature (*T* = 50 K) in a wavelength range of 300 to 900 nm.

Table 2. Elemental composition measured by ICP-OES and an element analyzer with the different 4H-SiC crystal samples (ppm: part per million, mg/kg).

Sample Condition	Element Composition		
	S1	S2	S3
SiC (wt%)	99.5 wt%	99.7 wt%	99.8 wt%
N (mg/kg)	2097	310	220
B (mg/kg)	9.25	1.93	0.17
Al (mg/kg)	56	0.12	0.10

Figure 6b shows the PL spectra of the three types of 4H-SiC crystal samples at extremely low temperatures (*T* = 50 K). Both N-Al and N-B DAP emissions were observed at 420 and 580 nm, respectively, at extremely low temperatures [30–32]. The peak intensity of the N-Al DAP emissions was observed to be the highest in S1. The peak intensities of the N-B DAP emissions for S2 were observed to be high, suggesting an increase in the N-B DAP density at extremely low temperatures [30–32]. The peak intensity of the N-B DAP emissions increased with the nitrogen concentration. As shown in the magnified image of Figure 6b, sample S2 showed a weak PL peak at 370 nm in the low-temperature PL spectra, which was caused by the emission of nitrogen from the nitrogen-bounce excitation [30–32]. Meanwhile, the N-B DAP emissions of S1 at 533 nm were the weakest due to the N-Al DAP emissions at 420 nm [30–32]. In the case of sample S1, the N-Al DAP emissions were dominant in the spectra at 420 nm, and they were affected by visible-green-light luminescence quenching at 533 nm.

To better understand the relationship between the impurities and PL properties, ICP-OES and an elemental analysis were performed to investigate the elemental composition of B, Al, and N in the three 4H-SiC crystal samples (S1, S2, and S3). The results are presented in Table 2.

From the obtained results, the concentration of each element was converted into units of atoms/cm^3 so that the recombination ratio of the donor–acceptor (RDA) (C_{D-A} and $2C_B/(C_N - C_B)$) could be calculated using Equation (1) [22,23]. The B, Al, and N concentrations were defined as C_B, C_{Al}, and C_N, respectively. The parameter C_{D-A} was calculated as a function of $C_B - (C_{Al} - C_N)$ [31–33]. The calculated concentrations are listed in Table 3.

$$\hat{C}_{D-A} = \ln C_B - (\ln C_{Al} - \ln C_N) \tag{1}$$

Table 3. Calculated concentrations of B, N, and Al using the ICP-OES results (atoms/cm^3).

Sample Condition	The Calculated Concentration (atoms/cm^3)		
	S1	S2	S3
C_N (atoms/cm^3)	2.10×10^{20}	3.10×10^{19}	2.20×10^{19}
C_B (atoms/cm^3)	1.20×10^{18}	2.50×10^{17}	2.20×10^{16}
C_{Al} (atoms/cm^3)	2.90×10^{18}	6.12×10^{15}	4.99×10^{15}
RDA of C_{D-A} (a.u.)	46	49	46
The ratio of $2C_B/(C_N - C_B)$	0.01	0.02	0.002

The DAP recombination rate is an efficient measure of emission luminescence [33–35]. The RDA is proportional to the donor concentration (C_D) and acceptor concentration (C_A). The correlation between C_D, C_A, and the PL properties in SiC samples has already been shown [22]. The increase in PL intensity of N-B DAP for S2 was shown to be due to an increase in C_N. The Aukerman and Millea model [33–35] suggests that the correlation between the DAP recombination and concentration causes an increase in PL intensity with the increase in the difference between C_N and C_B, which is called the N-B concentration gap. When the N-B concentration gap is larger than C_B, it becomes saturated. When the N-B concentration gap exceeds twice the value of C_B, the PL emission intensity decreases due to non-radiative defects (non-emission), as shown in the S3 sample [32].

To compare the correlation between the N-B concentrations and PL, we calculated the value of $2C_B/(C_N - C_B)$. When N-B emissions were observed, the N impurity—substituted for C in the SiC lattice—made it difficult for the electrons/holes between the impurity levels to transition to non-radiative emission at 533 nm, resulting in luminescence quenching. In the S3 sample, there was no emission, as the calculated ratio of $2C_B/(C_N - C_B)$ was 0.01 or less. In the S1 sample, Al and N existed in the hexagonal lattice of SiC, leading to a blue emission spectrum derived from the N-B DAP luminescence quenching. This shows that N-B DAP luminescence quenching can be extinguished depending on the concentrations of N and B impurities in 4H-SiC. When the calculated ratio of $2C_B/(C_N - C_B)$ was less than 0.01, the PL intensity was enhanced. When the Al concentration (C_{Al}) was high in 4H-SiC (S1), N-B DAP emissions could not be generated with the N-Al DAP emissions at 420 nm [30–32,35].

Based on the PL results, a schematic diagram of the proposed recombination paths of the main impurities in the 4H-SiC crystals is shown in Figure 7. When the wavefunction of an electron bound to a donor can interact with a hole that is bound to an acceptor, DAP recombination occurs. Typical DAP luminescence spectra were observed in 4H-SiC at room temperature (T = 298 K) and extremely low temperatures (T = 50 K). In Figure 7, the N-Al and N-B DAP PL spectra are shown. The N-Al DAP PL peaks were located in the relatively high-energy region, and the N-B DAP peaks appeared in the low-energy region because the Al acceptor levels were low, and the boron levels were relatively high. As described in Figure 7, the incorporation of B into 4H-SiC induced two boron-related levels, shallow boron and deep boron. In the N-B DAP PL, the deep boron centers were mainly involved in radiative recombination [30–32,35].

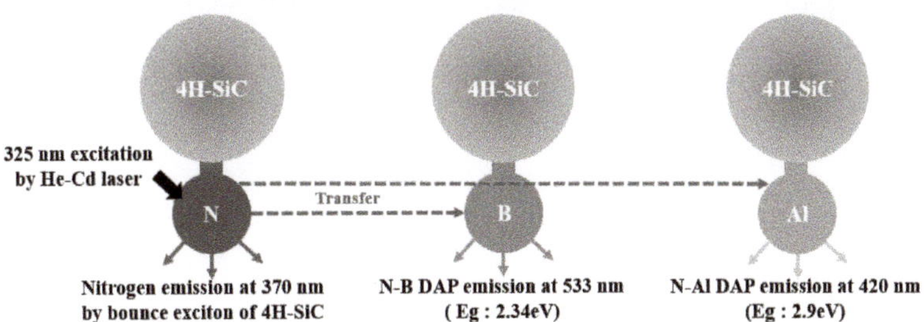

Figure 7. Schematic diagram of the proposed recombination paths with the elemental impurities incorporated into the 4H-SiC crystals.

The optical transmittance spectra of the 4H-SiC samples in the range of 200–1000 nm are shown in Figure 8a for samples S1 and S3. The optical transmittance of sample S1 was below 40%, and that of sample S3 was also above 40%. For the S1 sample, more N doping could decrease the optical transmissivity of the 4H-SiC crystal for wavelengths from 600 to 1000 nm. The absorption at 463 nm caused by nitrogen doping may weaken N-Al DAP emission extraction. Most importantly, Al and N co-doping caused the N donor and Al acceptor at hexagonal sites to be dominated in the SiC crystal, resulting in more non-radiative recombination and light absorption losses [22]. This absorption band was caused by the transition between the bottom levels of the conduction band and the top levels of the valence band, while the absorption band was caused by the induction of the energy level of the N impurities [35].

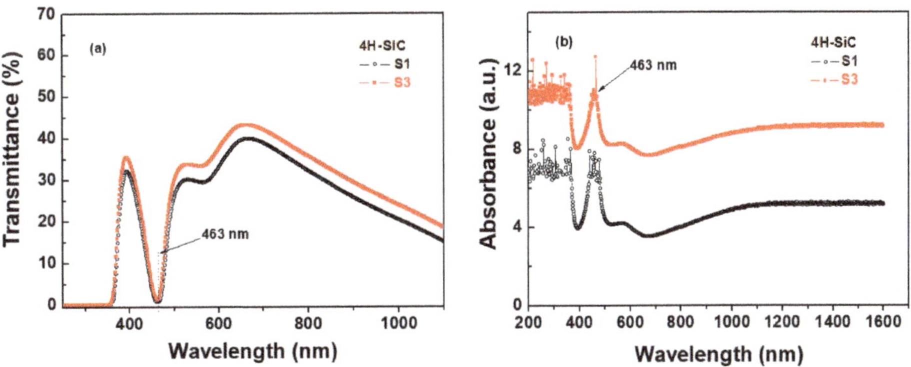

Figure 8. (**a**) Optical transmittance and (**b**) absorbance spectra of 4H-SiC crystal samples at room temperature (T = 298 K).

The absorption spectra of the 4H-SiC samples in the range of 200–1600 nm are shown in Figure 8b. The absorption peak at 463 nm caused by nitrogen doping may weaken the N-Al DAP emissions. These N-B DAP emissions of the 4H-SiC samples were responsible for the low optical transmissivity, and the light extraction of the N-B DAP emissions in the visible light range was weakened [22,23]. The aluminum and nitrogen co-doping led to the N donor and Al acceptor at the hexagonal sites dominating in the SiC crystal, resulting in more non-radiative recombination and light absorption losses [22,35].

The optical band gap (E_g) was calculated from the transmittance spectra using the Tauc Equation (2) [36,37].

$$\alpha h\upsilon = A \cdot (h\upsilon - E_g)^{0.5} \qquad (2)$$

where α is the absorption coefficient, A is a constant, and $h\upsilon$ is the photon energy [36,37]. E_g can be evaluated from the relation between $\alpha h\upsilon$ and the photon energy ($h\upsilon$), as shown when plotting $(\alpha h\upsilon)^2$ vs. energy in Figure 9. The value of E_g of the samples can then be determined by extrapolating a straight line to cross with the $h\upsilon$ axis at zero, as shown in Figure 9. The value of E_g of the 4H-SiC that was grown was found to be 3.29 eV for both S1 and S3 [36,37].

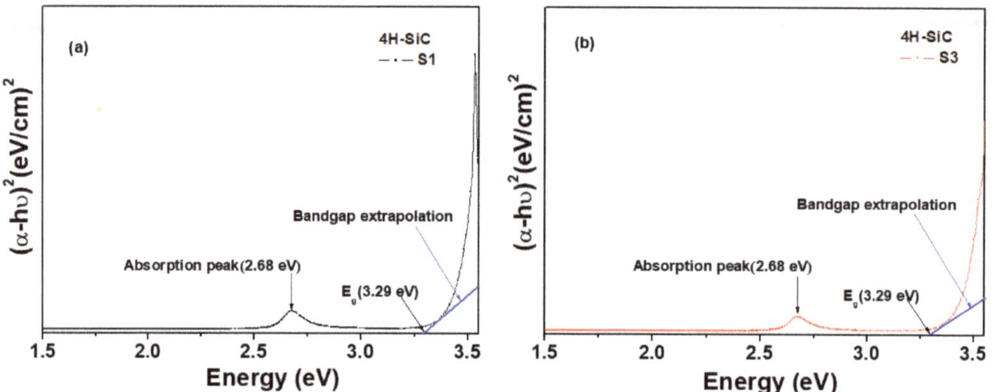

Figure 9. Plots of the variation in $(\alpha h\upsilon)^2$ versus photon energy ($h\upsilon$) of the 4H-SiC crystal samples grown with the PVT method with different 3C-SiC powders: (**a**) S1 sample and (**b**) S3 sample.

4. Conclusions

In summary, we investigated the correlation between the impurity concentrations and PL properties of single 4H-SiC crystals prepared with the PVT method. The single 4H-SiC crystals grown with the PVT process were evaluated by using Raman spectroscopy, XRD, ICP-OES, UV-Vis spectroscopy, and PL measurements. As a result, XRD peaks at two prominent peaks at 36° and 76°, which are reflections from the (0004) and (0008) planes, respectively, were found to correspond to 4H-SiC. After purification (S2 and S3), a few peaks of weak intensity were also separately observed at almost equal intervals, although the S1 sample had no small periodic peaks due to the large number of impurities. However, with XRD results, it is somewhat difficult to analyze the concentrations of trace impurities using peak shifts or weak minor peaks. Thus, we investigated the 4H-SiC crystal samples in terms of the trace impurity doping level by using the Raman and PL techniques. The Raman spectra of the 4H-SiC crystals that were grown were observed to have three characteristic peaks, which were detected at approximately 795, 800, and 970 cm^{-1}. In particular, the LO peak intensities of S2 and S3 decreased, and the LO peak shifts of the S2 and S3 samples occurred at lower wavenumbers than those of S1. It may be considered that reason was that the LO phonon mode also caused a shift in peaks toward lower frequencies, which could probably be attributed to the decrease in the grain size, internal stress from impurities, and the atomic size effect. For the PL spectra at a low temperature (50 K), two prominent PL emission peaks were observed at 420 and 580 nm. These DAP emission peaks were attributed to the impurity concentration caused by the doping of N and B in the single 4H-SiC crystals. The value of $2C_B/(C_N - C_B)$ was employed to evaluate the correlation between the N-B concentrations and PL. As the ratio of $2C_B/(C_N - C_B)$ was less than 0.01, there were no emissions in the S3 sample. Since Al and N existed in the S1 sample, N-B DAP emissions were not observed due to due to the luminescence quenching by the N-Al DAP emissions at 420 nm. These results show that N-B DAP luminescence quenching can be extinguished depending on the concentrations of N and B impurities in 4H-SiC. When the $2C_B/(C_N - C_B)$ ratio was 0.01 or less, the 4H-SiC samples with high C_{Al} values showed no N-B DAP emissions at 420 nm. Thus, the PL technique is a useful technique for detecting the lower trace impurity doping levels in 4H-SiC crystal samples.

Author Contributions: Conceptualization, S.-K.K., E.Y.J. and M.-H.L.; methodology, S.-K.K. and M.-H.L.; software, S.-K.K. and E.Y.J.; validation, S.-K.K., E.Y.J. and M.-H.L.; formal analysis, S.-K.K., E.Y.J. and M.-H.L.; investigation, S.-K.K., E.Y.J. and M.-H.L.; resources, S.-K.K. and M.-H.L.; data curation, S.-K.K., E.Y.J. and M.-H.L.; writing, original draft preparation, S.-K.K., E.Y.J. and M.-H.L.; writing—review and editing, S.-K.K., E.Y.J. and M.-H.L.; visualization, S.-K.K. and E.Y.J.; supervision, E.Y.J. and M.-H.L.; project administration, M.-H.L.; funding acquisition, M.-H.L. All authors have read and agreed to the published version of the manuscript.

Funding: This research was funded by the Technology Development Program (No. S2797799) from the Ministry of SMEs and Startups.

Institutional Review Board Statement: Not applicable.

Informed Consent Statement: Not applicable.

Data Availability Statement: Not applicable.

Acknowledgments: The authors would like to thank Sang-Geul Lee at the Korea Basic Science Institute (Daegu) for their useful discussions and for providing the XRD data. In addition, the authors would like to thank Hongtaek Kim at Horiba Korea Ltd. for his useful discussions and for providing the PL data.

Conflicts of Interest: The authors declare no conflict of interest.

References

1. Deng, L.; Wang, X.; Hua, X.; Lu, S.; Wang, J.; Wang, H.; Wang, B. Purification of β-SiC powders by heat treatment in vacuum. *Adv. Compos. Hybrid Mater.* **2021**, 00372. [CrossRef]
2. Pomaska, M.; Beyer, W.; Neumann, E.; Finger, F.; Ding, K. Impact of microcrystalline silicon carbide growth using hot-wire chemical vapor deposition on crystalline silicon surface passivation. *Thin Solid Film* **2015**, *595*, 217–220. [CrossRef]
3. Lee, K.-I.; Seok, D.C.; Jang, S.O.; Choi, Y.S. Development of silicon carbide atomic layer etching technology. *Thin Solid Film* **2020**, *707*, 138084. [CrossRef]
4. Yousefi, M.; Rahim-abadi, M.M. Improvement of the mechanical and oxidation resistance of pyrolytic carbon coatings by co-deposition synthesis of pyrolytic carbon-silicon carbide nanocomposite. *Thin Solid Film* **2020**, *713*, 138320. [CrossRef]
5. Mas'udah, K.W.; Diantoro, M.; Fuad, A. Synthesis and structural analysis of silicon carbide from silica rice husk and activated carbon using solid-state reaction. *IOP Conf. Ser. J. Phys. Conf. Ser.* **2018**, *1093*, 012033. [CrossRef]
6. Via, F.L.; Severino, A.; Anzalone, R.; Bongiorno, C.; Litrico, G.; Mauceri, M.; Schoeler, M.; Schuh, P.; Wellmann, P. From thin film to bulk 3C-SiC growth: Understanding the mechanism of defects reduction. *Mater. Sci. Semicond. Process.* **2018**, *78*, 57–68. [CrossRef]
7. Huseynov, E.M.; Naghiyev, T.G. Study of thermal parameters of nanocrystalline silicon carbide (3C-SiC) using DSC spectroscopy. *Appl. Phys. A* **2021**, *127*, 267. [CrossRef]
8. Lu, F.; Tarekegne, T.; Ou, Y.; Kamiyama, S.; Ou, H. Temperature-dependent photoluminescence properties of porous Fluorescen SiC. *Sci. Rep.* **2019**, *9*, 16333. [CrossRef]
9. Gadalla, M.N.; Greenspon, A.S.; Defo, R.K.; Zhang, X.; Hu, E.L. Enhanced cavity coupling to silicon vacancies in 4H silicon carbide using laser irradiation and thermal annealing. *Proc. Acad. Natl. Sci. USA* **2021**, *118*, e2021768118. [CrossRef]
10. Berwian, P.; Kaminzky, D.; Roßhirt, K.; Kallinger, B.; Friedrich, J.; Oppel, S.; Schneider, A.; Schütz, M. Imaging defect luminescence of 4H-SiC by ultraviolet-photoluminescence. *Solid State Phenom.* **2016**, *242*, 484–489. [CrossRef]
11. Hashemi, A.; Linderälv, C.; Krasheninnikov, A.V.; Ala-Nissila, T.; Erhart, P.; Komsa, H.P. Photoluminescence line shapes for color centers in silicon carbide from density functional theory calculations. *Phys. Rev. B* **2021**, *103*, 125203. [CrossRef]
12. Chen, B.-Y.; Chi, C.-C.; Hsu, W.K.; Ouyang, H. Synthesis of SiC/SiO$_2$ core–shell nanowires with good optical properties on Ni/SiO$_2$/Si substrate via ferrocene pyrolysis at low temperature. *Sci. Rep.* **2021**, *11*, 233. [CrossRef]
13. Hiller, D.; López-Vidrier, J.; Gutsch, S.; Zacharias, M.; Nomoto, K.; König, D. Defect-induced luminescence quenching vs. charge carrier generation of phosphorus incorporated in silicon nanocrystals as function of size. *Sci. Rep.* **2017**, *7*, 863. [CrossRef]
14. Cheng, D.J.; Shyy, W.J.; Kuo, D.H. Growth characteristics of CVD beta-silicon carbide. *J. Electrochem. Soc.* **1987**, *134*, 3145–3149. [CrossRef]
15. Wang, Z.; Dai, X.-Y.; Xu, S.-P.; Xu, M. Preparation of SiC powders by carbonthemal reduction method at low temperature. *MATEC Web Conf.* **2017**, *114*, 02015. [CrossRef]
16. Gao, P.; Xin, J.; Liu, X.; Zheng, Y.; Shi, E. Control of 4H polytype of SiC crystals by moving up the crucible to adjust the temperature field of the growth interface. *CrystEngComm* **2019**, *21*, 6964–6968. [CrossRef]
17. Shin, D.-G.; Kim, B.-S.; Son, H.-R.; Kim, M.-S. Study on the growth of 4H-SiC single crystal with high purity SiC fine powder. *J. Korean Cryst. Growth Cryst. Technol.* **2019**, *29*, 383–388.
18. Luo, X.; Ma, W.; Zhou, Y.; Liu, D.; Yang, B.; Dai, Y. Radiation synthesis and photoluminescence property of silicon carbide nanowires via carbothermic reduction of silica. *Nanoscale Res. Lett.* **2010**, *5*, 252. [CrossRef]

19. Ugraskan, V.; Isik, B.; Yazici, O.; Cakar, F. Surface characterization and synthesis of boron carbide and silicon carbide. *Solid State Sci.* **2021**, *118*, 106636. [CrossRef]

20. Irfan, M.; Ajmal, M.; Mazhar, M.E.; Usmani, M.N.; Ahmad, S.; Abbas, W.; Mahmood, M.; Hussain, M. Growth and characterization of 4H-SiC by thermal evaporation method. *Dig. J. Nanomater. Biostruct.* **2019**, *14*, 243–247.

21. Arora, A.; Pandey, A.; Patel, A.; Dalal, S.; Yadav, B.S.; Goyal, A.; Raman, R.; Thakur, O.P.; Tyagi, R. Polytype switching identification in 4H-SiC single crystal grown by PVT. *J. Mater. Sci. Mater. Electron.* **2020**, *31*, 16343–16351. [CrossRef]

22. Liu, X.; Zhuo, S.-Y.; Gao, P.; Huang, W.; Yan, C.-F.; Shi, E.-W. Donor acceptor pair emission in fluorescent 4H-SiC grown by PVT method. *AIP Adv.* **2015**, *5*, 047133. [CrossRef]

23. Khashan, K.S.; Ismail, R.A.; Mahdi, R.O. Synthesis of SiC nanoparticles by SHG 532 nm Nd:YAG laser ablation of silicon in ethanol. *Appl. Phys. A* **2018**, *124*, 443. [CrossRef]

24. Aldalbahi, A.; Li, E.; Rivera, M.; Velazquez, R.; Altalhi, T.; Peng, X.; Feng, P.X. A new approach for fabrications of SiC basedphoto detectors. *Sci. Rep.* **2016**, *6*, 23457. [CrossRef] [PubMed]

25. Nakashima, S.-I.; Mitani, T.; Tomobe, M.; Kato, T.; Okumura, H. Raman characterization of damaged layers of 4H-SiC induced by scratching. *AIP Adv.* **2016**, *6*, 015207. [CrossRef]

26. Peng, Y.; Hu, X.; Xu, X.; Chen, X.; Peng, J.; Han, J.; Dimitrijev, S. Temperature and doping dependence of the Raman scattering in 4H-SiC. *Opt. Mater. Express* **2016**, *6*, 2725–2733. [CrossRef]

27. Bauer, M.; Gigler, A.M.; Huber, A.J.; Hillenbrand, R.; Stark, R.W. Temperature-depending Raman line-shift of silicon carbide. *J. Raman Spectrosc.* **2009**, *40*, 1867–1874. [CrossRef]

28. Wan, L.; Zhao, D.; Wang, F.; Xu, G.; Lin, T.; Tin, C.-C.; Feng, Z.; Feng, Z.C. Quality evaluation of homoepitaxial 4H-SiC thin films by a Raman scattering study of forbidden modes. *Opt. Mater. Express* **2018**, *8*, 119–127. [CrossRef]

29. Kwasnicki, P. Evaluation of Doping in 4H-SiC by Optical Spectroscopies. Ph.D. Thesis, Université Montpellier II—Sciences et Techniques du Languedoc, Montpellier, France, 2014.

30. Zhuo, S.-Y.; Liu, X.-C.; Xu, T.-X.; Yan, C.-F.; Shi, E.-W. Strong correlation between B-Al-N doping concentration fluctuation and photoluminescence effects of f-SiC. *AIP Adv.* **2018**, *8*, 075130. [CrossRef]

31. Nagasawa, F.; Takamura, M.; Sekiguchi, H.; Miyamae, Y.; Oku, Y.; Nakahara, K. Prominent luminescence of silicon-vacancy defects created in bulk silicon carbide p–n junction diodes. *Sci. Rep.* **2021**, *11*, 1497. [CrossRef]

32. Liu, L.; Liu, A.; Bai, S.; Lv, L.; Jin, P.; Ouyang, X. Radiation resistance of silicon carbide Schottky diode detectors in D-T fusion neutron detection. *Sci. Rep.* **2017**, *7*, 13376. [CrossRef] [PubMed]

33. Allgaier, R.S. Extension of the aukerman-willardson two band hall coefficient analysis. *J. Appl. Phys.* **1965**, *36*, 2429. [CrossRef]

34. Wei, Y.; Tarekegne, A.T.; Ou, H. Influence of negative-U centers related carrier dynamics on donor-acceptor-pair emission in fluoescent SiC. *J. Appl. Phys.* **2018**, *124*, 054901. [CrossRef]

35. Zhuo, S.-Y.; Liu, X.-C.; Huang, W.; Kong, H.-K.; Xin, J.; Shi, E.-W. Photoluminescence in fluorescent 4H-SiC single crystal adjusted by B, Al, and N ternary dopants. *Chin. Phys. B* **2019**, *28*, 017101. [CrossRef]

36. Rasheed, M.N.; Maryam, A.; Fatima, K.; Iqbal, F.; Afzal, M.; Syvajarvi, M.; Murtaza, H.; Zhu, B.; Asghar, M. Enhanced electrical properties of nonstructural cubic silicon carbide with graphene contact for photovoltaic applications. *Dig. J. Nanomater. Biostruct.* **2020**, *15*, 963–972.

37. Alhusaiki-Alghamdi, H.M. Effect of silicon carbide (SiC) nanoparticles on the spectroscopic properties and performance of PMMA/PC Polymer Blend. *J. Mod. Phys.* **2019**, *10*, 487–499. [CrossRef]

 *compounds*

Review

The Nitrogen Bond, or the Nitrogen-Centered Pnictogen Bond: The Covalently Bound Nitrogen Atom in Molecular Entities and Crystals as a Pnictogen Bond Donor

Pradeep R. Varadwaj [1,2,*], **Arpita Varadwaj** [1], **Helder M. Marques** [2] and **Koichi Yamashita** [1]

[1]   Department of Chemical System Engineering, School of Engineering, The University of Tokyo 7-3-1, Tokyo 113-8656, Japan; varadwaj.arpita@gmail.com (A.V.); yamasita@chemsys.t.u-tokyo.ac.jp (K.Y.)
[2]   Molecular Sciences Institute, School of Chemistry, University of the Witwatersrand, Johannesburg 2050, South Africa; helder.marques@wits.ac.za
*   Correspondence: pradeep@t.okayama-u.ac.jp

Abstract: The nitrogen bond in chemical systems occurs when there is evidence of a net attractive interaction between the electrophilic region associated with a covalently or coordinately bound nitrogen atom in a molecular entity and a nucleophile in another, or the same molecular entity. It is the first member of the family of pnictogen bonds formed by the first atom of the pnictogen family, Group 15, of the periodic table, and is an inter- or intra-molecular non-covalent interaction. In this featured review, we present several illustrative crystal structures deposited in the Cambridge Structure Database (CSD) and the Inorganic Crystal Structure Databases (ICSD) to demonstrate that imide nitrogen is not the only instance where nitrogen can act as an electrophilic agent. Analysis of a set of carefully chosen illustrative crystal systems shows that a covalently bound nitrogen atom in a variety of molecular entities features a σ-hole or even a π-hole, and these have the ability to sustain attractive engagements with negative sites to form inter- and/or intramolecular interactions that drive, or assist, the formation of a crystalline phase.

Keywords: pnictogen bonding; nitrogen as pnictogen bond donor; geometries; crystal structure analysis; ICSD and CSD database analyses; MESP characterizations; sum of the van der Waals radii concept

Citation: Varadwaj, P.R.; Varadwaj, A.; Marques, H.M.; Yamashita, K. The Nitrogen Bond, or the Nitrogen-Centered Pnictogen Bond: The Covalently Bound Nitrogen Atom in Molecular Entities and Crystals as a Pnictogen Bond Donor. *Compounds* **2022**, *2*, 80–110. https://doi.org/10.3390/compounds2010007

Academic Editor: Juan Mejuto

Received: 18 February 2022
Accepted: 7 March 2022
Published: 15 March 2022

Publisher's Note: MDPI stays neutral with regard to jurisdictional claims in published maps and institutional affiliations.

1. Introduction

IUPAC definitions, features and characteristic properties that can be used to identify non-covalent interactions, such as hydrogen bonding, chalcogen bonding and halogen bonding in chemical systems have been promulgated in 2011 [1], 2013 [2] and 2019 [3], respectively. A task group has been charged with categorizing tetrel bonding, pnictogen bonding, and other non-covalent interactions, involving the elements of Groups 14–16 of the periodic table [4]. The proposed characteristics of these bonding interactions have emerged from observations of a variety of engineered chemical systems in the crystalline, liquid and gas phases, and examining signatures arising from IR [5,6], Raman [7,8], UV/vis [9,10], and NMR [10,11] methods, and from ab initio and density functional theory calculations [12–14]. The characteristic features vary from system to system because they are dependent on the nature of the electron density distribution associated with the electron donor and electron acceptor fragments driving the non-covalent interactions.

In this featured review, we show that nitrogen in molecular entities can act as an electrophile when covalently bonded with appropriate electron-withdrawing atomic domain(s). This electrophilic feature of N can form nitrogen-centered pnictogen bonding interactions (or simply nitrogen bonding interactions) when in close proximity with a negative site on the same or a neighboring molecule, and hence is responsible in part for the stability of the resulting supermolecular or intramolecular entity.

The nitrogen bond in chemical systems occurs when there is evidence of a net attractive interaction between the electrophilic region associated with a covalently or coordinately bound nitrogen atom in a molecular entity and a nucleophile in another, or the same molecular entity. It is the first member of the family of pnictogen bonds formed by the first atom of the pnictogen family, Group 15, of the periodic table, and is an inter- or intra-molecular non-covalent interaction.

A nitrogen bond in a molecular entity may be regarded as a σ-hole centered pnictogen bond, especially when the covalently bound nitrogen features a positive σ-hole along the R–N bond extension, where R is the remainder of the molecular entity and is in attractive engagement with a negative site in the same or a neighboring molecule. A σ-hole is defined as a charge density deficient region on an atom A that appears along the outer extension of the R–A covalent bond [15,16]. By contrast, a nitrogen bond may be regarded as a π-hole centered pnictogen bond when the nitrogen in molecular entities features a π-hole [17] on its electrostatic surface and has the ability to engage attractively with a negative site in a neighboring molecule, or a site that has an electron density different to that of the π-hole, thus providing stability to the geometry of the resulting structure. A σ-hole interaction in a chemical system is generally observed to be directional, whereas a π-hole is non-directional. There are many studies on σ-hole and π-hole [17] interactions in chemical systems, including similarities, differences, controversies and misconceptions [18–25].

Nevertheless, nitrogen is a crucial constituent in the development of high-density materials, and, of course, is crucial for ammonia synthesis [26–29]. Apart from its well-known use as a feedstock for the production of fertilizers [30], we may be on the brink of a viable ammonia economy [31]. "Green ammonia" is a carbon-free hydrogen-containing compound with immense interest due to its high density and high hydrogen storage capability [29,32–34].

We considered nitrogen because a fundamental understanding of its modes of interaction in forming complex chemical systems would surely elevate our current knowledge of nitrogen-centered pnictogen bonding (i.e., nitrogen as a pnictogen bond donor). The role of pnictogen bonding containing heavier members of the pnictogen family (viz. P, As, Sb and Bi) in catalysis has also been demonstrated on several occasions [35–38], and in other areas, such as anion transport and recognition chemistry [39], solution and gas-phase chemistry [40–42], computational chemistry [13], supramolecular chemistry [43], coordination chemistry [44,45], medicinal chemistry [46], crystallography [47,48] and crystal engineering [49,50], among other research fields.

Nitrogen is the lightest member of the pnictogen family (Group 15), and the third most electronegative element after fluorine and oxygen. When it is present in a molecular entity, it is well known to be a site for interaction with electrophiles [51]. HCN, N_2, NO, HCCCN, CH_3CN, and NO_3^- are a few examples where nitrogen is entirely negative, and hence, serves as a site to interact with electrophiles to form molecular complexes in the gas phase or adducts in the solid state. Nitrogen in molecules, such as NH_3 and in amines, is often found, both experimentally and theoretically [52,53], to be negative. This is a consequence of its strong electronegative character. It is also less polarizable than the other members of the pnictogen family. Its lone-pair electrons, if visualized from a Lewis structure viewpoint, dominate along the extension of the R–N covalent bonds in compounds in which it is covalently bonded and is an electron density donor towards electrophiles, forming hydrogen bonds, halogen bonds or chalcogen bonds, and acts as a Lewis base to metal ions [54]. This is the probable reason why fluorine, oxygen and nitrogen—being highly electronegative but with low polarizability—often have a negative σ-hole [21]. However, under certain circumstances, nitrogen can act as an acceptor of electron density, for example, in hypervalent non-covalent interactions, as demonstrated by Chandra and coworkers [55]. They observed an intermolecular interaction between nitromethane, CH_3NO_2 and NH_3 at low temperature within an inert gas matrix, which was characterized by IR spectroscopy and supported by first-principles calculations. The nitrogen in CH_3NO_2 acts as an electron density acceptor from the nitrogen in NH_3, forming a $CH_3O_2N\cdots NH_3$ dimer and is stabilized by an $N\cdots N$ pnictogen bond. The directional prevalence of this interaction over the C–H···N and N–H···O hydrogen bonding interactions was shown by ab initio calculations, demonstrating that σ-hole/π-hole driven interactions lead to the formation of the dimer, despite nitrogen's low polarizability.

Presented in this review are a set of illustrative crystal systems, demonstrating that nitrogen in the molecular entities that comprise these crystal systems serves as a nitrogen bond donor (electrophilic), and hence is responsible (at least in part) for their geometric stability and functionality. It should be noted that the definition and characteristic features of pnictogen bonding are yet to be formalized, and the term "pnictogen bonding" has come into use only relatively recently [56–59]. The origin and etymology of the terms pnictogen (also sometimes referred to as pnigogen or pnicogen) and pnictide can be traced back to a suggestion by the Dutch chemist Anton Eduard van Arkel (1893–1976) in the early 1950s [60]. Unsurprisingly, therefore, the characteristic (geometric) features of pnictogen bonding in several crystal structures re-examined in this work were not formally identified and characterized by the original workers who had reported the crystals. Our search of the Cambridge Structure Database (CSD) [61] and the Inorganic Chemistry Structure Database (ICSD) [62] produced hundreds—if not thousands—of structures in which nitrogen could conceivably feature a positive site. This was speculated upon based on the directional feature, together with the intermolecular distance of separation. It is worth mentioning that pnictogen bonds have been known since the middle of last century—if not formally identified and named as such—as have halogen and hydrogen bonds, and were certainly not first described only in 2011, as has been asserted [63].

The characterization of non-covalent interactions in many crystal structures has been undertaken based on the "less than the sum of the van der Waals (vdW) radii" concept [64], a widespread concept that has been invoked, for example, in structural and supramolecular chemistry [64–66], biological and medicinal chemistry [67,68], and crystallography [69–74]. According to this concept, when the inter- or intramolecular distance of separation is associated with a structural motif, for example, Pn···D (Pn = the pnictogen atom; D = an electron donor, such as O, N, a halogen anion, etc.), is less than the sum of the vdW radii of Pn and D, then it possible that Pn and D atomic basins are bonded to each other by an attractive engagement. This bonding has been referred to as a "close contact", or simply "a non-covalent interaction" [75]. The criterion, however, rejects those interactions that have intermolecular bond distances slightly larger than the sum of vdW radii of Pn and D. This could be misleading because the proposed vdW radii [76,77] are not necessarily accurate [78] (within an uncertainty of ±0.2 Å) because a spherical symmetry of atoms in molecules was generally assumed for their determination [75], and that there have been many systems reported theoretically and experimentally that fail to strictly obey the criterion [79–83]. Since the charge density profile of atoms in molecules is anisotropic, the vdW "radius" of an atom in a molecular entity is likely to vary between molecular entities. As several have suggested previously [66,75,77,84], we recognize a potential non-covalent interaction between interacting atomic basins even when the interaction distance is greater by several tenths of an Ångstrom than the sum of the respective vdW radii of bonded atomic basins. While doing so, directional features, together with chemical intuition and interpretations provided by the original authors, were considered. In addition, the underlying concepts of molecular electrostatic surface potentials (MESP) [79–83] were utilized in several instances to verify the putative interactions that emerged from the "less than the sum of the vdW radii" concept.

Given the sheer number of candidate structures available in the CSD and ICSD, we decided not to attempt a statistical analysis to delineate the possible occurrence and range of bond distances and angles of approach featuring intermolecular contracts formed between a nitrogen atom and other interacting atomic domains. To do so, one would have to carefully inspect each and every structure to observe whether nitrogen in those crystals does indeed have a positive site; chemical intuition alone does not suffice, and the results could be misleading without extensive computational work. Our search of, for example, the CSD, resulted in thousands of crystals, especially when the intermolecular distance and the angle associated with the motif R–N⋯X (R = any atom of the periodic table; X = O, N, C, F, Cl, Br, and I, etc.) were constrained to the range of 2.5–3.9 Å and 150°–180°. When we inspected the structures, it was found that (1) a large body of them comprises primary, secondary and tertiary interactions, and (2) many of them also comprise N that feature a negative site and is engaged in hydrogen bonding, (3) and many of them contain structures with missing H atoms or overlapping molecular fragments, such that it was very difficult to draw any conclusion about whether N could be a pnictogen bond donor in the crystals above. However, when the distance and angle criteria associated with the motif were restricted to the ranges, 2.8–3.9 Å and 170°–180° (or 3.0–3.5 Å and 170°–180°) (or 3.3–3.7 Å and 175°–180°), the number of hits in the CSD search was greatly reduced to a few hundred. This enabled us to visually scan most of the crystal structures and to select those where nitrogen-centered pnictogen bonding was likely to occur. We, therefore, emphasize that the examples given below are not exhaustive, but illustrative, since our ultimate aim is to inform the reader of the potential significance of nitrogen-centered pnictogen bonding in crystals so that this can be borne in mind when new materials are designed in silico.

2. Computational Details

We energy-minimized several monomeric entities responsible for the illustrative crystal systems chosen for this review, using density functional theory at the ωB97XD [85] and second-order Møller–Plesset (MP2) levels of theory [86]. For reasons described below, basis sets, such as Jorge-ATZP (where ATZP is the augmented triple-ζ plus polarization basis set), Aug-CC-pVTZ and def2-TZVPD were chosen.

The MESP calculations were performed on the fully relaxed geometries of the monomeric entities in the gas phase. This was done to provide insight into the possible ability of these entities—when in close proximity to another similar (or different) entity—to interact attractively with the partner species, causing or assisting in the formation of a supermolecular assembly. Specifically, we computed the local most minimum and maximum of potential, $V_{S,min}$ and $V_{S,max}$, respectively, on their electrostatic surfaces. The 0.001 a.u. (electrons bohr^{-3}) isoelectron density envelope that arbitrarily defines the van der Waals surface of a molecular entity was used on which to compute the electrostatic potential [87–89]. We used the sign and magnitude of these potentials to infer whether specific regions on the surfaces of these molecular entities are electrophilic or nucleophilic [51,89–91].

Accordingly, we utilized the following concepts to provide insight into the way a specific site in a molecular entity makes an attractive engagement with another site in another similar, or different, molecular entity, contributing to the formation of the crystal lattice. A specific region on an atom in a molecular entity was considered to be electrophilic when the sign of either $V_{S,min}$ or $V_{S,max}$ was positive, i.e., $V_{S,min} > 0$ or $V_{S,max} > 0$. When they are negative, $V_{S,min} < 0$ or $V_{S,max} < 0$, then the region was considered to be nucleophilic [92–94].

An intermolecular, or intramolecular interaction, generally occurs (but not always) when a region of an atom or fragment with a positive $V_{S,min}$ (or $V_{S,max}$) is in close proximity to that with a negative $V_{S,min}$ (or $V_{S,max}$). For instance, σ-holes on an atom A in a molecular entity along an extension of the bond R–A are generally associated with $V_{S,max}$ and lone-pair, and π-holes with $V_{S,min}$. Clearly, the nucleophilic portions of a given entity interact attractively with the regions of the most positive electrostatic potential on a neighboring (or the same) entity, leading to σ⋯lone-pair, σ⋯π, σ⋯σ and π⋯π interactions. This is the case for interactions that are coulombic in origin (a positive site attracting a negative one [54,80,95]). However, there are examples known in which anti-electrostatic interactions have been shown to occur [96,97], as well as instances where a positive site attracts a positive site [98], and a negative site attracts a negative site [51] when they are in close proximity. Although this has been suggested to be a consequence of electrostatic polarization [99–103], dispersion plays a significant role in stabilizing such interactions [25,89,98]. Other researchers also used a similar methodology for theoretical studies of non-covalent interactions in other chemical systems [104–108].

We employed the concepts of Type-I, Type-II and Type-III geometric topologies of bonding (Scheme 1) to characterize nitrogen-centered pnictogen bonds in the crystals illustrated in this overview [24,94].

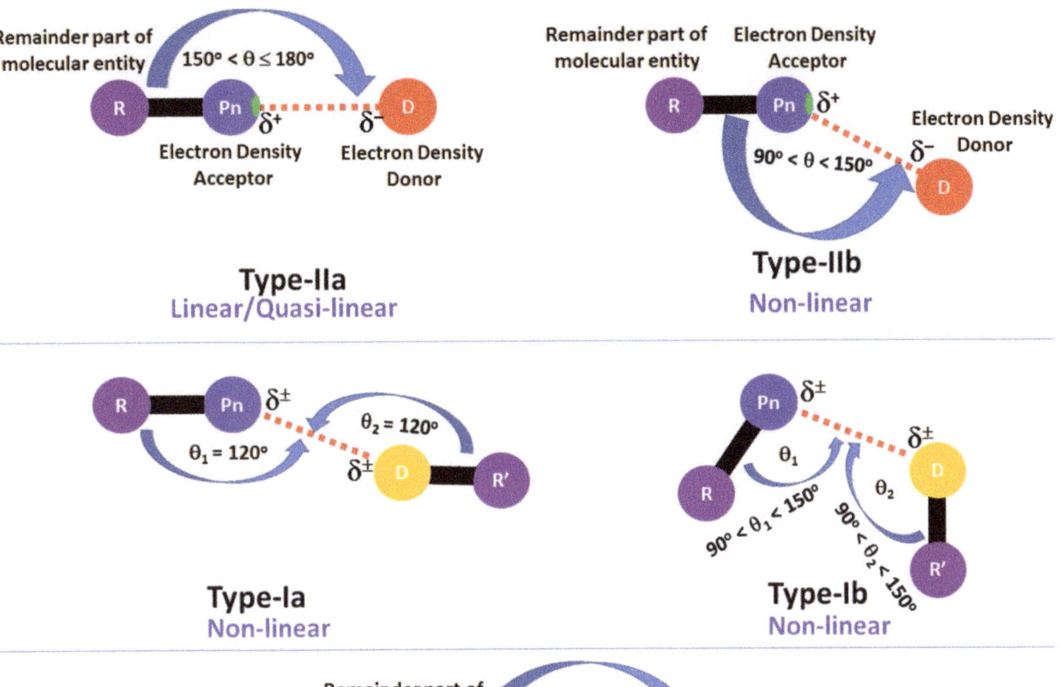

Scheme 1. Geometric topology of chemical bonding interactions formed by the pnictogen atom in molecular complexes and crystals, where Pn, D, R and R' refer to the covalently bound pnictogen atom, the interacting atomic domain (generally nucleophilic), and the remaining part of the molecular entities associated with Pn and D atomic basins, respectively. $\delta^\pm$ signifies the local polarity (positive or negative), and the small region on atom Pn along R–Pn bond extension colored in green indicates a σ-hole.

The electronic structure calculations were performed using the Gaussian 16 program package [109]. The Mercury 4.0 [110], Gaussview 5.0 [111], AIMAll [112], Multiwfn [113], and VMD [114] suite of programs were utilized for the analysis and drawing.

3. Illustrative Chemical Systems in the Crystalline Phase

3.1. The Solid-State Structure of Dinitrogen, N_2

We begin by looking at dinitrogen itself. Solid N_2 displays exceptional polymorphism under extreme conditions. Some of its known phases include α, β, γ, δ, ε, ι and θ [115–117]. Because the phases of the crystal depend on an external agency (temperature and pressure), the packing is different in each phase. Consequently, the intermolecular interactions between the N_2 molecules in these crystals are different.

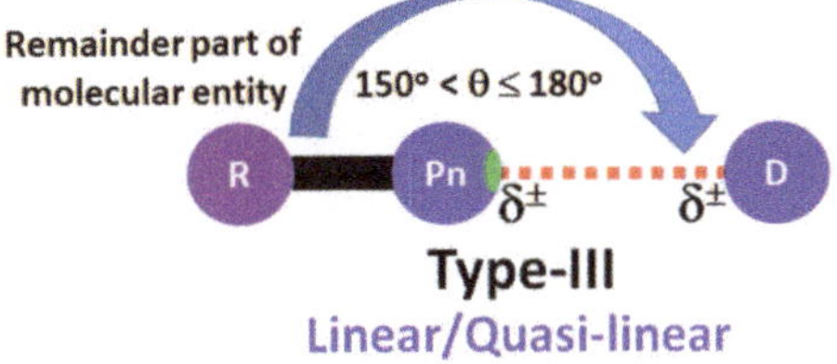

The α-phase (third allotrope) of solid N_2, with $Z = 4$, space group *Pa-3*, Figure 1a, determined using electron diffraction measurements [118], is mainly stabilized by $\pi\cdots N$ (lone-pair) interactions between the N_2 molecules. The centers of the N_2 molecules are on an f.c.c. lattice with each molecule pointing in a different direction. The intermolecular distance between N in one molecule and the midpoint of the π bond in a neighboring molecule is 3.565 Å. We attribute this interaction to π-centered pnictogen/nitrogen bonding.

The β-phase of hexagonal N_2 ($Z = 2$), Figure 1b [119], comprises two types of intermolecular bonding modes, including $N\cdots N$ and $\pi\cdots\pi$. The first of these contacts occur between the N_2 molecules along the crystallographic c-axis, as shown in Figure 1b, with $r(N\cdots N) = 3.015$ Å and $\angle N\equiv N\cdots N = 132.3°$. It is perhaps a Type-Ib topology of pnictogen bonding. The second type of contact is the result of a slip parallel arrangement between the N_2 molecules in the crystal along the crystallographic a and b axes, with an intermolecular distance of 3.861 Å between the centroids of a pair of triple bonds on two neighboring N_2 molecules ($r(\pi\cdots\pi) = 3.861$ Å).

The γ-phase of tetragonal N_2 ($Z = 2$), Figure 1c [119], consists of $N\cdots N$ and $\pi\cdots N$(lone-pair) interactions. The former ones are longer than the latter. Similar to the β-phase, the $N\cdots N$ contacts follow a Type-Ia topology of bonding, with $r(N\cdots N) = 3.292$ Å and $\angle N\equiv N\cdots N = 121.8°$, but are weaker. The $\pi\cdots N$ (lone-pair) pnictogen bonded interactions have $r(N\cdots\pi) = 3.415$ Å and are stronger than the $\pi\cdots\pi$ interactions in β-N_2. Regardless of the nature of the intermolecular interactions, they are all marginally shorter or somewhat longer than twice the vdW radius of N, 3.32 Å ($r_{vdW}(N) = 1.66$ Å [77]).

The ε-phase of rhombohedral N_2 ($Z = 24$), Figure 1d [120], has two types of $N\equiv N\cdots N$ bond distances, $r(N\cdots N) = 2.727$ Å and 2.801 Å, corresponding to $\angle N\equiv N\cdots N$ of 133.1° and 120.9°, respectively, and are less than twice the vdW radius of N, 3.32 Å. Additionally, N(lone-pair)$\cdots\pi$ interactions may be present between the N_2 molecules in the crystal, with $r(N\cdots\pi) = 2.958$ Å.

The crystal structures of the high-pressure δ and $\delta*$ phases of nitrogen were also investigated using single-crystal X-ray diffraction (not shown) [121]. The structure of the δ phase is isostructurally similar to that of γ-O_2. Thus, it comprises spherically disordered molecules, with a preference for avoiding pointing along the cubic $\langle 100 \rangle$ directions, and disk-like molecules with a uniform distribution of orientations. The structure of the $\delta*$ phase is tetragonal, and the space group was identified unambiguously as $P4_2/ncm$ at 14.5 GPa.

We did not observe any potentially directional interactions in either of the pressure-induced phases of N_2. In all cases, the $N\cdots N$ interactions follow a Type-Ia/Type-Ib topology of bonding and occur between interacting sites of dissimilar electron density. The possibility of the $N\cdots N$ and $\pi\cdots N$ (lone-pair) interaction can be inferred from the MESP model of an isolated molecule, Figure 1e. The delocalized bonding region in $N\equiv N$ is equipped with a belt of positive potential characterized by near equivalent local most maxima (tiny red spheres). In contrast to what might be expected on the surface of a covalent bound halogen in a molecular entity, such as, for example, in HBr and HCl, we observed that each N along the $N\equiv N$ bond extension is accompanied by a local most minima of potential (tiny blue spheres in Figure 1e) and is very negative. This signifies that there is a buildup on charge density on the surface of N along the outer $N\equiv N$ and the buildup is significant compared to the lateral sites on the same atoms.

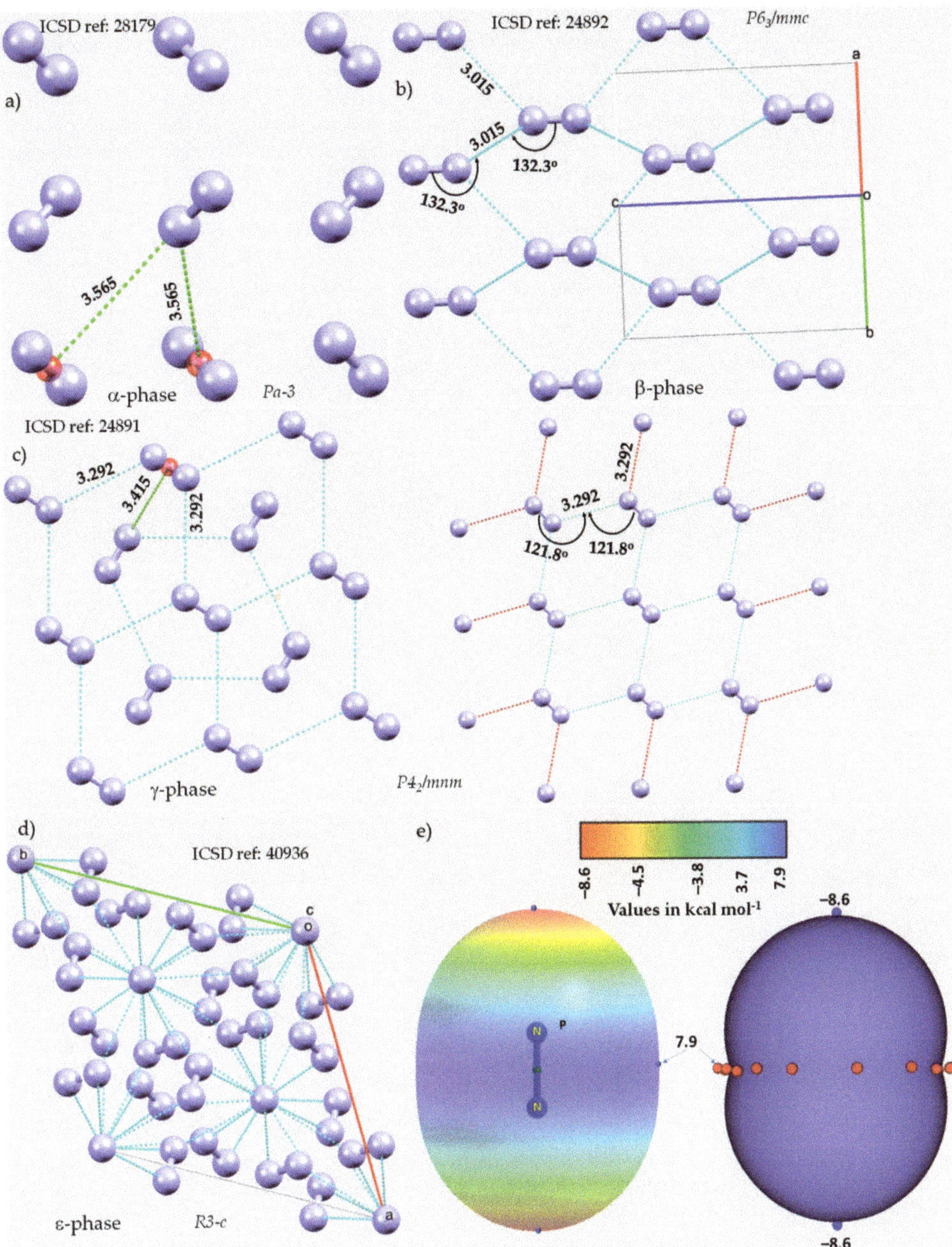

Figure 1. The nature of bonding interactions between the N$_2$ molecules in the crystals of (**a**) α-N$_2$, (**b**) β-N$_2$, (**c**) γ-N$_2$, and (**d**) ε-N$_2$. (**e**) The 0.001 a.u. isodensity mapped MESP plot of an isolated N$_2$ molecule, obtained at the MP2(full)/aug-cc-pVTZ level of theory. The local most minima and maxima of potential are marked by tiny circles in blue and red, respectively. Selected bond lengths and bond angles are in Å and degree, respectively. Crystallographic axes are not shown in (**a,c**) for clarity. The crystal symmetry is shown for each case, together with ICSD references.

3.2. The Nitrogen Trihalides, NX₃ and Their Crystal Structures

We now discuss a set of other examples in which covalently bound nitrogen acts as an electrophile, as observed in crystalline systems reported over many years. In these systems, covalently bound nitrogen, as in NX_3 (X = F, Cl, Br, I), has a positive region on the R–N bond extension that is capable of attracting a negative site on an identical, or in a different molecule. This can be appreciated by looking at the MESPs shown in Figure 2, obtained using ωB97XD/Jorge-ATZP. The top view of Figure 2 shows nitrogen having a positive electrostatic potential along the X–N bond extensions, regardless of the nature of the halogen derivative attached to it. It is largest (16.6 kcal·mol⁻¹) along the F–N bond extensions in NF_3. As one passes from NF_3 through NCl_3 to NBr_3, there is a decrease in $V_{S,max}$ from 16.6 kcal·mol⁻¹ to 5.4 kcal·mol⁻¹ to 2.7 kcal·mol⁻¹, and the stability of the σ-hole on N in NX_3 occurs in the order X = F > Cl > Br. By contrast, the lone-pair dominated region on N is negative, and the negativity increases in the order NF_3 (−4.1 kcal·mol⁻¹) > NCl_3 (−11.8 kcal·mol⁻¹) > NBr_3 (−14.2 kcal·mol⁻¹). These results demonstrate the amphoteric character of N in NX_3.

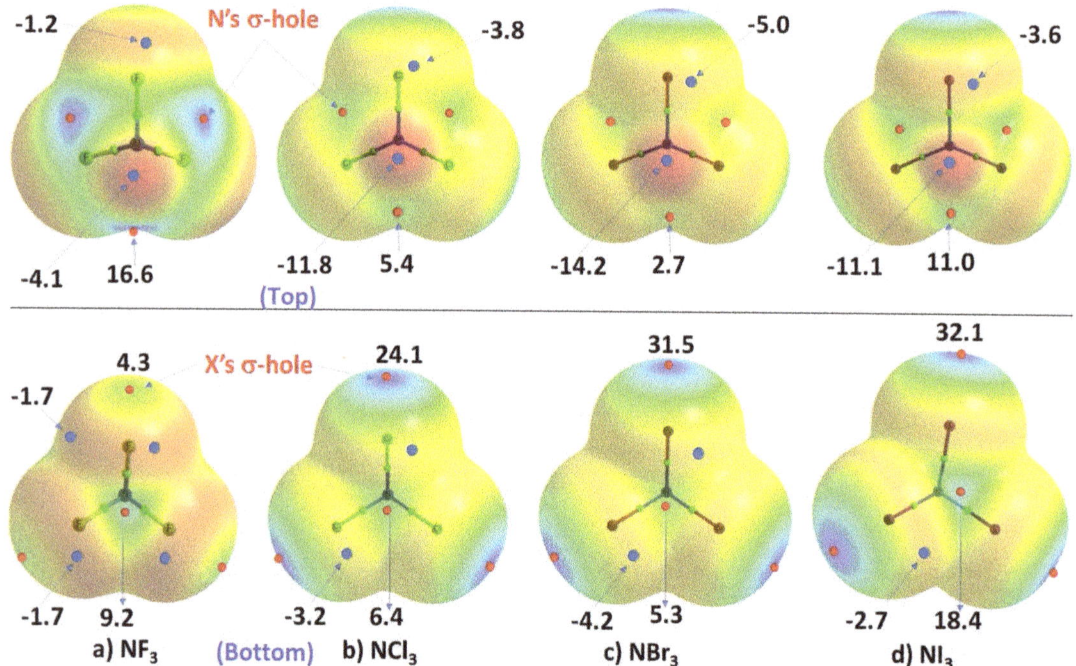

Figure 2. Comparison of the ωB97XD/Jorge-ATZP calculated 0.001 a.u. isodensity envelope mapped potential on the electrostatic surfaces of NX_3 (X = F, Cl, Br, I) molecules. Selected $V_{S,max}$ and $V_{S,min}$ values in kcal·mol⁻¹ are shown, which are the local most maximum and minimum of potential (red and blue circles), respectively. Two views of each MESP graph are shown. In the top view, bonded N faces the reader, whereas, in the bottom view, it is the three halogens that face the reader. The Quantum Theory of Atoms in Molecules (QTAIM [122])-based molecular graph is shown for each entity, with the circles in green representing bond critical points and bonds in atom color.

The MESP results for NI_3 are also included in Figure 2, but clearly do not follow the same trend noted above. This may be an artifact of the basis set we used for these calculations, and we clarify this further below. Whilst the numerical values of potential may be out of line, the reactive nature of the N site along and around the I–N bond extensions in NI_3 behaves qualitatively similar to that found for the other three NX_3 entities.

As shown in the bottom view of Figure 2, $V_{S,max}$ in NX_3 increases monotonically from X = F through I along the N–X bond extension, indicating that the magnitude of the σ-holes associated with these potentials increases in the same order and the σ-holes are surrounded by belt-like negative potentials. In other words, the least polarizable F has the smallest σ-hole on N–F bond extensions and the most polarizable I has the largest σ-hole on the N–I bond extensions. The difference in $V_{S,max}$ between NBr_3 and NI_3 is not large and, as we commented above, the MESP values for NI_3 with the chosen basis set used may not be very reliable.

In all cases, the central region of the triangular face formed by the three X atoms in each NX_3 is found to be positive. The positive and negative nature of $V_{S,max}$ and $V_{S,min}$ on the surface of each constituent atomic domain in NX_3 suggest that each of them not only has the ability to form complexes with another identical (or different) molecule but also has the ability to act both as an acid and a base.

In order to determine whether or not the inconsistency in the nature of the negative and positive regions on the surface of specific atoms in the series of four molecules NX_3 is a basis set artifact, we examined the same properties using another basis set, def2-TZVPD, available in the basis set exchange library [123]. We have also adopted a higher level of theory, MP2, for the same calculation as it is one of the simplest and most useful levels of theory beyond Hartree–Fock that accounts for effects arising from electron–electron correlation. The results are summarized in Table 1.

Table 1. ωB97XD/def2-TZVPD and MP2(full)/def2-TZVPD computed 0.001 a.u. isodensity envelope mapped the local most minima and maxima of potential on the electrostatic surfaces of NX_3 (X = F, Cl, Br, I) molecules. $V_{S,max}$ and $V_{S,min}$ values in kcal·mol^{-1}.

Species	NF_3		NCl_3		NBr_3		NI_3	
Method	ωB97XD	MP2 (Full)	ωB97XD	MP2 (Full)	ωB97XD	MP2 (Full)	ωB97XD	MP2 (Full)
Extrema of potential	$V_{s,max}$							
N–X (one on each X)	4.7	5.2	25.0	24.5	29.7	28.2	36.3	32.1
X–N (one on each extension)	16.5	16.5	5.2	4.6	3.5	3.5	2.1	2.3
Center of the triangular X_3 face	8.8	7.5	5.6	4.9	5.4	4.3	6.4	4.9
	$V_{S,min}$							
On lone-pair region on N	−4.1	−3.7	−11.9	−12.6	−14.1	−13.8	−17.2	−15.2
On F (within the X_3 face)	−1.8	−1.9	−3.7	−3.4	−4.3	−3.9	−4.9	−4.0
On X (opposite of the X_3 face)	−1.3	−1.3	−4.0	−4.1	−4.8	−4.6	−5.3	−4.8

As can be seen from the data in Table 1, and regardless of the correlated method used in conjunction with the def2-TZVPD basis set, the N has a positive σ-hole along the X–N bond extensions. These σ-holes become less positive as the size of the halogen in NX_3 increases from F down to I, and there are three equivalent holes on N in each NX_3. By contrast, the halogen derivative has a single σ-hole on its surface along the N–X bond extension, so there are also three positive σ-holes on the surfaces of the three X atoms that are also equivalent (only one is listed in Table 1). They systematically increase with the increasing size of the halogen in NX_3, in agreement with what might be expected from their polarizabilities.

3.2.1. Nitrogen Trifluoride, NF_3

The crystal structure of NF_3 was reported only recently, and corresponds to the low-temperature α-phase [124]. Since powder neutron diffraction measurements were performed, it is expected that the intermolecular geometry is more accurate than might have been obtained from X-ray diffraction measurements.

The α-phase of NF_3 crystallizes in the orthorhombic space group *Pnma*, with lattice parameters $a = 6.71457(13)$ Å, $b = 7.30913(14)$ Å, $c = 4.55189(8)$ Å, cell volume $(V) = 223.396(7)$ Å^3, and $Z = 4$ at $T = 6$ K. The β-phase of NF_3 corresponds to a high-temperature phase and was observed to be a plastic crystal (space group $P4/mnm$) with lattice parameters $a = 15.334(6)$ Å, $c = 7.820(3)$ Å, $V = 1838.6(12)$ Å^3, and $Z = 30$ at $T = 60$ K. It was suggested that the crystal structure of this latter phase is closely related to that of the Frank–Kasper sigma phase, but the one deposited in the ICSD (ref. code 1891641) does not contain the geometry of β-NF$_3$ (the fluorine atoms are missing).

The unit-cell of the α-phase comprises four units of NF$_3$, as shown in Figure 3a. It shows the intermolecular bonding modes between the building blocks responsible for the unit-cell of the NF$_3$ crystal; these are consistent with the attraction between the positive and negative sites of electrostatic potential localized on different atomic domains (Figure 3b). In particular, Type-IIa N–F···F and F–N···F non-covalent interactions are observed and are inferred from the intermolecular distances and angles of interaction. The former is less directional but shorter than the latter (Figure 3a).

The real nature of the intermolecular interactions between the molecular units in the crystal may not be apparent in inspecting the unit-cell alone; a periodic extension of the unit-cell is necessary. The 2×2 supercell structure of α-NF$_3$ is shown in Figure 3c. Although there are many more interactions between the molecules of NF$_3$ in the crystal, we have highlighted in Figure 3c only the prominent bonding modes formed by the N site. As revealed by the MESP model (Figure 2a), the three positive sites along the F–N bond extensions do indeed donate pnictogen-centered σ-hole bonds to the lateral portions of the F sites in the nearest NF$_3$ units (Figure 3a). Of the three, two are equivalent and the other is longer (3.169 Å vs. 3.368 Å). They are all directional since ∠F–N···F for each of the two equivalent interactions is 172.3° and that for the longer bond is 174.5° (Figure 3a).

In crystalline α-NF$_3$, the nitrogen in the NF$_3$ molecule acts as a hexa-furcated center in donating σ-hole bonds. It donates three highly directional σ-hole bonds (vide supra) and two relatively less directional equivalent σ-hole bonds ($r(N···F) = 3.301$ Å and ∠F–N···F $= 143.4°$) that are caused by the attraction between the positive site on the central region of the surface formed by the three fluorine atoms in one molecule and the negative region dominated by the lone-pair of a neighboring molecule, thereby forming five pnictogen bonds (Figure 4a,b). Because nitrogen has a negative site (readily appreciated in its Lewis structure) evidenced by the $V_{S,min} = -3.6$ kcal·mol^{-1} shown in Figure 2a, it shows a tendency to accept a σ-hole bond from the covalently bound fluorine of another interacting NF$_3$ molecule, responsible for the formation of an F···N halogen bond (Figure 4b). The back-to-back arrangement between the NF$_3$ molecules, Figure 4a, causing the additional intermolecular interactions noted above, is a result of a $\pi(N)···\pi(N)$ interaction. The behavior of N in forming so many non-covalent interactions appears to be unique to this system.

Our investigation shows that fluorine has the ability to act both as a hexa- and tetra-furcated center, as shown in Figure 4c,d, respectively. It is evident from Figure 4c that fluorine is an acceptor of three σ-hole bonds, thus forming one F–N···F and two N–F···F halogen bonds; it is a donor of a single σ-hole bond, forming an N–F···N halogen bond; and it shows capacity to form two or more N–F···F–N halogen-bonded Type-Ia and/or Type-Ib interactions—all within a distance of 2.90–3.20 Å. The Type-Ia N–F···F–N halogen-bonded contacts, such as those marked in Figure 4d, are commonly observed in fully fluorinated compounds, for example, the crystal of C$_6$F$_6$ [125], and in similar compounds [126–128]. Clearly, the complex topology of bonding between the NF$_3$ molecules in the crystal structure requires a variety of theoretical studies to detail the nature and the strength of the interactions involved.

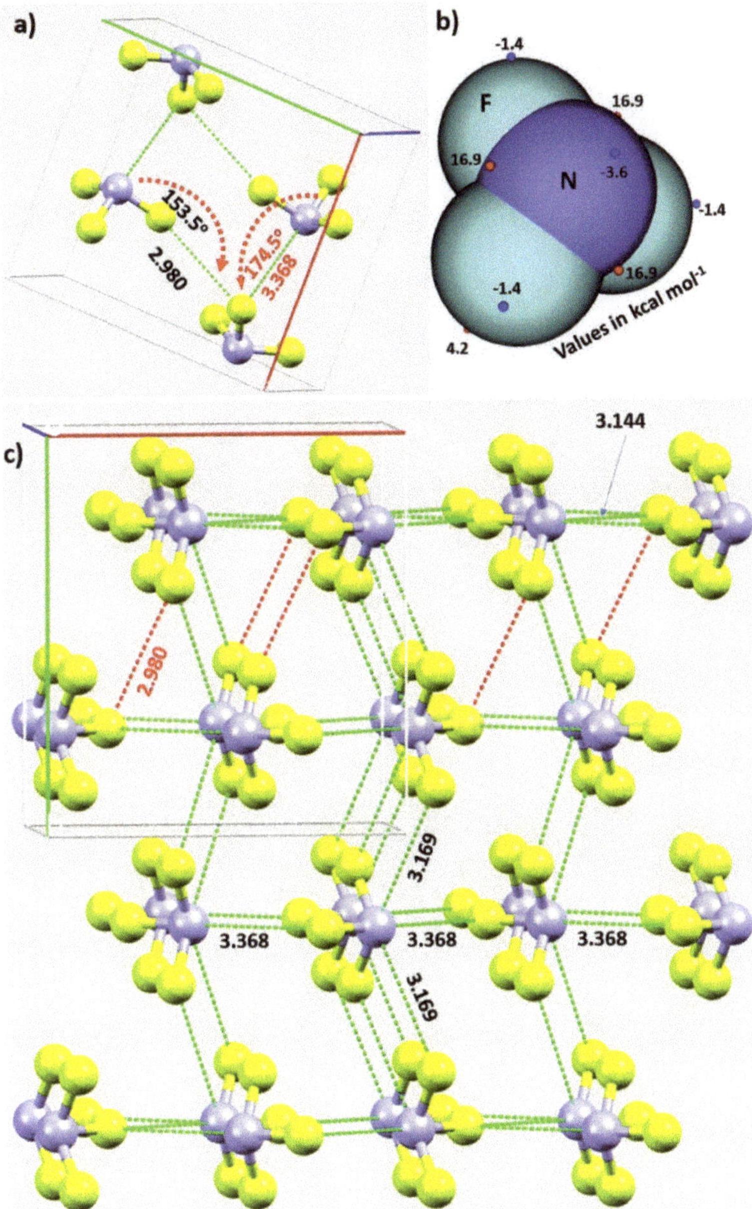

Figure 3. (**a**) The unit-cell of α-NF₃ (CSD ref. code 1891640), showing selected N⋯F and F⋯F σ-hole-centered intermolecular interactions and the intermolecular angles for the approach of the electrophile. (**b**) Illustration of selected local maximum and minimum of electrostatic potential (red and blue spheres, respectively) mapped on the 0.001 a.u. isoelectron density surface of NF₃ obtained at the ωB97XD/aug-cc-pVTZ level of theory. (**c**) The network of N⋯F and F⋯F σ-hole centered pnictogen and halogen bonding interactions in 2 × 2 supercell geometry of α-NF₃. Bond lengths and bond angles in Å and degrees, respectively. Dotted lines represent intermolecular interactions.

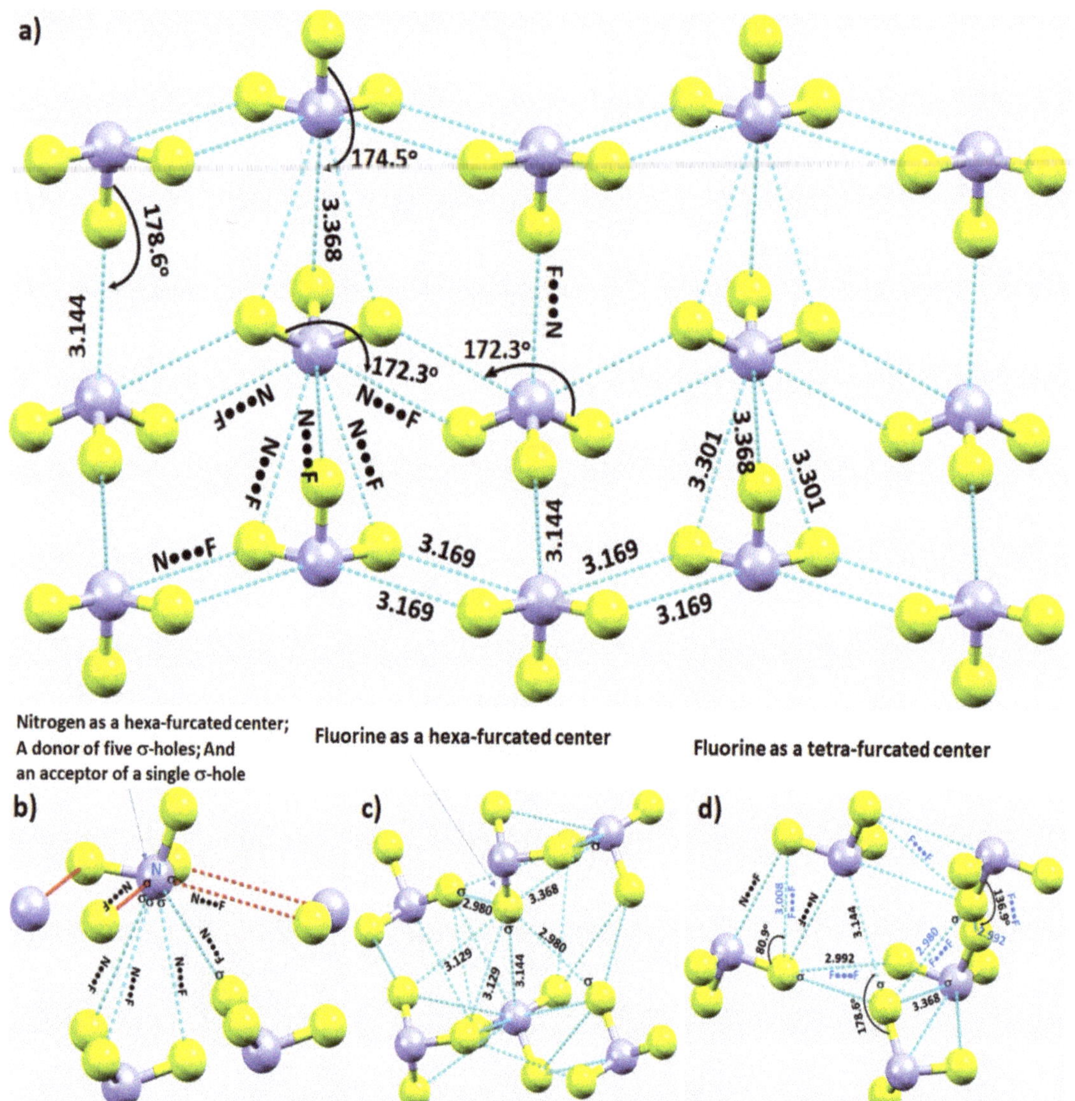

Figure 4. (**a**) Nature of N···F pnictogen bonds in the α-phase of the NF_3 crystal. (**b–d**) Nature of the local topology of intermolecular bonding interactions around N and F in NF_3. Bond lengths and bond angles in Å and degrees, respectively. The symbol "σ" on N/F in (**b–d**) refers to the covalently bonded atom donating the σ-hole.

3.2.2. Nitrogen Trichloride, NCl_3

The crystal structure of NCl_3 was reported in 1975 [129]. It crystalizes in the orthorhombic space group *Pnma*. There are 12 molecules in the unit-cell, as shown in Figure 5a. The mean N–Cl bond distance is 1.75(1) Å and the mean bond angle is 107(2)°. The insights gained from the values of $V_{S,max}$ and $V_{S,min}$ on the surface of an isolated NCl_3 molecule above (Figure 2b) can now be used to understand its chemical reactivity.

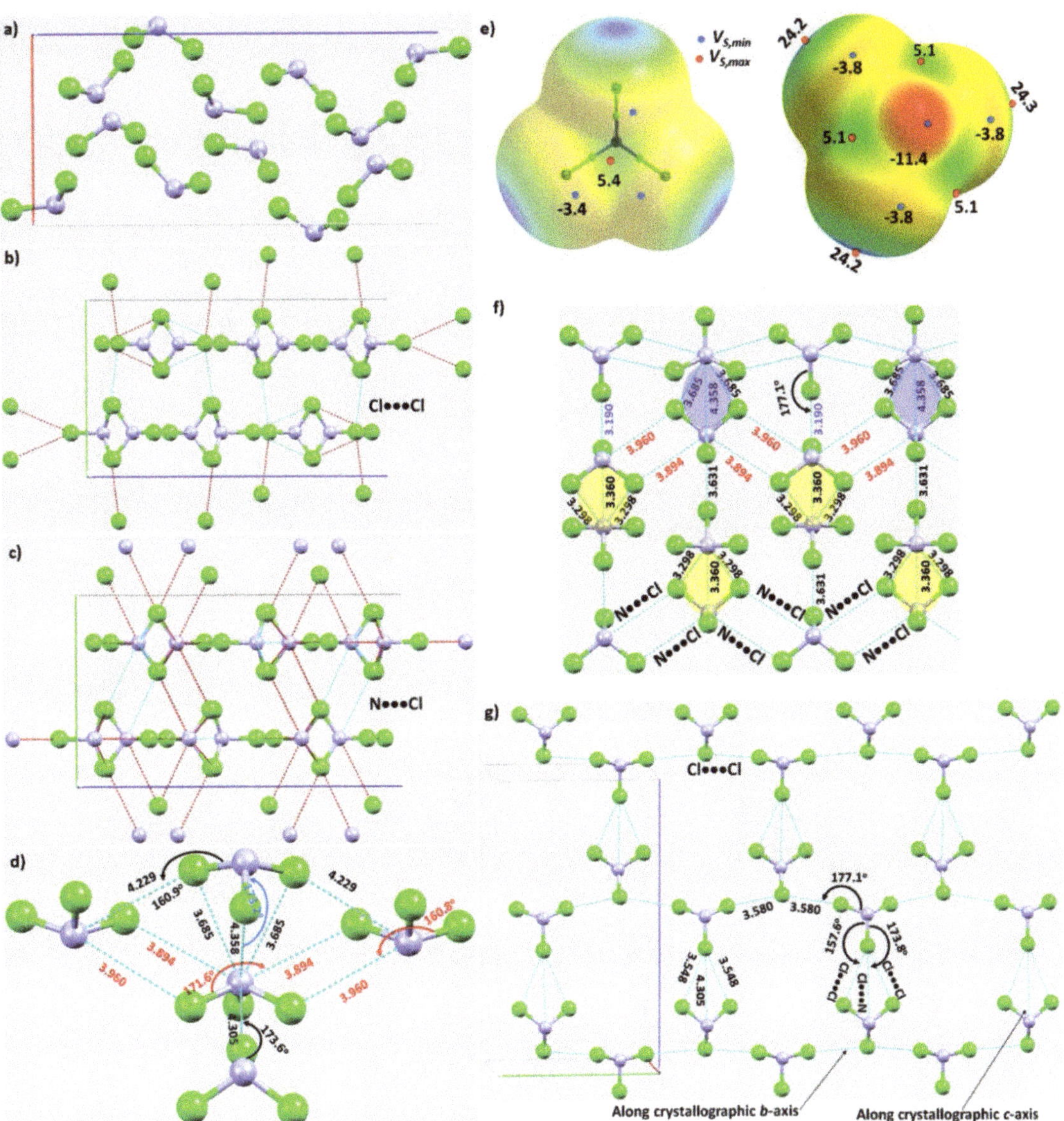

Figure 5. (**a**) The unit-cell of the NCl₃ crystal (ICSD ref. code: 4034). (**b**) Illustration of the nature of Cl···Cl intermolecular interactions between the molecular building blocks. (**c**) Illustration of the nature of the N···Cl interactions. (**d**) Nature of the local topology of bonding around N in NCl₃, showing N is a penta-furcated center. (**e**) Nature of the reactive sites on the electrostatic surface of the NCl₃ molecule responsible for the N···Cl, Cl···N and Cl···Cl interactions, computed with ωB97XD/aug-cc-pVTZ. (**f**) The nature of various N···Cl and Cl···N interactions in the crystal. (**g**) The nature of Cl···Cl and Cl···N halogen-bonded interactions in the crystal. Values of bond lengths and bond angles in Å and degrees, respectively.

The N–Cl···N and N–Cl···Cl intermolecular interactions holding the NCl₃ molecules together in the crystal can be rationalized by referring to Figure 5b,c, respectively. The extent to which covalently bound N in the NCl₃ molecule is engaged with the negative sites on neighboring NCl_3 molecules, and producing the observed solid-state structure, is illustrated in Figure 5d. The bonding features are as expected when positive and negative sites on molecular surfaces are in close proximity, inferred from the MESP of NCl_3 (Figure 5e).

The pattern of intermolecular interactions between bonded N and Cl in the NCl_3 crystal is illustrated in Figure 5f. It is evident that each N donates three σ-holes. Two of these, the attraction between a covalently bonded N in an NCl_3 molecule and chlorines on neighboring molecules, generate a zig-zag chain-like pattern. They are inequivalent, in contrast to the analogous interactions observed in the NF_3 crystal described above. The $r(N{\cdots}Cl)$ $(\angle Cl–N{\cdots}Cl)$ for these two contacts are 3.960 Å (160.8°) and 3.894 Å (171.6°), respectively, suggesting a Type-IIa nitrogen bonding topology. The remaining σ-hole bond formed by N, Figure 5d,f, is longer and quasi-linear ($r(N{\cdots}Cl)$ = 4.358 Å and $\angle Cl–N{\cdots}Cl$ = 173.2°), and is a result of the packing of molecules in the crystal. As shown in Figure 5d, each N site serves as a penta-furcated center in accepting and donating σ-hole bonds.

The three pnictogen bonds are augmented by σ-centered N–Cl$\cdots$N and N–Cl$\cdots$Cl halogen bonds in stabilizing the crystal. These Type-IIa N–Cl$\cdots$N bonds are highly directional ($r(Cl{\cdots}N)$ = 3.190 Å and $\angle N–Cl{\cdots}N$ = 177.1°), and are formed between the σ-hole on the N–Cl bond extension in one molecule and the lone-pair dominated region on N in another interacting molecule (Figure 5g).

The three N$\cdots$Cl contacts highlighted in yellow in Figure 5f are repeated throughout the crystal. Two of them are equivalent ($r(N{\cdots}Cl)$ = 3.298 Å) and the other slightly longer ($r(N{\cdots}Cl)$ = 3.360 Å). They are significantly bent, with $\angle Cl–N{\cdots}Cl$ of 140.3° and 140.1°, respectively. We assign them to be Type-IIb based on the classification of bonding topology provided in Scheme 1. They are enforced by the attraction between the positive site in the central region of the triangular face formed by three Cl atoms in the NCl_3 molecule and the lone-pair dominated negative site on the N of a neighboring molecule. One might characterize this as an $N(\pi){\cdots}N(\pi)$ pnictogen bond, consistent with the surface extrema revealed by the MESP model (Figure 5e).

Figure 5g shows the pattern of occurrence of Type-IIa N–Cl$\cdots$Cl halogen bonds. Those having a Cl$\cdots$Cl bond distance (and $\angle N–Cl{\cdots}Cl$) of 3.580 Å (168.4°) are more directional than those with the corresponding values of 3.548 Å (157.6°). Indeed, they are shorter and less directional than Cl$\cdots$N halogen bonds with bond distances (angles) of 4.305 Å (173.8°), Figure 5g. There are numerous Cl$\cdots$Cl contacts present in the crystal (not shown).

3.3. Halogen Azides, XN_3

Halogen azides XN_3 (X = F, Cl, Br, I), crystallographically known for some time [130–134], have also been studied more recently [135,136]. ClN_3 forms a chain-like polymer in the solid state, and BrN_3, a helical structure, due to the formation of intermolecular Br$\cdots$N$_\alpha$ and $N_\beta{\cdots}N_\gamma$ interactions. The unit-cells of IN_3 yield structures of polymeric $(–I–N_3–)_n$ chains that are interlocked into layers, yet two modifications were reported for IN_3. Its crystal structure, first determined by X-ray diffraction in 1993 [134], corresponds to the α-phase of the crystal. The second modification of the crystal is called the β-phase [135,137]. Although most of the past studies, including the recent one published in 2021 [137], have determined the lattice constants and important covalent bonding features, which have enhanced our knowledge of covalent p-block azide chemistry, the intermolecular bonding interactions between the XN_3 molecules responsible for the crystals are yet to be explored computationally. The study of Schulz and coworkers [135] demonstrates that ClN_3 adopts a polymeric structure in the solid state with short intermolecular Cl$\cdots$Cl distances, as was observed for the elemental halogen. The crystal structures of ClN_3 [135], and BrN_3 [136], are illustrated in Figure 6a,b and Figure 7a, respectively.

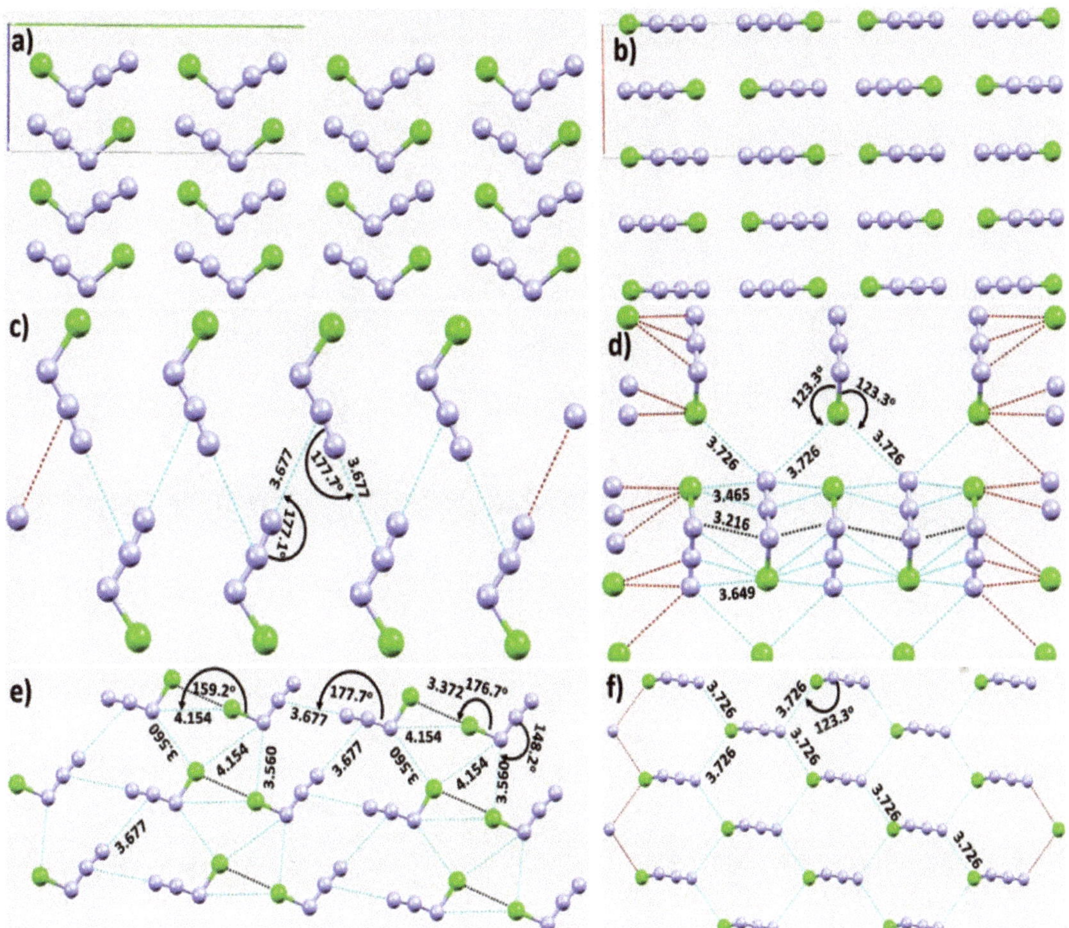

Figure 6. (**a,b**) Two different views of the crystal structure of chlorine azide (ICSD ref code: 424502). The nature of (**c**) $\pi(N)\cdots(N)$ contacts, and (**c,d**) a pattern of the $\pi(N)\cdots(N)$, $N\cdots Cl$, and $Cl\cdots Cl$ contacts in the crystal. (**e**) Illustration of $Cl\cdots Cl$ and $\pi\cdots N$ intermolecular halogen bonding and pnictogen bonding interactions in the crystal. (**f**) The nature of non-linear $Cl\cdots N$ halogens in the same crystal. Selected bond distances and bond angles are in Å and degree, respectively.

From the resonance structure, the central N atom of N_3^- in XN_3 is expected to be positive; this is indeed what we find with our MESP calculations (vide infra). The calculated $V_{S,max}$ and $V_{S,min}$ are both positive on the surface of this atom, and carry opposite signs for the terminal and covalently bonded N. This explains why the σ-hole on the X atoms in a given XN_3 ($X = Cl$, Br) is directed towards the negative N/X of a neighboring molecule, and why the negative tip on the terminal N atom in a given XN_3 is directed towards the positive region on the central N atom of the N_3^- in another XN_3 (see, for example, Figure 6c–f). Both the interaction types are short and significantly linear, with $r(\pi(N)\cdots N)$ and $\angle N{=}N\cdots N$ values of 3.677 Å and 177.7° in NCl_3 (Figure 6e), respectively, and of 3.093 Å and 163.3° in NBr_3 (Figure 7b), respectively. The nature of the intermolecular links between bonded atomic basins can be deduced from Figure 6c–f, including the potential involvement of Type-I and Type-II $\pi(N)\cdots N$ (Figure 6c–e), and $Cl\cdots N$ and $Cl\cdots Cl$ (Figure 6e,f) nitrogen and halogen bonding interactions. In some cases, a secondary interaction may arise as a consequence of the primary interaction. For instance, the primary Cl⋯Cl Type-IIa interaction ($r(\pi(N)\cdots N)$ = 3.372 Å) shown in Figure 6e is likely to be responsible for the development of the Cl⋯N secondary interactions ($r(Cl\cdots N)$ = 4.154 Å) that are probably very weak.

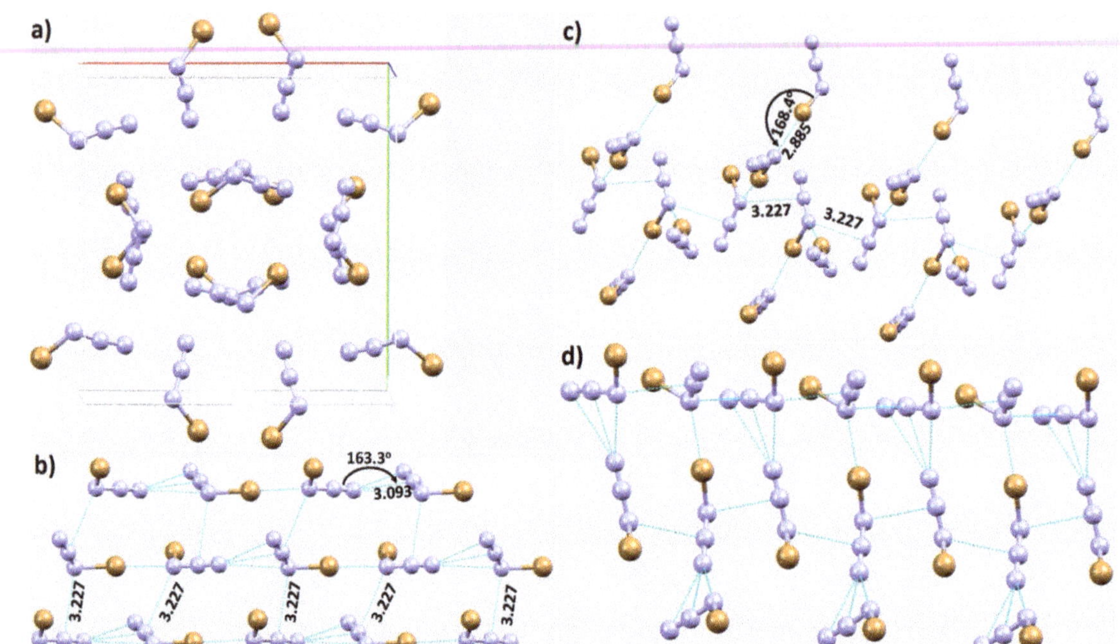

Figure 7. (**a**) The crystal structure of bromine azide (ICSD ref code: 423741). The nature of (**b**) $\pi(N)\cdots(N)$ and Br⋯N contacts, and (**c**) a pattern of the $\pi(N)\cdots(N)$, and Br⋯N contacts in the crystal. (**d**) The nature of non-linear $\pi(N)\cdots(N)$ pnictogen bonds in the same crystal. Selected bond distances and bond angles are in Å and degree, respectively.

The packing between the molecular units in ClN_3, or BrN_3 (Figure 7a), is not very different, and the pattern of intermolecular interactions found in ClN_3 is also evident in BrN_3. In addition to the observed $\pi(N)\cdots(N)$ and Br⋯N pnictogen- and halogen-bonded interactions between the BrN_3 units (Figure 7b,c), the slip parallel arrangement between the NBr_3 units (Figure 7d), which is also seen in the crystal of ClN_3, and which is caused by the attraction between π electron densities, is seen in Figure 7d. Clearly, N–X⋯N halogen bonding and N=N⋯N pnictogen bonding interactions play a vital role in the assembly of the molecular units, and hence, in the stability of these azido-based crystals.

The conclusions reached about the reactivity of the surfaces of XN_3 molecules above are consistent with the positive and negative signs of the local most minima and maxima of potential, summarized in Table 2. As can be seen from the data, the most positive region is identified on the surface of the central N atom of N_3^- in XN_3 in XN_3 compared to that of the terminal and halogen-bonded nitrogen atoms. The largest values of $V_{S,max}$ are 30.2 and 30.8 kcal·mol^{-1} in FN_3 with ωB97XD and MP2 (full), respectively. This becomes less positive as the size of the halogen increases from F down to I in XN_3, a trend that is independent of the correlated method employed. On the other hand, the axial portion of the covalently bonded halogen atom is negative in FN_3 but positive in XN_3 (X = Cl, Br, I). This is consistent with F in FN_3 being significantly less polarizable than the heavier halogens in XN_3 (X = Cl, Br, I), as expected from their polarizabilities (F < Cl < Br < I), and explains why covalently bound Cl and Br atoms make σ-hole centered halogen bonding interactions with the negative site centered on the covalently bonded N in a neighboring molecule (see Figures 6e,f and 7b,c for the ClN_3 and BrN_3 systems, respectively).

Table 2. ωB97XD/def2-TZVPD and MP2(full)/def2-TZVPD computed 0.001 a.u. isodensity envelope mapped local most minima and maxima of potential on the electrostatic surfaces of XN_3 (X = F, Cl, Br, I) molecules. $V_{S,max}$ and $V_{S,min}$ values are in kcal·mol^{-1}. Atom numbering of the N_3^- moiety in all members of the series is as in FN_3.

Species	FN$_3$		ClN$_3$		BrN$_3$		IN$_3$	
Method	ωB97XD	MP2 (Full)	ωB97XD	MP2 (Full)	ωB97XD	MP2 (Full)	ωB97XD	MP2 (Full)
Extrema				$V_{S,min}$				
On N3 of N2=N3	−4.5	−4.3	−8.4	−9.1	−10.3	−11.2	−13.8	−13.8
On N1 of N2=N1	−12.5	−11.6	−14.6	−13.4	−15.0	−13.4	−15.9	−14.5
On X of N1–X	−14.5	−13.4	−5.6	−4.1	−3.7	−2.1	−1.1	0.02
				$V_{S,max}$				
On X of N–X	−3.4	−1.7	23.5	25.3	31.7	33.1	42.8	41.2
On N1 of N2=N1	−6.4	−5.3	−2.2	−0.8	−0.5	1.0	2.5	3.4
On N2	25.8	24.6	23.7	21.5	22.4	20.1	21.0	18.3
On N2	30.2	30.8	24.6	23.4	22.3	20.7	17.8	16.4

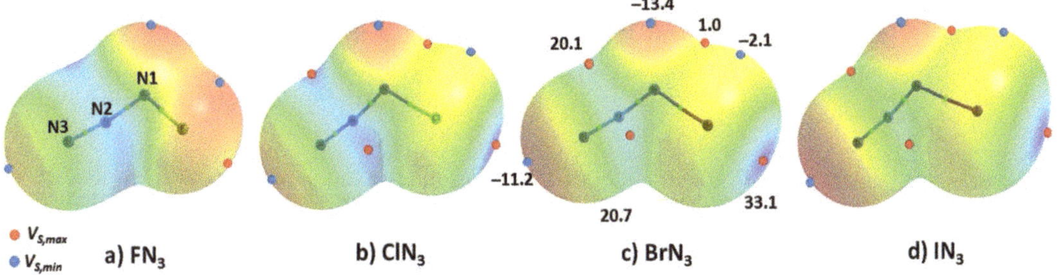

The MP2(full) and ωB97XD predicted electrostatic potentials are consistent with the conclusions reached above for most regions on the surfaces of XN_3, although MP2(full) produces lower values for most sites. A regular trend in a minimum or maximum on the surface of a given molecule is maintained across the series. For instance, the $V_{S,min}$ on the surface of the terminal N along the N=N bond extension in FN_3 is calculated with MP2(full) to be negative (−4.3 kcal·mol^{-1}) and is increasingly larger on the same atom in the other three systems of the family as one passes from FN_3 to ClN_3 to BrN_3 to IN_3. An anomaly is found on the nature of the lateral portion of the covalently bonded iodine in IN_3, for which MP2(full) and ωB97XD have predicted the $V_{S,min}$ associated with it to be slightly positive or close to neutral (0.02 kcal·mol^{-1}) and fully negative (−1.1 kcal·mol^{-1}), respectively. Which one of these is more reliable can only be cross-checked with a computational exploration of the interactions of this site with the negative site on other molecules. Nevertheless, this result unequivocally suggests that the magnitude of electrostatic potential (and sometimes its sign) is dependent on the nature of electron correlation [23,25,89], and hence care should be taken when computing potentials on the electrostatic surfaces of chemical systems. Another way to overcome this is to use a slightly larger isodensity envelope, such as 0.0015 a.u. or 0.0020 a.u., on which to compute the potential. We have discussed the usefulness of these envelopes elsewhere [22,25,87,89].

3.4. Other Azides

Our conclusions about the nature of intermolecular bonding in the halogen azides are consistent with the views of Bursch and coworkers [138] for analogous systems. Specifically, these authors have reported the properties of non-covalent interactions between azides and oxygen-containing moieties in some chemical systems. Representative examples of molecules reported in that study are shown in Figure 8. Each shows close azide–oxygen contacts in the solid state. It was observed that the intramolecular $N{\cdots}O$ nitrogen bonded contacts are much shorter than the sum of the van der Waals radii of the respective atomic basins. For the $N_{terminal}{-}N_{central}{\cdots}X$ angle (where X is a pnictogen, chalcogen or halogen) in the systems examined, there was a strong accumulation of data points between 85° and 130°. As stated by the authors, caution needs to be exercised in interpreting the angular nature of these interactions because every structure was not individually examined. In the representative structures shown in Figure 8, the $N_{terminal}{-}N_{central}{\cdots}O$ angles are 105.4°, 103,7°, 88.4° and 93.9° for Figure 8a–d, respectively. Although the intramolecular interactions in the representative structures were observed to be unusually bent (perhaps not unexpectedly as they are developed intramolecularly), they are characteristic of pnictogen bonding since the central N atom of the azide moiety in the molecular entities is positive, and is involved in an attractive interaction with the negative sites on covalently bonded O. One may attribute the $N{\cdots}O$ interaction to a $\pi{\cdots}$lone-pair type pnictogen bond since a p-type orbital on N is responsible for driving this interaction. This may also explain why the directionality of the π-centered pnictogen bond does not follow any specific (linear) topology of bonding. The theoretically calculated association energies for the systems above were reported to be ranged from -1.0 to -5.5 kcal·mol^{-1}.

The final azide we consider is phosphorus azide, P_3N_{21} [139]. The unit-cell contains two molecules of P_3N_{21}. They are bonded to each other through $\pi(N_3){\cdots}N$ and $N{\cdots}N$ noncovalent interactions, as shown in Figure 9a,b. The $N{\cdots}N$ intermolecular distances range between 2.90 and 3.30 Å, and are all less than twice the van der Waals radius of N, 3.32 Å. Clearly, intermolecular pnictogen bonding is one of the principal features responsible for the formation of the crystalline phase of the P_3N_{21}.

3.5. Miscellaneous Examples

We end this section by looking at a number of other examples to illustrate the importance of the involvement of N-centered pnictogen bonding in crystalline materials.

The first is the crystal structure of nitrosium nitrate ($NO^+NO_3^-$), Figure 10a, reported only recently [140], but it is a system that has been studied many times, dating back to 1948 [141–146]. The crystalline state, as determined by XRD measurements at high pressure, consists of $N{\equiv}O^+$ cations and sp^2-trigonal planar NO_3^- anions stabilized via charge-assisted $N{\cdots}\pi$ and $N{\cdots}O$ nitrogen bonding interactions. Our analysis suggests that both N and O in NO^+ in the crystal are involved in several intermolecular interactions with the surrounding NO_3^- units. The N of each nitrosonium group has five nearest neighbor O sites from three surrounding NO_3^- units within its inner coordination sphere, with $r(N{\cdots}O)$ between 2.15 and 2.30 Å (some are shown in Figure 10b,c). Similarly, the O atom of NO^+ has five oxygen atoms from four closest NO_3^- units, as its first intermolecular neighbors. The $O{\cdots}O$ distances vary between 2.27 and 2.36 Å, slightly longer than the $N{\cdots}O$ bond distances (not shown), as expected since O is slightly larger than N (r_{vdW} (O) = 1.50 Å and $r_{vdW}(N)$ = 1.66 Å) [77]. This analysis is consistent with that of the authors of the study [140]. However, we identified that π-hole(N)${\cdots}$N interactions occur in the crystal at an intermolecular distance of 2.539 Å (see Figure 10b), and are undoubtedly π-hole-centered pnictogen bonds.

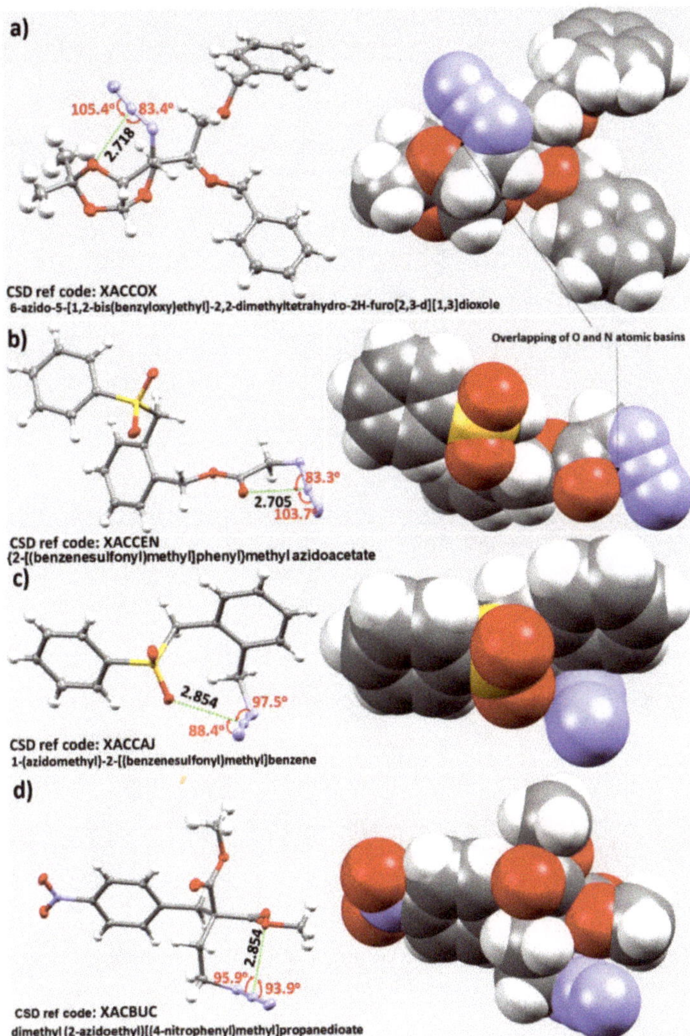

Figure 8. Representative examples of compounds with close intramolecular oxygen–azide contacts were observed in some structures reported in the solid state [138]; (**a**) 6-azido-5-[1,2-bis(benzyloxy)ethyl]-2,2-dimethyltetrahydro-2H-furo[2,3-d][1,3]dioxole; (**b**) {2-[(benzenesulfonyl)methyl]phenyl}methyl azidoacetate; (**c**) 1-(azidomethyl)-2-[(benzenesulfonyl)methyl]benzene; (**d**) dimethyl (2-azidoethyl)[(4-nitrophenyl)methyl]propanedioate. Nitrogen is depicted in blue, oxygen in red, sulfur in yellow, and carbon in gray. The interatomic distances between the central N of the azide and the closest oxygen atom are given in Å, and bond angles in degrees. Thermal ellipsoids are shown at the 50% probability level. The names of the molecular entities and CSD ref. codes are shown. The space-filling model is depicted in each case.

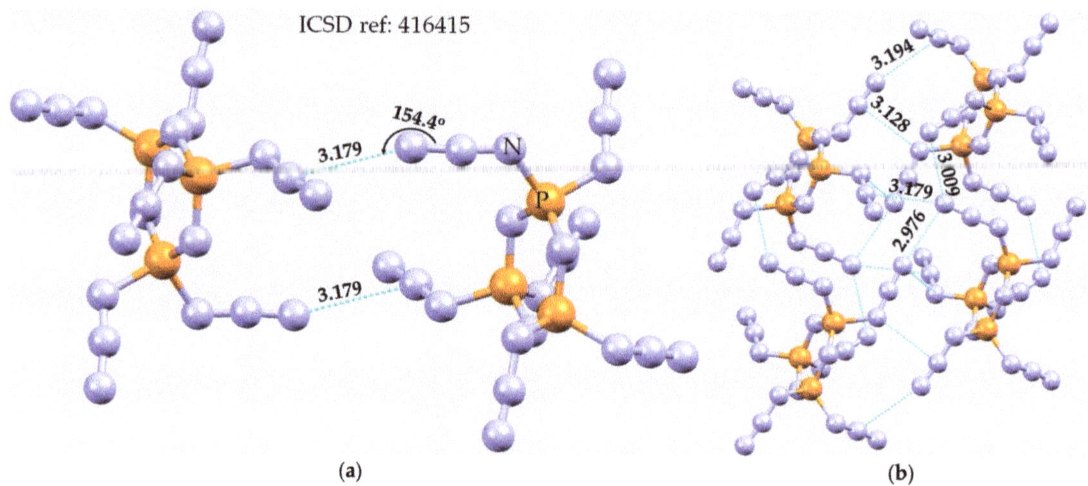

Figure 9. (**a**) The crystal structure of 2,2,4,4,6,6-hexa-azido-2,4,6-triphospha-1,3,5-triazine (P_3N_{21}) showing $\pi(N_3)\cdots N$ pnictogen bonded interactions. (**b**) Illustration of $N\cdots N$ and $\pi(N_3)\cdots N$ non-covalent interactions between four molecular units of P_3N_{21}. Selected bond lengths and bond angles are in Å and degree, respectively. Nitrogen is depicted in blue, and phosphorous in orange.

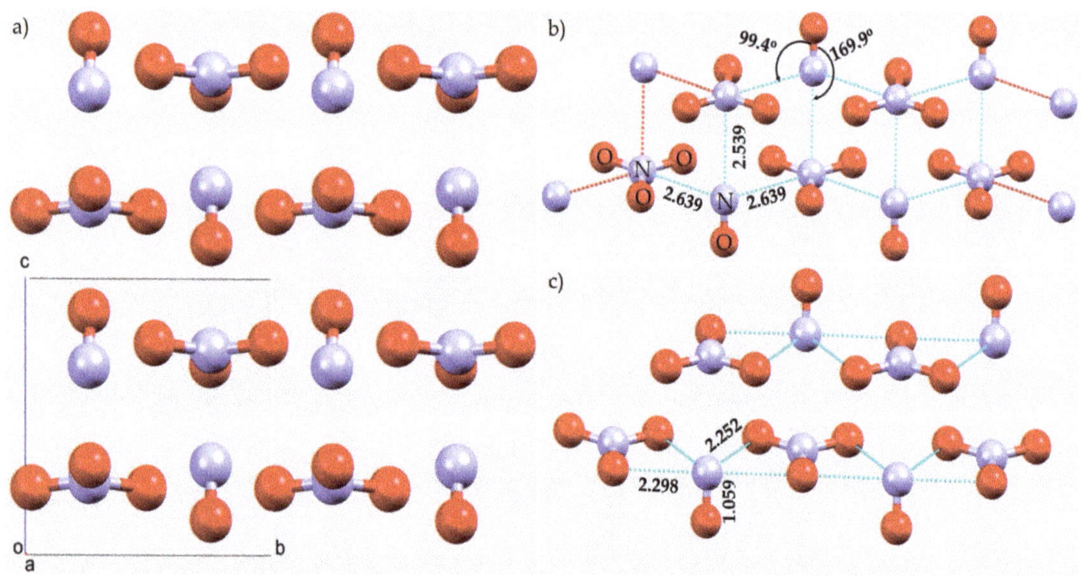

Figure 10. (**a**) The crystal structure of monoclinic ($P2_1/m$) nitrosium nitrate $NO^+NO_3^-$, CSD ref. code IMABAC. (**b,c**) Illustration of the nature of π-hole(N)$\cdots$N and N$\cdots$O bonding interactions in the crystal, respectively. Selected bond distances and bond angles are in Å and degree, respectively. Atom type is given in (**b**).

In crystalline tetramethylpyrazinium pentaiodide monohydrate [147], Figure 11, the feasibility of N$\cdots$I interactions is evident. They are charge-assisted pnictogen bonds, with bond distances in the range 3.7–4.1 Å, and are close to, or longer than, the sum of the vdW radii of N and I, 3.70 Å. They are π-centered since the π-hole on N accepts electron density from the iodine atom in I_3^-. In addition, there are numerous H$\cdots$I and (C)$\pi\cdots$I type intermolecular interactions that are also responsible for the overall structure of the crystalline material.

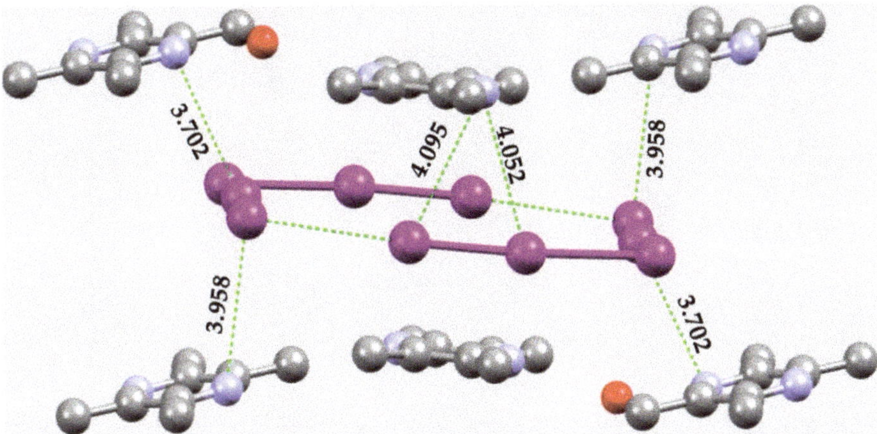

Figure 11. Illustration of N⋯I pnictogen bonds in the unit-cell of the crystal of tetramethylpyrazinium pentaiodide (CSD ref. code ZEPBIF). Nitrogen is depicted in blue, oxygen in red, iodine in purple, and carbon in gray. H atoms are not shown for clarity. Selected bond distances are in Å.

Many structures of compounds containing Group 15 elements deposited in the CSD owe their stability to a variety of intermolecular interactions, including pnictogen bonding. Group 15 elements in molecules are often trivalent or tetravalent and hence they may contain more than one σ-hole along the R–Pn bond extensions (viz. as in NX_3 shown above). Recently, for instance, Kumar and coworkers [148] suggested that tetrel and pnictogen bonds complement hydrogen and halogen bonds in framing the interactional landscape of barbituric acids. In particular, and by means of a CSD search, these workers have argued that the imide nitrogen atoms in some fluoro-, chloro- and bromo-substituted derivatives of barbituric acid and indandione can act as effective pnictogen bond donors, thereby featuring N⋯O nitrogen bonds. These contacts appear simultaneously with C⋯O tetrel bonds, and other interactions, and hence provide stability to the backbone of the overall crystal lattice.

As observed in the crystal structure of $NO^+NO_3^-$ (vide supra), one might speculate that N⋯O nitrogen bonding interactions occur in the crystal structures of ammonium 1,3,4,6-tetranitro-2,5-diazapentalene, $(NH_4)_2C_6N_6O_8$ [149], Figure 12a, and 1,3,4-trinitro-7,8-diazapentalene, $C_6H_3N_5O_6$ [150], Figure 12b, and that N⋯F pnictogen bonding occurs in the crystal of 3,3-bis(difluoroaminomethyl)oxetane, $C_5H_8F_4N_2O$ [151], Figure 12c. In the first two, the N⋯O interactions are π-centered; the π electron density delocalized on the surface of N of the NO_2 fragment in $[C_6N_6O_8]^{2-}$ is in an attractive engagement with the O's lone-pair in another unit of $[C_6N_6O_8]^{2-}$. The occurrence of the pnictogen bonded interaction in the crystal is facilitated by charge-assisted H⋯N hydrogen bonds between NH_4^+ and the dianion. The N(π)⋯O interactions are equivalent, with r(N(π)⋯O) = 2.945 Å, and are less than vdW radii of N and O, 3.16 Å. In the case of the crystal structure of 1,3,4-trinitro-7,8-diazapentalene, $C_6H_3N_5O_6$, there is no ion pair, yet the N(π)⋯O pnictogen bonded interactions still occur between the $C_6H_3N_5O_6$ units. However, in this case, primary/secondary interactions (viz. H⋯O hydrogen bonds, and C(π)⋯O interactions) are expected to play a crucial role in the overall stability of the crystal.

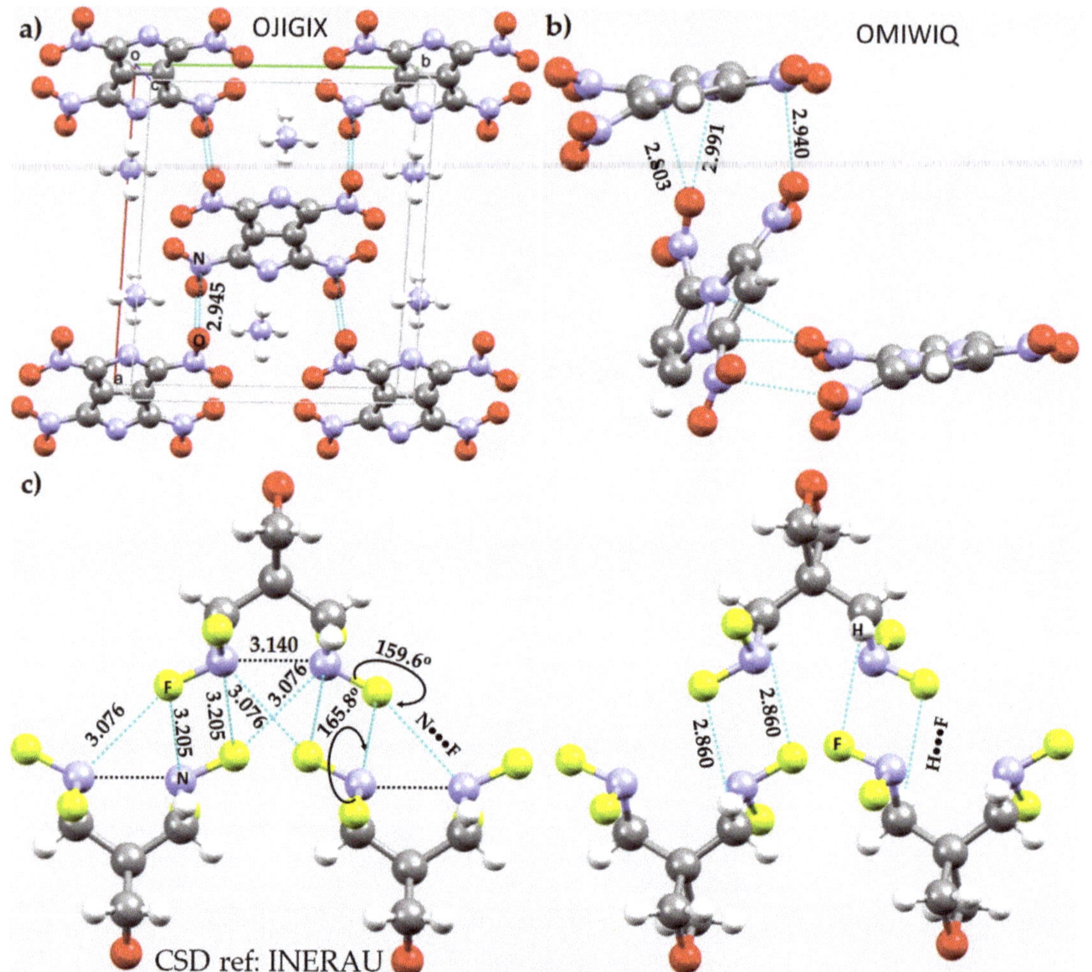

Figure 12. The nature of nitrogen-centered pnictogen bonding in the crystals of (**a**) $(NH_4)_2C_6N_6O_8$ [149], (**b**) $C_6H_3N_5O_6$ [150], and (**c**) $C_5H_8F_4N_2O$ [151]. Selected bond distances are shown in Å and degree, respectively. Selected atom types are shown in (**a,c**) that are involved in making non-covalent interactions. The CSD ref. code is given in each case.

In the crystal of $C_5H_8F_4N_2O$ [151] the interacting molecules are bonded to each other through F–N···F nitrogen bonds, Figure 12c (left). They appear to show up in two different flavors: one with an intermolecular bond distance of 3.076 and the other with an intermolecular bond distance of 3.205 Å. The second is directional ($\angle$F–N···F = 165.8°) and follows the Type-IIa topology of bonding and appears along the extension of the F–N covalent bonds. The first is non-linear and is taken to be a π-hole centered interaction ($r(N(\pi)\cdots O) = 3.076$ Å). In addition, each monomer unit in the crystal has an $N(\pi)\cdots N(\pi)$ pnictogen bonding interaction, with $r(N(\pi)\cdots N(\pi)) = 3.140$ Å. There is a distinct possibility of intermolecular pnictogen bonding in the crystal being driven by H···F hydrogen bonds (Figure 12c, right) that are shorter than the pnictogen bonds. Therefore, the hydrogen bonds are probably the primary intermolecular interaction, leading to the formation of secondary pnictogen interactions in this crystalline material.

The crystal structure of a perhydropyrazolopyrazole is shown in Figure 13a [152]. The N atom in the N–N covalent bond of pyrazole seems to be positive along its outer extension, and the lateral site is negative. This explains why the electropositive H site of the hydroxyl group in an included methanol solvent molecule is able to form a hydrogen bond, leading to the formation of a reasonably strong O–H···N hydrogen bond (r(H···N) = 2.043 Å and ∠O–H···N = 170.9°; Figure 13b). We also observed Type-IIa quasi-linear N···Cl nitrogen bonds in the crystal with r(N···Cl) = 3.958 Å and ∠N–N···Cl = 176.4°, Figure 13b. This emerges from the fact that the lateral site in the covalently bound Cl, which is nucleophilic, interacts with the axial site of N along the N–N bond extensions, and that the interaction is directional. Apart from these interactions, there are numerous π···π stacking interactions, Cl···Cl halogen bonds, and H···Cl hydrogen bonds in the crystal (not explicitly shown).

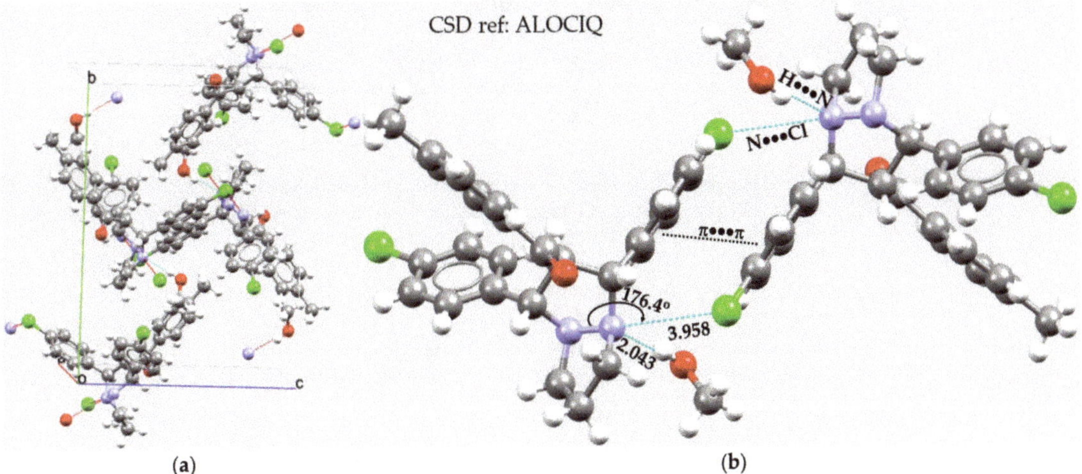

Figure 13. (**a**) The nature of N···Cl pnictogen bonding in the unit-cell of the crystal of a perhydropyrazolopyrazole, [1,3-bis(4-chlorophenyl)tetrahydro-1H,5H-pyrazolo[1,2-a]pyrazol-2-yl](4-methylphenyl)methanone [152]. There is a methanol solvate in the structure. (**b**) The local nature of intermolecular interactions between the monomeric units in the same crystal. Selected bond lengths and bond angles are in Å and degrees, respectively. The dotted lines represent intermolecular interactions, and those in red represent hanging contacts.

Olejniczak and coworkers [153] recently reported crystalline materials formed as a consequence of CH···N and N···N interactions in the high-pressure structures of some pyridazines, viz. 6-azido-1,2,4-triazolo[4,3-b]pyridazine ($C_5H_3N_7$) (as an anhydrate) and its 3-methyl derivative ($C_6H_5N_7$) (as a hydrated clathrate). Short C–H···N and N···N intermolecular contacts could be observed in the structures; the latter involves attraction exclusively between the azide groups. Figure 14a shows the unit-cell of the crystal of $C_5H_3N_7$, and Figure 14b shows the binary arrangement between the $C_5H_3N_7$ molecular units in the crystal. As shown in the latter, there is a pair each of N···N and H···N close contacts between the two interacting units. The first is more directional than the second and follows a Type-IIa pattern of nitrogen bonding. In addition to these interactions, nitrogen-centered π···π stacking interactions appear to play a very important role in the stability of the crystal, as may be inferred from Figure 14a. The intermolecular distances of the N···N contacts are different, 3.369 and 3.165 Å. The latter is shorter, and the former is longer than twice the vdW radius of N, 3.32 Å, and both are directional.

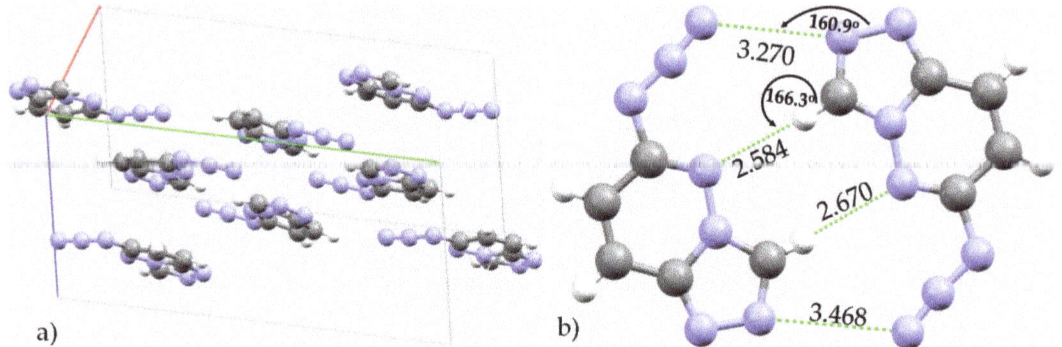

Figure 14. (**a**) The unit-cell of the crystal structure of 6-azido[1,2,4]triazolo[4,3-b]pyridazine ($C_5H_3N_7$), showing possible N···N close contacts between the interacting monomer units. (**b**) Illustration of N···N and H···N pnictogen and hydrogen bonds between the pair of two units of $C_5H_3N_7$. Bond lengths and bond angles in Å and degrees, respectively. The CSD ref. code of the crystal is NETJUR01.

Allcock coworkers [154] have reported the crystal structure of a series of short-chain and cyclic phosphazenes bearing *o*-dichloro- and *o*-dimethyl-phenoxy groups. Figure 15a shows one of them, $(O)PCl_2N=P(o\text{-}OC_6H_3Cl_2)_3$. Within the skeletal framework of the molecular entity, the lateral sites of the ortho-Cl atoms are negative. There is an overlap between the Cl and N atomic basins corresponding to an intermolecular distance of 3.264 Å (Figure 15b), which is less than the sum of the vdW radii of Cl and N atoms, 3.48 Å ($r_{vdW}(Cl)$ = 1.82 Å and $r_{vdW}(N)$ = 1.66 Å). This intramolecular N···Cl close contact is a π-hole centered nitrogen bond. The nitrogen site is also in an attractive engagement with the two nearest C and Cl atoms. The N···C close contact (r(N···C) = 2.901 Å) appears to be a π···π interaction, and the N(π)···Cl close contact with an intramolecular distance of 3.695 Å is a force consequence of the Cl–P···Cl pnictogen bond. The overall stability of the crystal is, however, a result of significant H···Cl and H···O, π···π stacking interactions (between arene moieties), and Cl···Cl halogen-bonded intermolecular interactions that may be deduced from the unit-cell of the crystal itself (Figure 15a).

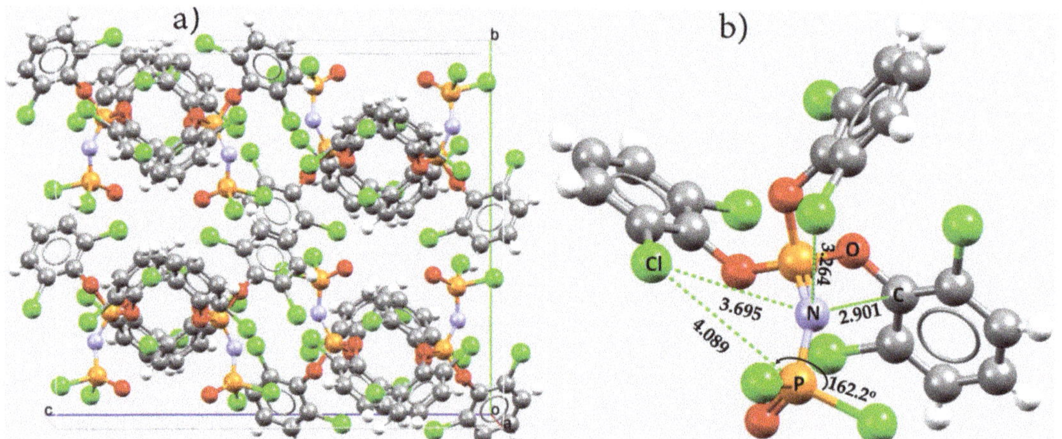

Figure 15. (**a**) The unit-cell of crystalline tris(2,6-dichlorophenoxy)phosphazeno-dichloro-oxo-phosphorus, $(O)PCl_2N=P(o\text{-}OC_6H_3Cl_2)_3$, (CSD ref. code PADJUZ). (**b**) Illustration of possible intramolecular pnictogen bonding within the framework of the molecular building block. Selected bond distances and bond angles in (**b**) are in Å and degree, respectively. The selected atom type is depicted.

4. Conclusions

In this featured review, we highlighted the importance and nature of the modes of nitrogen-centered pnictogen bonding (or, simply, the nitrogen bond) observed in many crystals deposited in the CSD and ICSD. Several of them have been known since the last century, even though the term "pnictogen bond", or even "nitrogen bond" was not coined during that time. They turn out to occur in four different flavors: $N(\sigma)\cdots$lone-pair, $N(\pi)\cdots$lone-pair, $N(\sigma)\cdots\pi$ and $N(\pi)\cdots\pi(N)$. The rationalization of these interacting modes was made possible by exploring the MESP of isolated monomers constituting the crystals in several instances, together with chemical intuition. We have shown that the singular occurrence of pnictogen bonding in crystals is very rare, and hence is accompanied or reinforced by other primary/secondary interactions in a great majority of crystals, such as halogen bonds, hydrogen bonds, and/or $\pi\cdots\pi$ stacking interactions. Whereas there have been few previous studies reported of pnictogen bonding with covalently bonded nitrogen as an electrophilic agent (viz. imide nitrogen [148]), as the nitrogen in the molecular entity is often observed to be negative, this overview highlighted the local modes of intra- and intermolecular bonding interactions in several examples containing nitrogen that has a positive site. This is expected to guide researchers on future studies of nitrogen-centered pnictogen bonding and the design in silico of novel materials. We did not carry out a statistical analysis of structures deposited in the CSD and ICSD of pnictogen bonding to reveal possible ranges of intermolecular distances and angles of the approach of the electrophile on N in molecular entities, since pnictogen bonding is heavily affected by other primary and secondary interactions. As pointed out by a reviewer, even though a single example would be sufficient proof of existence, stronger arguments can be made if such cases are definitely shown not to be outliers. Since no rigorous theoretical first-principles calculations have been performed on most of the systems highlighted in this study, it is expected that the illustrative crystal systems may be used as model systems to investigate the energy stability and other physio-chemical properties (viz. vibrational frequency shift and NMR chemical shifts) to reveal the characteristic features of pnictogen bonds. The attempt will no doubt help provide a reasonable description of the strength of nitrogen bonds in supermolecular entities.

Author Contributions: Conceptualization, project design, and project administration, P.R.V.; formal analysis and investigation, P.R.V. and A.V.; supervision, P.R.V.; writing—original draft, P.R.V. and A.V.; writing—review and editing, P.R.V., H.M.M., A.V. and K.Y. All authors have read and agreed to the published version of the manuscript.

Funding: This research received no external funding.

Institutional Review Board Statement: Not applicable.

Informed Consent Statement: Not applicable.

Data Availability Statement: This research did not report any data.

Acknowledgments: This work was entirely conducted using the various computation and laboratory facilities provided by the University of Tokyo and the Research Center for Computational Science of the Institute of Molecular Science (Okazaki, Japan). P.R.V. is currently affiliated with the University of the Witwatersrand (SA), and Nagoya University, Aichi 464-0814, Japan; A.V. is currently affiliated with AIST, Tsukuba 305-8560, Japan; K.Y. is currently affiliated with Kyoto University, ESICB, Kyoto, 615-8245, Japan. H.M.M. thanks the National Research Foundation, Pretoria, South Africa, and the University of the Witwatersrand for funding.

Conflicts of Interest: The authors declare no conflict of interest. The funders had absolutely no role in the design of the study; in the collection, analyses, or interpretation of data; in the writing of the manuscript, or in the decision to publish the results.

References

1. Arunan, E.; Desiraju, G.R.; Klein, R.A.; Sadlej, J.; Scheiner, S.; Alkorta, I.; Clary, D.C.; Crabtree, R.H.; Dannenberg, J.J.; Hobza, P.; et al. Definition of the hydrogen bond (IUPAC Recommendations 2011). *Pure Appl. Chem.* **2011**, *83*, 1637–1641. [CrossRef]
2. Desiraju, G.R.; Shing Ho, P.; Kloo, L.; Legon, A.C.; Marquardt, R.; Metrangolo, P.; Politzer, P.; Resnati, G.; Rissanen, K. Definition of the halogen bond (IUPAC Recommendations 2013). *Pure Appl. Chem.* **2013**, *85*, 1711–1713. [CrossRef]
3. Aakeroy, C.B.; Bryce, D.L.; Desiraju, R.G.; Frontera, A.; Legon, A.C.; Nicotra, F.; Rissanen, K.; Scheiner, S.; Terraneo, G.; Metrangolo, P.; et al. Definition of the chalcogen bond (IUPAC Recommendations 2019). *Pure Appl. Chem.* **2019**, *91*, 1889–1892. [CrossRef]
4. IUPAC Project: Categorizing Chalcogen, Pnictigen, and Tetrel Bonds, and Other Interactions Involving Group 14–16 Elements. Available online: https://iupac.org/projects/project-details/?project_nr=2016-001-2-300 (accessed on 31 January 2022).
5. Bakker, D.J.; Peters, A.; Yatsyna, V.; Zhaunerchyk, V.; Rijs, A.M. Far-Infrared Signatures of Hydrogen Bonding in Phenol Derivatives. *J. Phys. Chem. Lett.* **2016**, *7*, 1238–1243. [CrossRef] [PubMed]
6. Vasylyeva, V.; Catalano, L.; Nervi, C.; Gobetto, R.; Metrangolo, P.; Resnati, G. Characteristic redshift and intensity enhancement as far-IR fingerprints of the halogen bond involving aromatic donors. *CrystEngComm* **2016**, *18*, 2247–2250. [CrossRef]
7. Hardin, A.E.S.; Ellington, T.L.; Nguyen, S.T.; Rheingold, A.L.; Tschumper, G.S.; Watkins, D.L.; Hammer, N.I. A Raman Spectroscopic and Computational Study of New Aromatic Pyrimidine-Based Halogen Bond Acceptors. *Inorganics* **2019**, *7*, 119. [CrossRef]
8. Messina, M.T.; Metrangolo, P.; Navarrini, W.; Radice, S.; Resnati, G.; Zerbi, G. Infrared and Raman analyses of the halogen-bonded non-covalent adducts formed by α,ω-diiodoperfluoroalkanes with DABCO and other electron donors. *J. Mol. Struct.* **2000**, *524*, 87–94. [CrossRef]
9. Shen, Q.J.; Jin, W.J. Strong halogen bonding of 1,2-diiodoperfluoroethane and 1,6-diiodoperfluorohexane with halide anions revealed by UV-Vis, FT-IR, NMR spectroscopes and crystallography. *Phys. Chem. Chem. Phys.* **2011**, *13*, 13721–13729. [CrossRef] [PubMed]
10. Widner, D.L.; Robinson, E.R.; Perez, A.B.; Vang, H.G.; Thorson, R.A.; Driscoll, Z.L.; Giebel, S.M.; Berndt, C.W.; Bosch, E.; Speetzen, E.D.; et al. Comparing Strong and Weak Halogen Bonding in Solution: ^{13}C NMR, UV/Vis, Crystallographic, and Computational Studies of an Intramolecular Model. *Eur. J. Org. Chem.* **2017**, *2017*, 5739–5749. [CrossRef]
11. Hakkert, S.B.; Gräfenstein, J.; Erdelyi, M. The 15N NMR chemical shift in the characterization of weak halogen bonding in solution. *Faraday Discuss.* **2017**, *203*, 333–346. [CrossRef]
12. Gomila, R.M.; Frontera, A. Charge assisted halogen and pnictogen bonds: Insights from the Cambridge Structural Database and DFT calculations. *CrystEngComm* **2020**, *22*, 7162–7169. [CrossRef]
13. De Azevedo Santos, L.; Hamlin, T.A.; Ramalho, T.C.; Bickelhaupt, F.M. The pnictogen bond: A quantitative molecular orbital picture. *Phys. Chem. Chem. Phys.* **2021**, *23*, 13842–13852. [CrossRef] [PubMed]
14. Fanfrlík, J.; Zierkiewicz, W.; Švec, P.; Růžičková, Z.; Řezáč, J.; Michalczyk, M.; Růžička, A.; Michalska, D.; Hobza, P. Pnictogen bonding in pyrazine·PnX5 (Pn = P, As, Sb and X = F, Cl, Br) complexes. *J. Mol. Model.* **2017**, *23*, 328. [CrossRef] [PubMed]
15. Murray, J.S.; Lane, P.; Clark, T.; Riley, K.E.; Politzer, P. σ-Holes, π-holes and electrostatically-driven interactions. *J. Mol. Model.* **2012**, *18*, 541–548. [CrossRef]
16. Clark, T.; Hennemann, M.; Murray, J.S.; Politzer, P. Halogen bonding: The σ-hole. *J. Mol. Model.* **2007**, *13*, 291–296. [CrossRef] [PubMed]
17. Wang, H.; Wang, W.; Jin, W.J. σ-Hole Bond vs. π-Hole Bond: A Comparison Based on Halogen Bond. *Chem. Rev.* **2016**, *116*, 5072–5104. [CrossRef] [PubMed]
18. Zhang, Y.; Wang, W. The σ-hole···σ-hole stacking interaction: An unrecognized type of noncovalent interaction. *J. Chem. Phys.* **2020**, *153*, 214302. [CrossRef]
19. Politzer, P.; Murray, J.S. σ-Hole Interactions: Perspectives and Misconceptions. *Crystals* **2017**, *7*, 212. [CrossRef]
20. Mallada, B.; Gallardo, A.; Lamanec, M.; Torre, B.d.l.; Špirko, V.; Hobza, P.; Jelinek, P. Real-space imaging of anisotropic charge of σ-hole by means of Kelvin probe force microscopy. *Science* **2021**, *374*, 863–867. [CrossRef]
21. Politzer, P.; Murray, J.S. σ-holes and π-holes: Similarities and differences. *J. Comp. Chem.* **2018**, *39*, 464–471. [CrossRef]
22. Varadwaj, A.; Varadwaj, P.R.; Jin, B.-Y. Fluorines in tetrafluoromethane as halogen bond donors: Revisiting address the nature of the fluorine's σ-hole. *Int. J. Quantum Chem.* **2015**, *115*, 453–470. [CrossRef]
23. Varadwaj, P.R.; Varadwaj, A.; Jin, B.-Y. Halogen bonding interaction of chloromethane with several nitrogen donating molecules: Addressing the nature of the chlorine surface σ-hole. *Phys. Chem. Chem. Phys.* **2014**, *16*, 19573–19589. [CrossRef] [PubMed]
24. Ibrahim, M.A.A.; Moussa, N.A.M. Unconventional Type III Halogen···Halogen Interactions: A Quantum Mechanical Elucidation of σ-Hole···σ-Hole and Di-σ-Hole Interactions. *ACS Omega* **2020**, *5*, 21824–21835. [CrossRef] [PubMed]
25. Varadwaj, P.R.; Varadwaj, A.; Marques, H.M. Does Chlorine in CH3Cl Behave as a Genuine Halogen Bond Donor? *Crystals* **2020**, *10*, 146. [CrossRef]
26. Foster, S.L.; Bakovic, S.I.P.; Duda, R.D.; Maheshwari, S.; Milton, R.D.; Minteer, S.D.; Janik, M.J.; Renner, J.N.; Greenlee, L.F. Catalysts for nitrogen reduction to ammonia. *Nat. Catal.* **2018**, *1*, 490–500. [CrossRef]
27. Uyanik, M.; Kato, T.; Sahara, N.; Katade, O.; Ishihara, K. High-Performance Ammonium Hypoiodite/Oxone Catalysis for Enantioselective Oxidative Dearomatization of Arenols. *ACS Catal.* **2019**, *9*, 11619–11626. [CrossRef]
28. Hinokuma, S.; Sato, K. Ammonia Combustion Catalysts. *Chem. Lett.* **2021**, *50*, 752–759. [CrossRef]

29. Lamb, K.E.; Dolan, M.D.; Kennedy, D.F. Ammonia for hydrogen storage; A review of catalytic ammonia decomposition and hydrogen separation and purification. *Int. J. Hydrogen Energy* **2019**, *44*, 3580–3593. [CrossRef]

30. US Geological Survey. *Mineral Commodity Summaries 2020*; US Geological Survey: Reston, VA, USA, 2020. [CrossRef]

31. MacFarlane, D.R.; Cherepanov, P.V.; Choi, J.; Suryanto, B.H.R.; Hodgetts, R.Y.; Bakker, J.M.; Ferrero Vallana, F.M.; Simonov, A.N. A Roadmap to the Ammonia Economy. *Joule* **2020**, *4*, 1186–1205. [CrossRef]

32. Ogasawara, K.; Nakao, T.; Kishida, K.; Ye, T.-N.; Lu, Y.; Abe, H.; Niwa, Y.; Sasase, M.; Kitano, M.; Hosono, H. Ammonia Decomposition over CaNH-Supported Ni Catalysts via an $NH_2{}^-$-Vacancy-Mediated Mars–van Krevelen Mechanism. *ACS Catal.* **2021**, *11*, 11005–11015. [CrossRef]

33. Mukherjee, S.; Devaguptapu, S.V.; Sviripa, A.; Lund, C.R.F.; Wu, G. Low-temperature ammonia decomposition catalysts for hydrogen generation. *Appl. Catal. B Environ.* **2018**, *226*, 162–181. [CrossRef]

34. Rouwenhorst, K.H.R.; Engelmann, Y.; van 't Veer, K.; Postma, R.S.; Bogaerts, A.; Lefferts, L. Plasma-driven catalysis: Green ammonia synthesis with intermittent electricity. *Green Chem.* **2020**, *22*, 6258–6287. [CrossRef]

35. Gini, A.; Paraja, M.; Galmés, B.; Besnard, C.; Poblador-Bahamonde, A.I.; Sakai, N.; Frontera, A.; Matile, S. Pnictogen-bonding catalysis: Brevetoxin-type polyether cyclizations. *Chem. Sci.* **2020**, *11*, 7086–7091. [CrossRef] [PubMed]

36. Humeniuk, H.V.; Gini, A.; Hao, X.; Coelho, F.; Sakai, N.; Matile, S. Pnictogen-Bonding Catalysis and Transport Combined: Polyether Transporters Made In Situ. *JACS Au* **2021**, *1*, 1588–1593. [CrossRef]

37. Paraja, M.; Gini, A.; Sakai, N.; Matile, S. Pnictogen-Bonding Catalysis: An Interactive Tool to Uncover Unorthodox Mechanisms in Polyether Cascade Cyclizations. *Chem. A Eur. J.* **2020**, *26*, 15471–15476. [CrossRef]

38. Benz, S.; Poblador-Bahamonde, A.I.; Low-Ders, N.; Matile, S. Catalysis with Pnictogen, Chalcogen, and Halogen Bonds. *Angew. Chem. Int. Ed.* **2018**, *57*, 5408–5412. [CrossRef]

39. Taylor, M.S. Anion recognition based on halogen, chalcogen, pnictogen and tetrel bonding. *Coord. Chem. Rev.* **2020**, *413*, 213270. [CrossRef]

40. Qiu, J.; Song, B.; Li, X.; Cozzolino, A.F. Solution and gas phase evidence of anion binding through the secondary bonding interactions of a bidentate bis-antimony(iii) anion receptor. *Phys. Chem. Chem. Phys.* **2018**, *20*, 46–50. [CrossRef]

41. Legon, A.C. Tetrel, pnictogen and chalcogen bonds identified in the gas phase before they had names: A systematic look at non-covalent interactions. *Phys. Chem. Chem. Phys.* **2017**, *19*, 14884–14896. [CrossRef]

42. Lee, L.M.; Tsemperouli, M.; Poblador-Bahamonde, A.I.; Benz, S.; Sakai, N.; Sugihara, K.; Matile, S. Anion Transport with Pnictogen Bonds in Direct Comparison with Chalcogen and Halogen Bonds. *J. Am. Chem. Soc.* **2019**, *141*, 810–814. [CrossRef]

43. Moaven, S.; Andrews, M.C.; Polaske, T.J.; Karl, B.M.; Unruh, D.K.; Bosch, E.; Bowling, N.P.; Cozzolino, A.F. Triple-Pnictogen Bonding as a Tool for Supramolecular Assembly. *Inorg. Chem.* **2019**, *58*, 16227–16235. [CrossRef] [PubMed]

44. Mahmudov, K.T.; Gurbanov, A.V.; Aliyeva, V.A.; Resnati, G.; Pombeiro, A.J.L. Pnictogen bonding in coordination chemistry. *Coord. Chem. Rev.* **2020**, *418*, 213381. [CrossRef]

45. Mahmudov, K.T.; Huseynov, F.E.; Aliyeva, V.A.; Guedes da Silva, M.F.C.; Pombeiro, A.J.L. Noncovalent Interactions at Lanthanide Complexes. *Chem. Eur. J.* **2021**, *27*, 14370–14389. [CrossRef] [PubMed]

46. Liu, C.; Shin, J.; Son, S.; Choe, Y.; Farokhzad, N.; Tang, Z.; Xiao, Y.; Kong, N.; Xie, T.; Kim, J.S.; et al. Pnictogens in medicinal chemistry: Evolution from erstwhile drugs to emerging layered photonic nanomedicine. *Chem. Soc. Rev.* **2021**, *50*, 2260–2279. [CrossRef] [PubMed]

47. Fanfrlík, J.; Hnyk, D. Dihalogen and Pnictogen Bonding in Crystalline Icosahedral Phosphaboranes. *Crystals* **2018**, *8*, 390. [CrossRef]

48. Sobalev, S.; Matveychuk, Y.; Bartashevich, E. Features of the Pnictogen Bonds Formed by Neighboring Nitro Groups in Crystals. *Bull. South Ural State Univ. Chem.* **2019**, *11*, 66–75. [CrossRef]

49. Frontera, A.; Bauza, A. On the Importance of Pnictogen and Chalcogen Bonding Interactions in Supramolecular Catalysis. *Int. J. Mol. Sci.* **2021**, *22*, 12550. [CrossRef]

50. Shukla, R.; Chopra, D. Chalcogen and pnictogen bonds: Insights and relevance. *Curr. Sci.* **2021**, *120*, 1848–1853.

51. Varadwaj, A.; Varadwaj, P.R.; Jin, B.-Y. Can an entirely negative fluorine in a molecule, viz. perfluorobenzene, interact attractively with the entirely negative site (s) on another molecule (s)? Like liking like! *RSC Adv.* **2016**, *6*, 19098–19110. [CrossRef]

52. Varadwaj, P.R.; Cukrowski, I.; Marques, H.M. DFT-X3LYP Studies on the Coordination Chemistry of Ni^{2+}. Part 1: Six Coordinate $[Ni(NH_3)_n(H_2O)_{6-n}]^{2+}$ Complexes. *J. Phys. Chem. A* **2008**, *112*, 10657–10666. [CrossRef]

53. Varadwaj, P.R.; Marques, H.M. The physical chemistry of coordinated aqua-, ammine-, and mixed-ligand Co^{2+} complexes: DFT studies on the structure, energetics, and topological properties of the electron density. *Phys. Chem. Chem. Phys.* **2010**, *12*, 2126–2138. [CrossRef] [PubMed]

54. Murray, J.S.; Politzer, P. Can Counter-Intuitive Halogen Bonding Be Coulombic? *ChemPhysChem* **2021**, *22*, 1201–1207. [CrossRef] [PubMed]

55. Chandra, S.; Suryaprasad, B.; Ramanathan, N.; Sundararajan, K. Nitrogen as a pnicogen? Evidence for π-hole driven novel pnicogen bonding interactions in nitromethane–ammonia aggregates using matrix isolation infrared spectroscopy and ab initio computations. *Phys. Chem. Chem. Phys.* **2021**, *23*, 6286–6297. [CrossRef] [PubMed]

56. Thirumoorthi, R.; Chivers, T.; Vargas-Baca, I. S,C,S-Pnictogen bonding in pincer complexes of the methanediide $[C(Ph_2PS)_2]^{2-}$. *Dalton Trans.* **2011**, *40*, 8086–8088. [CrossRef]

57. Zahn, S.; Frank, R.; Hey-Hawkins, E.; Kirchner, B. Pnicogen Bonds: A New Molecular Linker? *Chem. Eur. J.* **2011**, *17*, 6034–6038. [CrossRef]

58. Scheiner, S. A new noncovalent force: Comparison of P···N interaction with hydrogen and halogen bonds. *J. Chem. Phys.* **2011**, *134*, 094315. [CrossRef]

59. Del Bene, J.E.; Alkorta, I.; Sanchez-Sanz, G.; Elguero, J. ^{31}P–^{31}P spin–spin coupling constants for pnicogen homodimers. *Chem. Phys. Lett.* **2011**, *512*, 184–187. [CrossRef]

60. Girolami, G.S. Origin of the Terms Pnictogen and Pnictide. *J. Chem. Ed.* **2009**, *86*, 1200. [CrossRef]

61. Groom, C.R.; Bruno, I.J.; Lightfoot, M.P.; Ward, S.C. The Cambridge Structural Database. *Acta Cryst.* **2016**, *B72*, 171–179. [CrossRef]

62. Hellenbrandt, M. The Inorganic Crystal Structure Database (ICSD)—Present and Future. *Crystallogr. Rev.* **2004**, *10*, 17–22. [CrossRef]

63. Alkorta, I.; Elguero, J.; Frontera, A. Not Only Hydrogen Bonds: Other Noncovalent Interactions. *Crystals* **2020**, *10*, 180. [CrossRef]

64. Cavallo, G.; Metrangolo, P.; Milani, R.; Pilati, T.; Priimagi, A.; Resnati, G.; Terraneo, G. The halogen bond. *Chem. Rev.* **2016**, *116*, 2478–2601. [CrossRef] [PubMed]

65. Roy, K.; Kar, S.; Das, R.N. Chapter 2—Chemical Information and Descriptors. In *Understanding the Basics of QSAR for Applications in Pharmaceutical Sciences and Risk Assessment*; Roy, K., Kar, S., Das, R.N., Eds.; Academic Press: Boston, MS, USA, 2015; pp. 47–80. [CrossRef]

66. Dance, I. Distance criteria for crystal packing analysis of supramolecular motifs. *New J. Chem.* **2003**, *27*, 22–27. [CrossRef]

67. Motiejunas, D.; Wade, R.C. 4.09—Structural, Energetic, and Dynamic Aspects of Ligand–Receptor Interactions. In *Comprehensive Medicinal Chemistry II*; Taylor, J.B., Triggle, D.J., Eds.; Elsevier: Oxford, UK, 2007; pp. 193–213. [CrossRef]

68. Tredwell, M.; Gouverneur, V. 1.5 Fluorine in Medicinal Chemistry: Importance of Chirality. In *Comprehensive Chirality*; Carreira, E.M., Yamamoto, H., Eds.; Elsevier: Amsterdam, The Netherlands, 2012; pp. 70–85. [CrossRef]

69. Desiraju, G.R. Crystal Engineering: From Molecule to Crystal. *J. Am. Chem. Soc.* **2013**, *135*, 9952–9967. [CrossRef] [PubMed]

70. Ganguly, P.; Desiraju, G.R. Van der Waals and Polar Intermolecular Contact Distances: Quantifying Supramolecular Synthons. *Chem. Asian J.* **2008**, *3*, 868–880. [CrossRef]

71. Desiraju, G.R. C–H ... O Hydrogen Bonding and the Deliberate Design of Organic Crystal Structures. *Mol. Cryst. Liq. Cryst. Sci. Technol. Sect. A Mol. Cryst. Liq. Cryst.* **1992**, *211*, 63–74. [CrossRef]

72. Rowland, R.S.; Taylor, R. Intermolecular Nonbonded Contact Distances in Organic Crystal Structures: Comparison with Distances Expected from van der Waals Radii. *J. Phys. Chem.* **1996**, *100*, 7384–7391. [CrossRef]

73. Gafner, G.; Herbstein, F.H.; Lee, C.M. Derivation of van der Waals radii from known crystal structures. *Acta Crystallogr. A* **1972**, *28*, 422–426. [CrossRef]

74. Hu, S.-Z.; Zhou, Z.-H.; Xie, Z.-X.; Robertson, B.E. A comparative study of crystallographic van der Waals radii. *Z. Kristall.—Cryst. Mater.* **2014**, *229*, 517–523. [CrossRef]

75. Politzer, P.; Murray, J.S. The use and misuse of van der Waals radii. *Struct. Chem.* **2021**, *32*, 623–629. [CrossRef]

76. Bondi, A. Van Der Waals Volumes and Radii. *J. Phys. Chem.* **1964**, *68*, 441–451. [CrossRef]

77. Alvarez, S. A cartography of the van der Waals territories. *Dalton Trans.* **2013**, *42*, 8617–8636. [CrossRef] [PubMed]

78. Schiemenz, G.P. The sum of van der Waals radii—A pitfall in the search for bonding. *Z. Naturforsch. B* **2007**, *62*, 235–243. [CrossRef]

79. Politzer, P.; Murray, J.S. Molecular Electrostatic Potentials and Chemical Reactivity. In *Reviews in Computational Chemistry*; John Wiley & Sons, Inc.: Hoboken, NJ, USA, 1991; Volume 2, pp. 273–312. [CrossRef]

80. Politzer, P.; Murray, J.S. Molecular Electrostatic Potentials: Significance and Applications. In *Chemical Reactivity in Confined Systems*; Chattaraj, P.K., Chakraborty, D., Eds.; John Wiley & Sons, Ltd.: Hoboken, NJ, USA, 2021; pp. 113–134. [CrossRef]

81. Mishra, P.C.; Kumar, A. Molecular electrostatic potentials and fields: Hydrogen bonding, recognition, reactivity and modelling. In *Theoretical and Computational Chemistry*; Murray, J.S., Sen, K., Eds.; Elsevier: Amsterdam, The Netherlands, 1996; Volume 3, pp. 257–296.

82. Liu, L.; Miao, L.; Li, L.; Li, F.; Lu, Y.; Shang, Z.; Chen, J. Molecular Electrostatic Potential: A New Tool to Predict the Lithiation Process of Organic Battery Materials. *J. Phys. Chem. Lett.* **2018**, *9*, 3573–3579. [CrossRef]

83. Alipour, M.; Mohajeri, A. Molecular Electrostatic Potential as a tool for Evaluating the Etherification Rate Constant. *J. Phys. Chem. A* **2010**, *114*, 7417–7422. [CrossRef] [PubMed]

84. Pauling, L. *The Nature of the Chemical Bond*, 3rd ed.; Cornell University Press: New York, NY, USA, 1960.

85. Chai, J.-D.; Head-Gordon, M. Long-range corrected hybrid density functionals with damped atom–atom dispersion corrections. *Phys. Chem. Chem. Phys.* **2008**, *10*, 6615–6620. [CrossRef] [PubMed]

86. Frisch, M.J.; Head-Gordon, M.; Pople, J.A. A direct MP2 gradient method. *Chem. Phys. Lett.* **1990**, *166*, 275–280. [CrossRef]

87. Varadwaj, P.R.; Varadwaj, A.; Marques, H.M.; Yamashita, K. The Phosphorous Bond, or the Phosphorous-Centered Pnictogen Bond: The Covalently Bound Phosphorous Atom in Molecular Entities and Crystals as a Pnictogen Bond Donor. *Molecules* **2022**, *27*, 1487. [CrossRef]

88. Murray, J.S.; Politzer, P. Molecular Surfaces, van der Waals Radii and Electrostatic Potentials in Relation to Noncovalent Interactions. *Croat. Chem. Acta* **2009**, *82*, 267–275.

89. Varadwaj, A.; Marques, H.M.; Varadwaj, P.R. Is the Fluorine in Molecules Dispersive? Is Molecular Electrostatic Potential a Valid Property to Explore Fluorine-Centered Non-Covalent Interactions? *Molecules* **2019**, *24*, 379. [CrossRef]

90. Varadwaj, P.R. Does Oxygen Feature Chalcogen Bonding? *Molecules* **2019**, *24*, 3166. [CrossRef] [PubMed]

91. Varadwaj, P.R.; Varadwaj, A.; Marques, H.M. Very strong chalcogen bonding: Is oxygen in molecule capable of forming it? A First Princiles Perspective. *Authorea*, 2020; *preprints*. [CrossRef]

92. Varadwaj, P.R.; Varadwaj, A.; Marques, H.M.; Yamashita, K. Chalcogen Bonding in the Molecular Dimers of WCh$_2$ (Ch = S, Se, Te): On the Basic Understanding of the Local Interfacial and Interlayer Bonding Environment in 2D Layered Tungsten Dichalcogenides. *Int. J. Mol. Sci.* **2022**, *23*, 1263. [CrossRef] [PubMed]

93. Varadwaj, P.R.; Marques, H.M.; Varadwaj, A.; Yamashita, K. Chalcogen···Chalcogen Bonding in Molybdenum Disulfide, Molybdenum Diselenide and Molybdenum Ditelluride Dimers as Prototypes for a Basic Understanding of the Local Interfacial Chemical Bonding Environment in 2D Layered Transition Metal Dichalcogenides. *Inorganics* **2022**, *10*, 11. [CrossRef]

94. Varadwaj, P.R.; Varadwaj, A.; Marques, H.M. Halogen Bonding: A Halogen-Centered Noncovalent Interaction Yet to Be Understood. *Inorganics* **2019**, *7*, 40. [CrossRef]

95. Murray, J.S.; Politzer, P. Hydrogen Bonding: A Coulombic σ-Hole Interaction. *J. Indian Inst. Sci.* **2020**, *100*, 21–30. [CrossRef]

96. Holthoff, J.M.; Engelage, E.; Weiss, R.; Huber, S.M. "Anti-Electrostatic" Halogen Bonding. *Angew. Chem. Int. Ed.* **2020**, *59*, 11150–11157. [CrossRef]

97. Loy, C.; Holthoff, J.M.; Weiss, R.; Huber, S.M.; Rosokha, S.V. "Anti-electrostatic" halogen bonding in solution. *Chem. Sci.* **2021**, *12*, 8246–8251. [CrossRef]

98. Varadwaj, A.; Varadwaj, P.R.; Yamashita, K. Do surfaces of positive electrostatic potential on different halogen derivatives in molecules attract? Like attracting like! *J. Comput. Chem.* **2018**, *39*, 343–350. [CrossRef]

99. Politzer, P.; Murray, J.S.; Clark, T.; Resnati, G. The s-hole revisited. *Phys. Chem. Chem. Phys.* **2017**, *19*, 32166–32178. [CrossRef]

100. Politzer, P.; Murray, J.S.; Clark, T. Halogen bonding and other σ-hole interactions: A perspective. *Phys. Chem. Chem. Phys.* **2013**, *15*, 11178–11189. [CrossRef]

101. Clark, T.; Politzer, P.; Murray, J.S. Correct electrostatic treatment of noncovalent interactions: The importance of polarization. *WIREs Comput. Mol. Sci.* **2015**, *5*, 169–177. [CrossRef]

102. Politzer, P.; Murray, J.S. An Overview of Strengths and Directionalities of Noncovalent Interactions: σ-Holes and π-Holes. *Crystals* **2019**, *9*, 165. [CrossRef]

103. Clark, T.; Murray, J.S.; Politzer, P. Role of Polarization in Halogen Bonds. *Aust. J. Chem.* **2014**, *67*, 451–456. [CrossRef]

104. Adonin, S.A.; Gorokh, I.D.; Novikov, A.S.; Samsonenko, D.G.; Plyusnin, P.E.; Sokolov, M.N.; Fedin, V.P. Bromine-rich complexes of bismuth: Experimental and theoretical studies. *Dalton Trans.* **2018**, *47*, 2683–2689. [CrossRef]

105. Rozhkov, A.V.; Krykova, M.A.; Ivanov, D.M.; Novikov, A.S.; Sinelshchikova, A.A.; Volostnykh, M.V.; Konovalov, M.A.; Grigoriev, M.S.; Gorbunova, Y.G.; Kukushkin, V.Y. Reverse Arene Sandwich Structures Based upon π-Hole···[M^{II}] (d^8 M = Pt, Pd) Interactions, where Positively Charged Metal Centers Play the Role of a Nucleophile. *Angew. Chem.* **2019**, *131*, 4208–4212. [CrossRef]

106. Dabranskaya, U.; Ivanov, D.M.; Novikov, A.S.; Matveychuk, Y.V.; Bokach, N.A.; Kukushkin, V.Y. Metal-Involving Bifurcated Halogen Bonding C–Br···η^2(Cl–Pt). *Cryst. Growth Des.* **2019**, *19*, 1364–1376. [CrossRef]

107. Kashina, M.V.; Kinzhalov, M.A.; Smirnov, A.S.; Ivanov, D.M.; Novikov, A.S.; Kukushkin, V.Y. Dihalomethanes as Bent Bifunctional XB/XB-Donating Building Blocks for Construction of Metal-involving Halogen Bonded Hexagons. *Chem. Asian J.* **2019**, *14*, 3915–3920. [CrossRef]

108. Katkova, S.A.; Mikherdov, A.S.; Kinzhalov, M.A.; Novikov, A.S.; Zolotarev, A.A.; Boyarskiy, V.P.; Kukushkin, V.Y. (Isocyano Group π-Hole)···[dz^2-M^{II}] Interactions of (Isocyanide)[M^{II}] Complexes, in which Positively Charged Metal Centers (d^8-M = Pt, Pd) Act as Nucleophiles. *Chem. Eur. J.* **2019**, *25*, 8590–8598. [CrossRef]

109. Frisch, M.J.; Trucks, G.W.; Schlegel, H.B.; Scuseria, G.E.; Robb, M.A.; Cheeseman, J.R.; Scalmani, G.; Barone, V.; Mennucci, B.; Petersson, G.A.; et al. *Gaussian 09*, revision C.01; Gaussian, Inc.: Wallinford, CT, USA, 2016.

110. Macrae, C.F.; Bruno, I.J.; Chisholm, J.A.; Edgington, P.R.; McCabe, P.; Pidcock, E.; Rodriguez-Monge, L.; Taylor, R.; van de Streek, J.; Wood, P.A. Mercury 4.0: From visualization to analysis, design and prediction. *J. Appl. Cryst.* **2008**, *41*, 466–470. [CrossRef]

111. Dennington, R.; Keith, T.; Millam, J. *GaussView*, version 5.0.9; Semichem, Inc.: Shawnee Mission, KS, USA, 2009.

112. Keith, T.A. *AIMAll*, version 19.10.12; TK Gristmill Software: Overland Park, KS, USA, 2019. Available online: https://aim.tkgristmill.com (accessed on 24 February 2022).

113. Lu, T.; Chen, F. Multiwfn: A multifunctional wavefunction analyzer. *J. Comp. Chem.* **2012**, *33*, 580–592. [CrossRef] [PubMed]

114. Humphrey, W.; Dalke, A.; Schulten, K. VMD—Visual Molecular Dynamics. *J. Mol. Graph.* **1996**, *14*, 33–38. [CrossRef]

115. Turnbull, R.; Hanfland, M.; Binns, J.; Martinez-Canales, M.; Frost, M.; Marqués, M.; Howie, R.T.; Gregoryanz, E. Unusually complex phase of dense nitrogen at extreme conditions. *Nat. Commun.* **2018**, *9*, 4717. [CrossRef] [PubMed]

116. Hanfland, M.; Lorenzen, M.; Wassilew-Reul, C.; Zontone, F. Structures of Molecular Nitrogen at High Pressures. *Rev. High Press. Sci. Tecnol.* **1998**, *7*, 787–789. [CrossRef]

117. Vos, W.L.; Finger, L.W.; Hemley, R.J.; Hu, J.Z.; Mao, H.K.; Schouten, J.A. A high-pressure van der Waals compound in solid nitrogen-helium mixtures. *Nature* **1992**, *358*, 46–48. [CrossRef]

118. Venables, J.A.; English, C.A. Electron diffraction and the structure of [alpha]-N$_2$. *Acta Cryst. B* **1974**, *30*, 929–935. [CrossRef]

119. Schuch, A.F.; Mills, R.L. Crystal Structures of the Three Modifications of Nitrogen 14 and Nitrogen 15 at High Pressure. *J. Chem. Phys.* **1970**, *52*, 6000–6008. [CrossRef]

120. Mills, R.L.; Olinger, B.; Cromer, D.T. Structures and phase diagrams of N$_2$ and CO to 13 GPa by X-ray diffraction. *J. Chem. Phys.* **1986**, *84*, 2837–2845. [CrossRef]

121. Stinton, G.W.; Loa, I.; Lundegaard, L.F.; McMahon, M.I. The crystal structures of δ and δ^* nitrogen. *J. Chem. Phys.* **2009**, *131*, 104511. [CrossRef]

122. Bader, R.F. *Atoms in Molecules: A Quantum Theory*; Oxford University Press: Oxford, UK, 1990.

123. Pritchard, B.P.; Altarawy, D.; Didier, B.; Gibson, T.D.; Windus, T.L. A New Basis Set Exchange: An Open, Up-to-date Resource for the Molecular Sciences Community. *J. Chem. Inf. Model.* **2019**, *59*, 4814–4820. [CrossRef]

124. Ivlev, S.I.; Conrad, M.; Hoelzel, M.; Karttunen, A.J.; Kraus, F. Crystal Structures of α- and β-Nitrogen Trifluoride. *Inorg. Chem.* **2019**, *58*, 6422–6430. [CrossRef] [PubMed]

125. Varadwaj, A.; Varadwaj, P.R.; Marques, H.M.; Yamashita, K. Revealing Factors Influencing the Fluorine-Centered Non-Covalent Interactions in Some Fluorine-substituted Molecular Complexes: Insights from First-Principles Studies. *ChemPhysChem* **2018**, *19*, 1486–1499. [CrossRef] [PubMed]

126. Kawai, S.; Sadeghi, A.; Xu, F.; Peng, L.; Orita, A.; Otera, J.; Goedecker, S.; Meyer, E. Extended halogen bonding between fully fluorinated aromatic molecules. *ACS Nano* **2015**, *9*, 2574–2583. [CrossRef] [PubMed]

127. Varadwaj, A.; Varadwaj, P.R.; Marques, H.M.; Yamashita, K. Comment on "Extended Halogen Bonding between Fully Fluorinated Aromatic Molecules: Kawai et al., ACS Nano, 2015, 9, 2574–2583". CondMat Public Archive. Available online: https://arxiv.org/ftp/arxiv/papers/1802/1802.09995.pdf (accessed on 12 October 2021).

128. Varadwaj, P.R.; Varadwaj, A.; Jin, B.-Y. Unusual bonding modes of perfluorobenzene in its polymeric (dimeric, trimeric and tetrameric) forms: Entirely negative fluorine interacting cooperatively with entirely negative fluorine. *Phys. Chem. Chem. Phys.* **2015**, *17*, 31624–31645. [CrossRef] [PubMed]

129. Hartl, H.; Schöner, J.; Jander, J.; Schulz, H. Die Struktur des festen Stickstofftrichlorids (−125 °C). *Z. Anorg. Allg. Chem.* **1975**, *413*, 61–71. [CrossRef]

130. Dehnicke, K. The Chemistry of Iodine Azide. *Angew. Chem. Int. Ed. Engl.* **1979**, *18*, 507–514. [CrossRef]

131. Dehnicke, K. The Chemistry of the Halogen Azides. *Adv. Inorg. Chem.* **1983**, *26*, 169–200.

132. Tornieporth-Oetting, I.C.; Klapötke, T.M. Covalent Inorganic Nonmetal Azides. In *Combustion Efficiency and Air Quality*; Hargitta, I., Vidóczy, T., Eds.; Springer: Boston, MA, USA, 1995.

133. Tornieporth-Oetting, I.C.; Klapötke, T.M. Recent developments in the chemistry of binary nitrogen-halogen species. *Comments Inorg. Chem.* **1994**, *15*, 137–169. [CrossRef]

134. Buzek, P.; Klapötke, T.M.; von Ragué Schleyer, P.; Tornieporth-Oetting, I.C.; White, P.S. Iodine Azide. *Angew. Chem. Int. Ed. Engl.* **1993**, *32*, 275–277. [CrossRef]

135. Lyhs, B.; Bläser, D.; Wölper, C.; Schulz, S.; Jansen, G. A Comparison of the Solid-State Structures of Halogen Azides XN3 (X = Cl, Br, I). *Angew. Chem. Int. Ed.* **2012**, *51*, 12859–12863. [CrossRef]

136. Lyhs, B.; Bläser, D.; Wölper, C.; Schulz, S.; Jansen, G. Solid-State Structure of Bromine Azide. *Angew. Chem. Int. Ed.* **2012**, *51*, 1970–1974. [CrossRef] [PubMed]

137. Müller, U.; Ivlev, S.; Schulz, S.; Wölper, C. Automated Crystal Structure Determination Has its Pitfalls: Correction to the Crystal Structures of Iodine Azide. *Angew. Chem. Int. Ed.* **2021**, *60*, 17452–17454. [CrossRef] [PubMed]

138. Bursch, M.; Kunze, L.; Vibhute, A.M.; Hansen, A.; Sureshan, K.M.; Jones, P.G.; Grimme, S.; Werz, D.B. Quantification of Noncovalent Interactions in Azide–Pnictogen, –Chalcogen, and –Halogen Contacts. *Chem. Eur. J.* **2021**, *27*, 4627–4639. [CrossRef] [PubMed]

139. Göbel, M.; Karaghiosoff, K.; Klapötke, T.M. The First Structural Characterization of a Binary P–N Molecule: The Highly Energetic Compound P_3N_{21}. *Angew. Chem. Int. Ed.* **2006**, *45*, 6037–6040. [CrossRef] [PubMed]

140. Laniel, D.; Winkler, B.; Koemets, E.; Fedotenko, T.; Chariton, S.; Milman, V.; Glazyrin, K.; Prakapenka, V.; Dubrovinsky, L.; Dubrovinskaia, N. Nitrosonium nitrate ($NO^+NO_3^-$) structure solution using in situ single-crystal X-ray diffraction in a diamond anvil cell. *IUCrJ* **2021**, *8*, 208–214. [CrossRef]

141. Addison, C.C.; Thompson, R. Ionic Reactions in Liquid Dinitrogen Tetroxide. *Nature* **1948**, *162*, 369–370. [CrossRef]

142. Parts, L.; Miller, J.T., Jr. Nitrosonium Nitrate. Isolation at 79°–205°K and Infrared Spectra of the Polymorphic Compound. *J. Chem. Phys.* **1965**, *43*, 136–139. [CrossRef]

143. Agnew, S.F.; Swanson, B.I.; Jones, L.H.; Mills, R.L.; Schiferl, D. Chemistry of nitrogen oxide (N_2O_4) at high pressure: Observation of a reversible transformation between molecular and ionic crystalline forms. *J. Phys. Chem.* **1983**, *87*, 5065–5068. [CrossRef]

144. Agnew, S.F.; Swanson, B.I.; Jones, L.H.; Mills, R.L. Disproportionation of nitric oxide at high pressure. *J. Phys. Chem.* **1985**, *89*, 1678–1682. [CrossRef]

145. Sihachakr, D.; Loubeyre, P. High-pressure transformation of N_2/O_2 mixtures into ionic compounds. *Phys. Rev. B* **2006**, *74*, 064113. [CrossRef]

146. Somayazulu, M.; Madduri, A.; Goncharov, A.F.; Tschauner, O.; McMillan, P.F.; Mao, H.-K.; Hemley, R.J. Novel Broken Symmetry Phase from N_2O at High Pressures and High Temperatures. *Phys. Rev. Lett.* **2001**, *87*, 135504. [CrossRef] [PubMed]

147. Bailey, R.D.; Pennington, W.T. Tetramethylpyrazinium polyiodides. *Acta Crystallogr. B* **1995**, *51*, 810–815. [CrossRef]

148. Kumar, V.; Scilabra, P.; Politzer, P.; Terraneo, G.; Daolio, A.; Fernandez-Palacio, F.; Murray, J.S.; Resnati, G. Tetrel and Pnictogen Bonds Complement Hydrogen and Halogen Bonds in Framing the Interactional Landscape of Barbituric Acids. *Cryst. Growth Des.* **2021**, *21*, 642–652. [CrossRef]

149. Butcher, R.J.; Bottaro, J.C.; Gilardi, R. Ammonium 1,3,4,6-tetranitro-2,5-diazapentalene. *Acta Crystallogr. E* **2003**, *59*, o1149–o1150. [CrossRef]

150. Butcher, R.J.; Bottaro, J.C.; Gilardi, R. 1,3,4-Trinitro-7,8-diazapentalene. *Acta Crystallogr. E* **2003**, *59*, o1780–o1782. [CrossRef]
151. Gilardi, R.; Evans, R.N.; Manser, G.E. 3,3-Bis(difluoroaminomethyl)oxetane, a promising new energetic material. *Acta Crystallogr. E* **2003**, *59*, o2032–o2034. [CrossRef]
152. Molchanov, A.P.; Efremova, M.M.; Kryukova, M.A.; Kuznetsov, M.A. Selective and reversible 1,3-dipolar cycloaddition of 6-aryl-1,5-diazabicyclo[3.1.0]hexanes with 1,3-diphenylprop-2-en-1-ones under microwave irradiation. *Beilstein J. Org. Chem.* **2020**, *16*, 2679–2686. [CrossRef]
153. Olejniczak, A.; Katrusiak, A.; Podsiadlo, M.; Katrusiak, A. Crystal design by CH . . . N and N . . . N interactions: High-pressure structures of high-nitrogen-content azido-triazolopyridazines compounds. *Acta Crystallogr. B* **2020**, *76*, 1136–1142. [CrossRef]
154. Allcock, H.R.; Ngo, D.C.; Parvez, M.; Visscher, K. Cyclic and short-chain linear phosphazenes with hindered aryloxy side groups. *J. Chem. Soc. Dalton Trans.* **1992**, *10*, 1687–1699. [CrossRef]

Communication

Volatile Organic Compounds in *Dactylorhiza* Species

Marisabel Mecca [1], **Rocco Racioppi** [1], **Vito Antonio Romano** [1], **Licia Viggiani** [1], **Richard Lorenz** [2] **and Maurizio D'Auria** [1,*]

[1] Dipartimento di Scienze, Università della Basilicata, V.le dell'Ateneo Lucano 10, 85100 Potenza, Italy; marisabelmecca@libero.it (M.M.); rocco.racioppi@unibas.it (R.R.); vitoantonio_romano@libero.it (V.A.R.); licia.viggiani@unibas.it (L.V.)

[2] AHO Baden-Württemberg, 64686 Lautertal, Germany; lorenz@orchids.de

* Correspondence: maurizio.dauria@unibas.it; Tel.: +39-0971-205-480

Abstract: HS-SPME-GC–MS analysis of the scent of *Dactylorhiza viridis* revealed the presence of verbenone (28.86%), caryophyllene (25.67%), β-terpineol (9.48%), and δ-cadinene (6.94%). In the scent of *Dactylorhiza romana* β-ocimene (18.69%), pentadecane (18.40%), α-farnesene (14.65%), and isopropyl 14-methylpentadecanoate (14.32%) were found. *Dactylorhiza incarnata* contained tetradecane (11.07%), pentadecane (28.40%), hexadecane (19.53%), heptadecane (17.33%), and α-cubenene (11.48%). Analysis of *Dactylorhiza saccifera* showed the presence of caryophyllene (17.38%), pentadecane (6.43%), hexadecane (6.13%), and heptadecane (5.08%). Finally, the aroma components found in *Dactylorhiza sambucina* were caryophyllene (12.90%), β-sesquiphellandrene (32.16%), 4,5-di-*epi*-aristolochene (10.18%).

Keywords: *Dactylorhiza*; volatile organic compounds; solid phase microextraction; gas chromatography; mass spectrometry

Citation: Mecca, M.; Racioppi, R.; Romano, V.A.; Viggiani, L.; Lorenz, R.; D'Auria, M. Volatile Organic Compounds in *Dactylorhiza* Species. *Compounds* 2022, 2, 121–130. https://doi.org/10.3390/compounds2020009

Academic Editor: Juan Mejuto

Received: 21 February 2022
Accepted: 29 March 2022
Published: 12 April 2022

Publisher's Note: MDPI stays neutral with regard to jurisdictional claims in published maps and institutional affiliations.

1. Introduction

The taxonomy of the genus *Dactylorhiza* Necker ex Nevski is one of the most studied in the Orchidaceae family [1–8], widespread in Eurasia, North Africa, Alaska [9,10], with a great variety of confused and difficult to classify forms in precise taxa. It includes a number of species that varies strongly between authors, from twelve [1] to seventy-five species [6], or to thirty-six species and forty-six subspecies [11]. Traditional classifications recognize four sections of the genus *Dactylorhiza*: Aristatae, Sambucinae, Iberanthus, and Dactylorhiza s.s. [6]. In the early 2000s, new genetic work led to the inclusion of *Coeloglossum viride*, the only species of the genus *Coeloglossum* Hartm. 1820, in the genus *Dactylorhiza* as *Dactylorhiza viridis* (L.) [12,13]. Several authors have used this new classification in recently published monographs [14–16], while others [17–19] have been reluctant to include *Coeloglossum viride* within *Dactylorhiza*, arguing that current evidence is insufficient.

In this work it was decided to consider *Coeloglossum* in *Dactylorhiza* recognizing *D. viride*, as reported in recent molecular biology [20–25] and molecular genetics studies [10,26–30]. The scents emitted by five species of *Dactylorhiza* present in Southern Italy, four present in Basilicata, *Dactylorhiza viridis* (L.) R. M. Bateman, Pridgeon, and M. W. Chase (Figure 1a), *Dactylorhiza romana* (Sebastiani) Soó (Figure 1b), *Dactylorhiza sambucina* (L.) Soó (Figure 2a), and one present in Campania, *Dactylorhiza saccifera* (Brongn.) Soó (Figure 2b), and *Dactylorhiza incarnata* (L.) Soó (Figure 2c) were analyzed.

D. viridis is an uncommon species in Basilicata; it always grows in small populations (from 1 to 10 plants) in open woods and pastures from 600 m a.s.l. at 2100 m a.s.l. It has a very variable morphology, due to its stature (plants more than 30 cm high in the open woods and meadows of the mid-hill, or a few centimeters low in altitude meadows), due to its color (from light green to reddish yellow), and the number of flowers. *D. viridis* is the only species of the group that has a rewarding strategy by rewarding pollinators with nectar, even if in minimal quantities [27], while, on the contrary, the other species of *Dactylorhiza* do not produce nectar by pursuing a strategy of food deception. Food-deceptive orchids are more common than sexually deceptive ones; their flowers provide signals of food gratification but do not provide re-compensation [31–33]. In Basilicata the populations of *D. romana* and *D. sambucina* are characterized by the simultaneous presence of flowers colored from yellow to red with more or less frequent various intermediate forms, while the populations of *D. saccifera*, and *D. incarnata*, have flowers with pinkish, white, or fleshy pinks often faded and attenuated in their features.

Figure 1. (**a**): *Dactylorhiza viridis*, Moliterno (Pz), 23 May 2018; (**b**): *Dactylorhiza romana*, Rifreddo (Pz), 10 April 2017. Photos of V. A. Romano.

The color, smell, size, and shape of flowers are key signals that attract pollinators in search of rewards [34–36]. Floral color is often included in studies on the evolutionary models of pollination systems while floral scents, whose characterization requires sophisticated and complex tools, have been little studied, even if they are often considered the main attraction for pollinators [37–40]. Different types of floral scent variations can be distinguished. The floral odor can vary in composition (when the flowers emit different compounds or distinct ratios of the same compounds) and/or emission speed (that is, when the flowers emit the same bouquet but in different quantities). In many cases, it has been reported that the chemical profiles of floral scents vary both between individuals of the same population and between different populations, through an almost infinite number of combinations of volatile compounds [41–47].

Comparative analysis of the emission of scents revealed a greater variation of the compounds among the same individuals in the deceptive orchids compared to the gratifying orchids [48,49]. The flower bouquet varies significantly between food deceptive orchids, even within the same genus, and can affect pollinator species and their behaviors in multiple ways [31,41]. The analysis of floral odors in orchids has been mainly carried out to characterize olfactory signals presumably attractive to pollinators, but only in a few cases have these studies been coupled, with the analysis of the pollinator response, to the volatile compounds emitted using electrophysiological techniques [48,50,51] and behavioral tests [52–55]. Therefore, there is an increasing need for a better understanding of the evolution of floral scents in order to obtain a more complete view of the mechanisms and patterns of evolution of pollination systems.

Food-deceptive orchids are pollinated by generalist pollinators [56], mainly bumblebee queens and various other bees [57–60], just emerged from the soil, without experience which, after the first attempts at pollination, abandon the plant, moving away in search of other inflorescences [61]. It is thought that the wide variability of the morphology of flowers and scents in deceptive species for food favors cross-pollination by reducing the learning in the recognition of flowers by newly emerged, inexperienced pollinators, thus limiting self-pollination and geitonogamy and favoring outbreeding.

Figure 2. (**a**): *Dactylorhiza sambucina*, Moliterno (Pz), 21 May 2018; (**b**): *Dactylorhiza saccifera* Abriola (Pz), 30 June 2017; (**c**): *Dactylorhiza incarnata*, Mandranello, Padula (Sa), 6 June 2018. Photos of V. A. Romano.

While *D. romana*, *sambucina*, and *saccifera* are widespread in the territory of Basilicata, also forming large populations, *D. incarnata* has been identified in a single station on the Maddalena Mountains (Padula. Salerno) on the border between Basilicata and Campania where it forms a large population mixed with *D. saccifera* and where it is possible to find many hybrid forms between the two species and also many plants of *D. incarnata* var. *ochroleuca*. This population of *D. incarnata* is the southernmost station reported in Italy.

The purpose of this work is to complete the collection of data on the various perfumes emitted by the species of orchids in Basilicata by using the same technique, solid phase microextraction performed on intact plats followed by gas chromatography–mass spectrometry. Until now, the volatile organic compounds constituents of the scent of the orchid species have been determined on several orchid species [62–69].

2. Materials and Methods

2.1. Plant Material

The sample of *D. viridis* was collected at Moliterno (Pz), 1100 m a.s.l., on 23 May 2018. The sample of *D. romana* was collected at Rifreddo (Pz), 1120 m a.s.l., on 10 April 2017. The sample of *D. sambucina* (yellow color) was collected at Moliterno (Pz), 1200 m a.s.l., on 21 May 2018. The sample of *D. saccifera* was collected at Abriola (Pz), 1200 m a.s.l., on 30 June 2017. The sample of *D. incarnata* was collected at Mandranello, Padula (Sa), 1200 m a.s.l., on 3 June 2018. The plants were collected by Vito Antonio Romano.

The plants were harvested about two weeks before flowering by taking all the clod of earth, taking care not to damage the root system, planted in special pots in the gardens of the University of Basilicata (Potenza 650 m a.s.l.), in waiting for their full bloom. Two days before the tests the plants were transferred to an air-conditioned room at 22 °C. The plants were tested whole without being damaged under a cylindrical glass bell (12 cm × 45 cm) in which only the inflorescence and the SPME probe were inserted.

To avoid contamination, the interior of the bell was isolated from the external environment with appropriate closing and sealing systems during the 24 h of the test (from 8 in the morning to 8 the following day).

In order to be sure that the internal environment of the bell was isolated from the external environment, various blank tests were carried out.

The plants were successively used for further studies on pollination, fertility, and germination of the plants. After these studies the plants were not in condition to be collected in an herbarium. However, these species can be recognized without ambiguities on the basis of their properties, well documented by the Figures 1 and 2. In view of the fact that the investigated taxa are rare wild plants, in order to preserve the species, we have chosen to use a single plant for our analysis.

2.2. Analysis of Volatile Organic Compounds

The SPME analysis of five different samples of *Dactylorhiza* has been performed. This way, the identified plants were collected and inserted in glass jars for 24 h where the fiber (DVB/CAR/PDMS) and SPME syringe were also placed. After this time the fiber was desorbed in a gas chromatographic apparatus equipped with a quadrupole mass spectrometer detector. A 50/30-μm DVB/CAR/PDMS module with 1 cm fiber (57328-U, Supelco, Milan, Italy) was employed to determine VOCs. SPME fiber was maintained in the bell jar for 24 h. The analytes were desorbed in the splitless injector at 250 °C for 2 min. Analyses were accomplished with an HP 6890 Plus gas chromatograph equipped with a Phenomenex Zebron ZB-5 MS capillary column (30-m × 0.25-mm i.d. × 0.25 μm FT) (Agilent, Milan, Italy). An HP 5973 mass selective detector (Agilent, Milan, Italy) in the range 0–800 *m/z* (Agilent) was utilized with helium at 0.8 mL/min as the carrier gas. The EI source was used at 70 eV. The analyses were performed by using a splitless injector. The splitless injector was maintained at 250 °C and the detector at 230 °C. The oven was held at 40 °C for 2 min, then gradually warmed, 8 °C/min, up to 250 °C and held for 10 min. Tentative identification of aroma components was based on mass spectra and Wiley 11 and NIST 14 library comparison. Single VOC peak was considered as identified when its experimental spectrum matched with a score over 90% that present in the library. All the analyses were performed in triplicate.

3. Results and Discussion

The scent of *Dactylorhiza* orchids has been the object of some studies in the past. Analysis of *D. sambucina* showed the presence of limonene, β-myrcene, α-pinene, α-bergamotene, β-bisabolene, caryophyllene, and β-selinene [70]. Dichloromethane extracts of flowers of *D. incarnata* showed the presence of 4-hydroxybenzaldehyde (16.88%), 4-hydrozybenzyl alcohol (30.60%), and methyl 4-hydroxyphenylacetate (34.37%) [71]. Pentane, diethyl ether extraction of flowers of *D. incarnata* showed the presence of nonanal, 9-(*Z*)-heptacosene, 9-(*Z*)-nonacosene, tricosane, pentacosane, and heptacosane [72] The analysis of the scent of *D. romana* (red flower), obtained by head space analysis of the scent absorbed on Porapak Q, showed the presence of several components, including nonanal (9.19%), sabinene (10.89%), (*E*)-ocimene (15.63%), and linalool (7.99%) [31]. Such different results, obtained by using different analytical techniques, allowed us to analyze the scent of these species by using SPME (solid phase microextraction) technique. The results are reported in Table 1.

The never studied before *D. viridis* contains, as its main components, verbenone (28.86%) and caryophyllene (25.67%), while other compounds present in relevant amounts are β-terpineol (9.48%) and δ-cadinene (6.94%) (Table 1 and Figure 3).

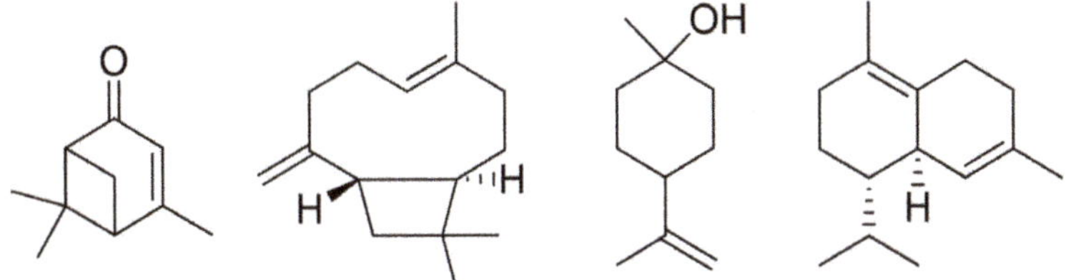

Figure 3. Main components of the scent of *D. viridis*.

In the scent of *D. romana* several compounds were found: β-ocimene (18.69%, found also in [31]), pentadecane (18.40%), α-farnesene (14.65%), and isopropyl 14-methylpentadecanoate (14.32%) (Table 1). Previous works published on the aroma of *D. incarnata* found phenolic compounds, in one case, while long chain alkanes and alkenes, were found in another case [71,72]. SPME analysis of the scent detected the presence of a mixture of hydrocarbons, but with lower molecular weight than those found in [72]. Tetradecane (11.07%), pentadecane (28.40%), hexadecane (19.53%), and heptadecane (17.33%) were the main components, together with α-cubenene (11.48%) (Table 1).

D. saccifera contained, as a main component, caryophyllene (17.38%), while other significant compounds were hydrocarbons, pentadecane (6.43%), hexadecane (6.13%), and heptadecane (5.08%) (Table 1). Finally, the aroma components found in *D. sambucina* were caryophyllene (12.90%, found also in [70]), β-sesquiphellandrene (32.16%, Figure 4), 4,5-di-*epi*-aristolochene (10.18%, Figure 4) (Table 1).

β-sesquiphellandrene   4,5-di-*epi*-aristolochene

Figure 4. Main components of the scent of *D. sambucina*.

Table 1. SPME-GC–MS analysis of *Dactylorhiza* species.

Compound	r.t. [a] (min.)	KI [b]	D. viridis	D. romana	D. incarnata	D. saccifera	D. sambucina
					Area % ± 0.03		
Mesityl oxide	4.94	782				2.03	
α-Pinene	7.78	933				0.20	0.30
Sabinene	8.77	972	2.27				2.44
β-Pinene	8.86	979					0.79
β-Myrcene	9.12	989	2.42				
2,2,4,6,6-pentamethyl-3-heptene	9.29	1018				0.60	
Limonene	9.79	1028				1.88	
Eucalyptol	10.00	1032	2.65				
β-Ocimene	10.33	1044		18.69			
β-Terpineol	10.74	1085	9.48				
Linalool	11.22	1100				0.85	
Lilac aldehyde A	12.09	1145				0.20	
Lilac aldehyde B	12.25	1154				0.90	
Dodecane	13.08	1200				0.45	
α-Terpineol	13.16	1209	2.07				
Lilac alcohol A	13.52	1221				2.08	
Verbenone	13.55	1223	28.86				
Lilac alcohol B	13.84	1235				4.00	
Citral	14.55	1265	4.21				
Tridecane	14.88	1300				2.88	0.85
2,4,4,6,6,8,8-Heptamethyl-2-nonene	16.02	1343				1.48	
α-Cubebene	16.42	1360			11.48		0.34
Tetradecane	16.64	1400			11.07	3.90	2.13
β-Elemene	16.69	1403					
Caryophyllene	17.22	1420	25.67			17.38	12.90
Methyl 2-phenylcyclopropanecarboxylate	17.31	1435				0.90	
cis-α-Bergamotene	17.38	1440					7.61
β-Farnesene	17.48	1458					0.68
Geranylacetone	17.61	1468		6.62			
Humulene	17.76	1473					0.42
4,5-Di-*epi*-aristolochene	18.23	1485					10.18

Table 1. *Cont.*

Compound	r.t. (a) (min.)	KI (b)	D. viridis	D. romana	D. incarnata	D. saccifera	D. sambucina
					Area % ± 0.03		
Pentadecane	18.25	1500	2.09	18.40	28.40	6.43	
β-Selinene	18.40	1511					1.28
α-Farnesene	18.45	1518		14.65		3.30	
β-Bisabolene	18.50	1523					2.64
δ-Cadinene	18.78	1530	6.94			1.00	
7-Hexadecenal	18.64	1555					
β-Sesquiphellandrene	18.76	1557					32.16
Elemicin	19.07	1566				3.30	
Hexadecane	19.71	1600			19.53	6.13	1.06
Methyl dihydrojasmonate	20.32	1650	2.55				1.23
Cyclotetradecane	20.47	1673				1.58	
Heptadecane	21.14	1700			17.33	5.08	0.64
3,5-Di-*t*-butyl-4-hydroxybenzaldehyde	22.13	1772				4.38	
Octadecane	22.48	1800			8.24	3.40	0.25
Phytone	23.14	1848					0.58
Nonadecane	23.70	1900				2.93	
Isopropyl 14-methylpentadecanoate	25.35	1915		14.32			1.94
7,9-Di-*t*-butyl-1-oxaspiro[4.5]deca-6,9-diene-2,8-dione	24.11	1929				1.88	
Eicosane	24.93	2000				1.83	
13-*epi*-Manoyl oxide	25.49	2015			3.42		
Docosane	27.22	2200				1.93	
Tricosane	28.31	2300				2.28	
Bis(1-phenylethyl)phenol	30.98	2426				2.08	

(a) R.T. retention time; (b) KI Kovats index.

4. Conclusions

The results described above can give us some useful information. First, the results of our analysis, performed by using SPME technique, are not in agreement with previous described scent composition [31,70–72]. This difference may depend on several factors: first of all, the different location of the plants under study. The different environmental conditions could induce plants to adopt different strategies for pollination. Second, the different analysis methodology could play a significant role. Furthermore, we can observe that every species adopts a different strategy. *D. viridis* has a scent where terpenes are the main components. This statement is applicable also to *D. romana*; however, it is noteworthy that the terpenes involved in the scent are different from those observed in the other species. The scent of *D. incarnata* only included hydrocarbons as components; hydrocarbons are present in the scent of *D. saccifera* but, in this case, caryophyllene is also present in relevant amount. Finally, terpenes were detected in the scent of *D. sambucina*, as in *D. viridis* and in *D. romana*, but the compounds involved in the scent, with the exception of caryophyllene, are different from those observed in the other species.

Author Contributions: Conceptualization, M.D. and V.A.R.; methodology, R.R.; investigation, R.R., L.V. and M.M.; data curation, R.L.; writing—original draft preparation, V.A.R. and M.D.; writing—review and editing, M.D., V.A.R. and R.L. All authors have read and agreed to the published version of the manuscript.

Funding: This research received no external funding.

Institutional Review Board Statement: Not applicable.

Informed Consent Statement: Not applicable.

Data Availability Statement: Not applicable.

Conflicts of Interest: The authors declare no conflict of interest.

References

1. Klinge, J. Dactylorchids, orchids subgeneris, monographiae prodromus. *Acta Horti Petropol.* **1898**, *17*, 145–201.
2. Vermeulen, P. *Studies on Dactylorchids*; Schotanus & Jens: Utrecht, The Netherlands, 1947.
3. Soó, R. Synopsis generis *Dactylorhiza* (Dactylorchis). *Ann. Univ. Sci. Bp. Biol.* **1960**, *3*, 335–357.
4. Senghas, K. Taxonomische Übersicht der Gattung Dactylorhiza Necker ex Nevski. *Jahresber. Naturwiss. Ver. Wupp.* **1968**, *21–22*, 32–67.
5. Nelson, E. *Monographie und Ikonographie der Orchidaceen-Gattung Dactylorhiza*; Speich: Zürich, Switzerland, 1976.
6. Averyanov, L.V. A review of the genus Dactylorhiza. In *Orchid Biology. Reviews and Perspectives*; Arditti, J., Ed.; Timber Press: Portland, OR, USA, 1990; pp. 159–206.
7. Pedersen, H.Æ. Species concept and guidelines for infraspecific taxonomic ranking in *Dactylorhiza* (Orchidaceae). *Nord. J. Bot.* **1998**, *18*, 289–311. [CrossRef]
8. Hedrén, M. Systematics of the *Dactylorhiza euxina/incarnata/maculata* polyploid complex (Orchidaceae) in Turkey: Evidence from allozyme data. *Plant Syst. Evol.* **2001**, *229*, 23–44. [CrossRef]
9. Delforge, P. *Guide des Orchidees d'Europe, d'Afrique du Nord et du Proche-Orient*, 2nd ed.; Delachaux et Niestle: Lausanne, Switzerland, 2001.
10. Efimov, P.G.; Philippov, E.G.; Krivenko, D.A. Allopolyploid speciation in Siberian *Dactylorhiza* (Orchidaceae, Orchidoideae). *Phytotaxa* **2016**, *258*, 101–120. [CrossRef]
11. Eccarius, W. *Die Orchideengattung Dactylorhiza*; Eisenach: Bürgel, Germany, 2016.
12. Bateman, R.M.; Pridgeon, A.M.; Chase, M.W. Phylogenetics of subtribe Orchidinae (Orchidoideae, Orchidaceae) based on nuclear ITS sequences. 2. Infrageneric relationships and taxonomic revision to achieve monophyly of Orchis sensu stricto. *Lindleyana* **1997**, *12*, 113–141.
13. Cribb, P.J.; Chase, M.W. Proposal to conserve the name *Dactylorhiza* Necker ex Nevski over *Coeloglossum* Hartm. (*Orchidaceae*). *Taxon* **2001**, *50*, 581–582. [CrossRef]
14. Dusak, F.; Pernot, P. *Les Orchidées Sauvages d'Île-de-France*; Biotope: Mèze, Germany, 2002.
15. Jacquet, P.; Scappaticci, G. *Une Répartition des Orchidées Sauvages de France*, 3rd ed.; Société Francaise d'Orchidophilie: Paris, France, 2003.
16. Foley, M.; Clarke, S. *Orchids of the British Isles*; Griffin Press: Cheltenham, UK, 2005.

17. Rossi, W.; Eldredge Maury, A. *Iconography of Italian Orchids*; Istituto Nazionale per la Fauna Selvatica 'Alessandro Ghigi': Bologna, Italy, 2002.

18. Delforge, P. *Guide des Orchidées d'Europe, d'Afrique du Nordiska et du Proche-Orient*, 3rd ed.; Delachaux et Niestlé: Lausanne, Switzerland, 2005.

19. Baumann, H.; Blatt, H.; Dierssen, K.; Dietrich, H.; Dostmann, H.; Eccarius, W.; Kretzschmar, H.; Kühn, H.-D.; Möller, O.; Paulus, H.F.; et al. *Die Orchideen Deutschlands*; Arbeitskreise Heimische Orchideen: Uhlstädt-Kirchhasel, Germany, 2005.

20. Bateman, R.M.; Hollingsworth, P.M.; Preston, J.; Yi-Bo, L.; Pridgeon, A.M.; Chase, M.W. Molecular phylogenetics and evolution of Orchidinae and selected Habenariinae (Orchidaceae). *Bot. J. Linn. Soc.* **2003**, *142*, 1–40. [CrossRef]

21. Shipunov, A.B.; Fay, M.F.; Pillon, Y.; Bateman, R.M.; Chase, M.W. *Dactylorhiza* (Orchidaceae) in European Russia: Combined molecular and morphological analysis. *Am. J. Bot.* **2004**, *91*, 1419–1426. [CrossRef]

22. Devos, N.; Raspé, O.; Oh, S.H.; Tyteca, D.; Jacquemart, A.L. The evolution of *Dactylorhiza* (Orchidaceae) allotetraploid complex: Insights from nrDNA sequences and cpDNA PCR-RFLP data. *Mol. Phylogenet. Evol.* **2006**, *38*, 767–778. [CrossRef]

23. Pillon, Y.; Fay, M.F.; Hedrén, M.; Bateman, R.M.; Devey, D.S.; Shipunov, A.B.; Van Ver Bank, M.; Chase, M.W. Evolution and temporal diversification of western European polyploid species complexes in *Dactylorhiza* (Orchidaceae). *Taxon* **2007**, *56*, 1185–1208. [CrossRef]

24. Inda, L.A.; Pimentel, M.; Chase, M.W. Chalcone synthase variation and phylogenetic relationships in *Dactylorhiza* (Orchidaceae). *Bot. J. Linn. Soc.* **2010**, *163*, 155–165. [CrossRef]

25. Balao, F.; Trucchi, E.; Wolfe, T.M.; Hao, B.-H.; Lorenzo, M.T.; Baar, J.; Sedman, L.; Kosiol, C.; Amman, F.; Chase, M.W.; et al. Adaptive sequence evolution is driven by biotic stress in a pair of orchid species (*Dactylorhiza*) with distinct ecological optima. *Mol. Ecol.* **2017**, *26*, 3649–3662. [CrossRef]

26. Givnish, T.J.; Spalink, D.; Ames, M.; Lyon, S.P.; Hunter, S.J.; Zuluaga, A.; Iles, W.J.D.; Clements, M.A.; Arroyo, M.T.K.; Leebens-Mack, J.; et al. Orchid phylogenomics and multiple drivers of their extraordinary diversification. *Proc. R. Soc. B Biol. Sci.* **2015**, *282*, 20151553. [CrossRef]

27. Bateman, R.M.; Rudall, P.J. Clarified relationship between *Dactylorhiza viridis* and *Dactylorhiza iberica* renders obsolete the former genus *Coeloglossum* (Orchidaceae: Orchidinae). *Kew Bull.* **2018**, *73*, 4. [CrossRef]

28. Bateman, R.M.; Murphy, A.R.; Hollingsworth, P.M.; Hart, M.L.; Denholm, I.; Rudall, P.J. Molecular and morphological phylogenetics of the digitate-tubered clade within subtribe *Orchidinae ss* (Orchidaceae: Orchideae). *Kew Bull.* **2018**, *73*, 54. [CrossRef]

29. Kaki, A.; Vafaee, Y.; Khadivi, A. Genetic variation of *Anacamptis coriophora, Dactylorhiza umbrosa, Himantoglossum affine, Orchis mascula*, and *Ophrys schulzei* in the western parts of Iran. *Ind. Crops Prod.* **2020**, *156*, 112854. [CrossRef]

30. Brandrud, M.K.; Baar, J.; Lorenzo, M.T.; Bateman, R.M.; Chase, M.W.; Hedrén, M.; Paun, O. Phylogenomic relationships of diploids and the origins of allotetraploids in *Dactylorhiza* (*Orchidaceae*): RADseq data track reticulate evolution. *Syst. Biol.* **2019**, *69*, 91–109. [CrossRef]

31. Salzmann, C.C.; Schiestl, F.P. Odour and colour polymorphism in the food deceptive orchid *Dactylorhiza romana*. *Plant Syst. Evol.* **2007**, *267*, 37–45. [CrossRef]

32. Jersáková, J.; Johnson, S.D.; Jürgens, A. Deceptive behaviour in plants. II. Food deception by plants: From generalized systems to specialized floral mimicry. In *Plant–Environment Interactions*; Baluska, F., Ed.; Springer: Berlin/Heidelberg, Germany, 2009; pp. 223–246.

33. Walsh, R.P.; Michaels, H.J. When it pays to cheat: Examining how generalized food deception increases male and female fitness in a terrestrial orchid. *PLoS ONE* **2017**, *12*, e0171286. [CrossRef]

34. Chittka, L.; Raine, N.E. Recognition of flowers by pollinators. *Curr. Opin. Plant Biol.* **2006**, *9*, 428–435. [CrossRef]

35. Sun, M.; Schlüter, P.M.; Gross, K.; Schiestl, F.P. Floral isolation is the major reproductive barrier between a pair of rewarding orchid sister species. *J. Evol. Biol.* **2015**, *28*, 117–129. [CrossRef]

36. Valenta, K.; Nevo, O.; Martel, C.; Chapman, C.A. Plant attractants: Integrating insights from seed dispersal and pollination ecology. *Evol. Ecol.* **2017**, *31*, 249–267. [CrossRef]

37. Dicke, M. Chemical ecology: From genes to communities. In *Chemical Ecology: From Genes to Ecosystem*; Dicke, M., Taken, W., Eds.; Springer: Dordrecht, The Netherlands, 2006; pp. 175–189.

38. Raguso, R.A. Wake up and smell the roses: The ecology and evolution of floral scent. *Ann. Rev. Ecol. Evol. Syst.* **2008**, *39*, 549–569. [CrossRef]

39. Raguso, R.A. Start making scents: The challenge of integrating chemistry into pollination ecology. *Entomol. Exp. Appl.* **2008**, *128*, 196–207. [CrossRef]

40. Whitehead, M.R.; Peakall, R. Integrating floral scent, pollination ecology and population genetics. *Funct. Ecol.* **2009**, *23*, 863–874. [CrossRef]

41. Raguso, R.A.; Levin, R.A.; Fooze, S.E.; Holmberg, M.W.; McDade, L.A. Fragrance chemistry, nocturnal rhythms and pollination "syndromes" in Nicotiana. *Phytochemistry* **2003**, *63*, 265–284. [CrossRef]

42. Hoballah, M.E.; Stuurman, J.; Turlings, T.C.J.; Guerin, P.M.; Connétable, S.; Kuhlemeier, C. The composition and timing of flower odour emission by wild Petunia axillaris coincide with the antennal perception and nocturnal activity of the pollinator Manduca sexta. *Planta* **2005**, *222*, 141–150. [CrossRef]

43. Knudsen, J.T.; Eriksson, R.; Gershenzon, J.; Stahl, B. Diversity and distribution of floral scent. *Bot. Rev.* **2006**, *72*, 1–120. [CrossRef]

44. Theis, N.; Lerdau, M.; Raguso, R.A. The challenge of attracting pollinators while evading floral herbivores: Patterns of fragrance emission in *Cirsium arvense* and *Cirsium repandum* (Asteraceae). *Int. J. Plant. Sci.* **2007**, *168*, 587–601. [CrossRef]
45. Dötterl, S.; Jareiß, K.; Salma Jhumur, U.; Jürgens, A. Temporal variation of flower scent in *Silene otitis* (Caryophyllaceae): A species with a mixed pollination system. *Bot. J. Linn. Soc.* **2012**, *169*, 447–460. [CrossRef]
46. Parachnowitsch, A.L.; Raguso, R.A.; Kessler, A. Phenotypic selection to increase floral scent emission, but not flower size or colour in bee-pollinated Penstemon digitalis. *New Phytol.* **2012**, *195*, 667–675. [CrossRef] [PubMed]
47. Delle Vedove, R.; Schatz, B.; Dufay, M. Understanding intraspecific variation of floral scent in light of evolutionary ecology. *Ann. Bot.* **2017**, *120*, 1–20. [CrossRef]
48. Salzmann, C.C.; Cozzolino, S.; Schiestl, F.P. Floral scent in food deceptive orchids: Species specificity and sources of variability. *Plant Biol.* **2007**, *9*, 720–729. [CrossRef]
49. Salzmann, C.C.; Nardella, A.M.; Cozzolino, S.; Schiestl, F.P. Variability in floral scent in rewarding and deceptive orchids: The signature of pollinator-imposed selection? *Ann. Bot.* **2007**, *100*, 757–765. [CrossRef]
50. Galizia, C.G.; Kunze, J.; Gumbert, A.; Borg-Karlson, A.K.; Sachse, S.; Markl, C.; Menzel, R. Relationship of visual and olfactory signal parameters in a food-deceptive flower mimicry system. *Behav. Ecol.* **2005**, *16*, 159–168. [CrossRef]
51. Stökl, J.; Twele, R.; Erdmann, D.; Francke, W.; Ayasse, M. Comparison of the flower scent of the sexually deceptive orchid *Ophrys iricolor* and the female sex pheromone of its pollinator *Andrena morio*. *Chemoecology* **2007**, *17*, 231–233. [CrossRef]
52. Dodson, C.H.; Dressler, R.L.; Hills, H.G.; Adams, R.M.; Williams, N.H. Biologically active compounds in orchid fragrances. *Science* **1969**, *164*, 1243–1249. [CrossRef]
53. Ayasse, M.; Schiestl, F.P.; Paulus, H.F.; Ibarra, F.; Francke, W. Pollinator attraction in a sexually deceptive orchid by means of unconventional chemicals. *Proc. R. Soc. B* **2003**, *270*, 517–522. [CrossRef]
54. Jersáková, J.; Jurgens, A.; Smilauer, P.; Johnson, S.D. The evolution of floral mimicry: Identifying traits that visually attract pollinators. *Funct. Ecol.* **2012**, *26*, 1381–1389. [CrossRef]
55. Peter, C.I.; Johnson, S.D. A pollinator shift explains floral divergence in an orchid species complex in South Africa. *Ann. Bot.* **2014**, *113*, 277–288. [CrossRef]
56. Dafni, A. Mimicry and deception in pollination. *Ann. Rev. Ecol. Syst.* **1984**, *15*, 259–278. [CrossRef]
57. Nilsson, L.A. Anthecology of *Orchis morio* (Orchidaceae) at its outpost in the north. *Nova Acta Regiae Soc. Sci. Ups.* **1984**, *5*, 166–179.
58. Van Der Cingel, N.A. *An Atlas of Orchid Pollination: European Orchids*; CRC Press: Boca Raton, FL, USA, 2001.
59. Cozzolino, S.; Schiestl, F.; Müller, A.; De Castro, O.; Nardella, A.M.; Widmer, A. Evidence for pollinator sharing in Mediterranean nectar mimic orchids: Absence of premating barriers? *Proc. R. Soc. B* **2005**, *272*, 1271–1278. [CrossRef]
60. Schiestl, F.P. On the success of a swindle: Pollination by deception in orchids. *Naturwissenschaften* **2005**, *92*, 255–264. [CrossRef]
61. Peter, C. Pollination, Floral Deception and Evolutionary Processes in *Eulophia* (Orchidaceae) and Its Allies. Ph.D. Thesis, University of KwaZulu-Natal, Durban, South Africa, 2009.
62. D'Auria, M.; Lorenz, R.; Racioppi, R.; Romano, V.A. Fragrance components of *Platanthera bifolia* subsp. osca. *Nat. Prod. Res.* **2017**, *31*, 1612–1619. [CrossRef]
63. D'Auria, M.; Lorenz, R.; Mecca, M.; Racioppi, R.; Romano, V.A.; Viggiani, L. Fragrance components of *Platanthera bifolia* subsp. osca and *Platanthera chlorantha* collected in several sites in Italy. *Nat. Prod. Res.* **2020**, *34*, 2857–3861. [CrossRef]
64. D'Auria, M.; Fascetti, S.; Racioppi, R.; Romano, V.A.; Rosati, L. Orchids from Basilicata: The Scent. In *Orchids Phytochemistry, Biology and Horticulture*; Merillon, J.-M., Kodja, H., Eds.; Springer: Berlin/Heidelberg, Germany, 2022; pp. 627–648.
65. D'Auria, M.; Lorenz, R.; Mecca, M.; Racioppi, R.; Romano, V.A. Aroma components of *Cephalanthera* orchids. *Nat. Prod. Res.* **2021**, *35*, 174–177. [CrossRef]
66. Mecca, M.; Racioppi, R.; Romano, V.A.; Viggiani, L.; Lorenz, R.; D'Auria, M. Volatile organic compounds from *Orchis* species found in Basilicata (Southern Italy). *Compounds* **2021**, *1*, 83–93. [CrossRef]
67. D'Auria, M.; Lorenz, R.; Mecca, M.; Racioppi, R.; Romano, V.A. The composition of the aroma of *Serapias* orchids in Basilicata (Southern Italy). *Nat. Prod. Res.* **2021**, *35*, 4068–4072. [CrossRef]
68. Mecca, M.; Racioppi, R.; Romano, V.A.; Viggiani, L.; Lorenz, R.; D'Auria, M. The scent of *Himantoglossum* species found in Basilicata (Southern Italy). *Compounds* **2021**, *1*, 164–173. [CrossRef]
69. Romano, V.A.; Rosati, L.; Fascetti, S.; Cittadini, A.M.R.; Racioppi, R.; Lorenz, R.; D'Auria, M. Spatial and temporal Variability of the floral scent emitted by *Barlia robertiana* (Loisel.) Greuter, a Mediterranean food-deceptive orchid. *Compounds* **2022**, *2*, 37–53. [CrossRef]
70. Nilsson, L.A. The pollination ecology of *Dactylorhiza sambucina* (Orchidaceae). *Bot. Not.* **1980**, *133*, 367–385.
71. Naczk, A.M.; Kowalkowska, A.K.; Wisniewska, N.; Halinski, D.P.; Kapusta, M.; Czerwicka, M. Floral anatomy, ultrastructure and chemical analysis in *Dactylorhiza incarnata/maculate* complex (Orchidaceae). *Bot. J. Linn. Soc.* **2018**, *187*, 512–536. [CrossRef]
72. Wróblewska, A.; Szczepaniak, L.; Bajguz, A.; Jedrzejczyk, I.; Talalaj, I.; Ostrowiecka, B.; Brzosko, E.; Jermakowicz, E.; Mirski, P. Deceptice strategy in *Dactylorhiza* orchids: Multidirectional evolution of floral chemistry. *Ann. Bot.* **2019**, *123*, 1005–1016. [CrossRef]

Article

The Scent of *Himantoglossum* Species Found in Basilicata (Southern Italy)

Marisabel Mecca [1], Rocco Racioppi [1], Vito Antonio Romano [1], Licia Viggiani [1], Richard Lorenz [2] and Maurizio D'Auria [1,*]

1 Dipartimento di Scienze, Università della Basilicata, V.le dell'Ateneo Lucano 10, 85100 Potenza, Italy; marisabelmecca@libero.it (M.M.); rocco.racioppi@unibas.it (R.R.); vitoantonio.romano@libero.it (V.A.R.); licia.viggiani@unibas.it (L.V.)
2 Arbeitskreis Heimische Orchideen Baden-Württemberg, D-69469 Weinheim, Germany; lorenz@orchids.de
* Correspondence: maurizio.dauria@unibas.it; Tel.: +39-0971-205-480

Abstract: The SPME (Solid Phase Microextraction) analysis of the scent of *H. hircinum* showed the presence of elemicin in the presence of a relevant amount of eugenol. The scent of the sample of *H. adriaticum* collected in Abruzzo showed the presence 4-amino-5-(4-morpholinylmethyl)-2-oxazolidinone, β-ocimene, decyl decanoate, and 9-tricosene as main components. The sample of *H. adriaticum* collected at Marsico Nuovo has an aroma where the main component was pentadecyl hexanoate, 9-tricosene, methyleugenol, tetradecane, pentadecane, and elemicin. The samples of *H. adriaticum* collected at Viggianello showed some similarities in the scent: the main components were 9-tricosene and methyleugenol.

Keywords: *Himantoglossum*; scent; gas chromatography; mass spectrometry; solid phase microextraction

Citation: Mecca, M.; Racioppi, R.; Romano, V.A.; Viggiani, L.; Lorenz, R.; D'Auria, M. The Scent of *Himantoglossum* Species Found in Basilicata (Southern Italy). *Compounds* **2021**, *1*, 164–173. https://doi.org/10.3390/compounds1030015

Academic Editor: Juan Mejuto

Received: 9 November 2021
Accepted: 7 December 2021
Published: 20 December 2021

Publisher's Note: MDPI stays neutral with regard to jurisdictional claims in published maps and institutional affiliations.

1. Introduction

The determination of the volatile organic compounds emitted from a natural source is smell and taste are the oldest of our senses. They probably developed in very primitive organisms as means of obtaining information about chemical changes in the organism's environment. Animals use smell and taste to find food and to assess its quality. The smell of food has a powerful effect on animals.

Living organisms use the chemical sense as a means of communications. If the communication is between different parts of the same organism, the messenger is referred to as a hormone. Chemicals used to carry signals from one organism to another are known as semiochemicals. In the case of flowers, the aroma components are mainly devoted to attracting pollinator insects.

In recent years, *Himantoglossum* s.l. has included other taxa of considerable interest and conservation [1,2]. Currently, the expanded genus *Himantoglossum* is composed of the subgenus *Himantoglossum* including all the species of the former genus *Himantoglossum*, the subgenus *Barlia*, consisting of the two species of the genus *Barlia*, and the subgenus *Comperia* consisting only of the former species *Comperia comperiana* (Table 1).

The attraction of the pollinator in orchids generally occurs first through the air diffusion of scents, then through sight, as the pollinator approaches its target inflorescence, and finally through tactile signals and hardly volatile, extractable compounds when it lands on the chosen flower.

H. hircinum and *H. adriaticum* have large, showy flowers with a long, captivating lip, often adorned with showy tufts of colored papillae that provide footholds acting as a guide for pollinators. Vöth (1990) speculates that these papillae could also be the seat of osmophores responsible for most of the volatile organic compounds emitted by the plant [3]. The scent emitted by these two species can be strong, unpleasant, or sweet. They are allogamous species that do not offer food reward; in fact, their short and sack-like spur does not contain nectar [4].

Teschner (1980) has shown that the spur of *H. hircinum* and *H. adriaticum* may contain small amounts of glucose in some populations [5]. Kropf and Renner (2008) chemically demonstrated the presence of nectar in *H. hircinum* [6]. Little is known about the pollination of the various species of the *H. hircinum* group; pollinators are thought to differ locally, with Teschner (1980) suggesting that solitary bees are the true pollinators [5].

In this work, we have dealt with the only two species present in Basilicata belonging to the old genus *Himantoglossum* Spreng. 1826, *Himantoglossum hircinum* (L.) Sprengel 1826 and *Himantoglossum adriaticum* H. Bauman 1978, while, in previous works we analyzed the perfumes of *Barlia robertiana* [7], a species until a few years ago considered to belong to the monospecific genus *Barlia* and now merged into the new clade *Himantoglossum*.

H. hircinum (Figure 1) has a Mediterranean-Atlantic distribution, from southern Great Britain to northern Africa. Present in Italy, the species is reported to be widespread in Sicily between 150 and 1750 m [8], sporadically in the southern regions, also reported in Tuscany, Liguria, and southern Piedmont [9].

Table 1. Taxonomy of the genus *Himantoglossum* s.l. generated by integrating the results of the present study with those of Sramkó [10].

Genus	Subgenus	Clade	Section	Species
Himantoglossum	*Comperia*	-	-	*comperianum*
	Barlia	-	-	*metlesicsianum*
				robertianum
	Himantoglossum	Hircinum-caprinum	Formosum	*formosum*
			Caprinum	*caprinum*
				montistauri
				calcaratum rumelicum
				calcaratum calcaratum
			Hircinum	*adriaticum*
				hircinum

Figure 1. *Himantoglossum hircinum.* Photo of V. A. Romano.

From the observations of one of the authors of the following work (VAR), we report a wide diffusion for Basilicata of this species, so much so that it is very common in the hilly and mountainous area of the Province of Potenza from 400 to 1500 m where it forms large populations with dozens of plants. It blooms from early May to early June.

H. adriaticum (Figure 2) has a Euro-Mediterranean distribution, present in southern Italy up to the Alpine regions, Slovenia, and Croatia; its northeastern limit also touches Austria, Hungary, and Slovakia [9].

Figure 2. *Himantoglossum adriaticum.* Photo of V. A. Romano.

In Basilicata it was reported by Gölz and Reinhard (1982), Conti et al. (2005), Fascetti et al. (2008), Romano et al. (2013) [11–14]. It blooms from mid-June to mid-July from 1300 to 1600 m. Many plants have never been observed on the same site, maximum 10 plants on an area of 0.5 ha. Generally, there are 1–3 isolated plants distant from other single specimens even a few km. The richest area is that of the Pollino National Park between Basilicata and Calabria [15].

For the perfume tests, *H. hircinum* plants from Basilicata and *H. adriaticum* plants from Basilicata and Abruzzo were used. All the plants were collected before anthesis and planted in the gardens of the University of Basilicata and tested, many days later, when they were in full bloom.

The scent of a flower can be an important factor determining the pollination of a plant. The study of the scent of the orchids was the object of several works in the past [16]. Unfortunately, several different approaches have been used in order to determine the composition of the aroma of an orchid. Extraction, headspace analysis, and SPME have used. Often, different chemical procedures allowed to obtain different results. On the basis of these considerations, a research project started devoted to the determination of the scent of the orchids found in Basilicata using the same procedure, solid phase microextraction coupled with gas chromatography and mass spectrometry. This way, the composition of the scent of *Platanthera bifolia* subsp. *osca* [17,18], *Platanthera chlorantha* [18], *Cephalanthera* orchids [19], *Serapias* orchids [20], *Gymnadenia* orchids [21], *Barlia robertiana* [7], *Neotinea* orchids [22], and *Orchis* species [23] has been investigated.

Some studies report some data on the scent of *H. hircinum*. (*E*)-Ocimene, elemicin, (*E*)-3-methyl-4-decenoic acid, (*Z*)-4-decenoic acid, and lauric acid were considered as the main components of the scent after absorption on charcoal [24], (*E,Z*)-2,6-dimethyl-3,5,7-octatrien-2-ol and the (*E,E*) isomer were claimed as major constituents of the aroma [25,26]. Finally, hexane extraction of the labella showed the presence of high molecular weight alkanes such as pentacosane, heptacosane, and nonacosane [27].

2. Experimental Section

2.1. Plant Material

The samples of *H. adriaticum* were collected at Comune di Cocullo, Prov dell'Aquila (Abruzzo), 1070 m. a.s.l., on 20 May 2017 (Sample 1), at Fontana delle Brecce, Marsico Nuovo (Pz), 1439 m. a.s.l., on 10 June 2017 (Sample 2), at Piano Visitone, Viggianello (Pz), 1500 m. a.s.l., on 12 June 2018 (samples 3 and 4). The sample of *H. hircinum* was collected at Contrada Manta, Potenza, 1000 m. a.s.l., on 1 May 2017. The plants were collected by Vito Antonio Romano.

The plants were harvested about two weeks before flowering by taking all the clod of earth, taking care not to damage the root system, planted in special pots in the gardens of the University of Basilicata (Potenza 650 m. a.s.l.), in waiting for their full bloom. Two days before the tests the plants were transferred to an air-conditioned room at 22 °C. The plants were tested, whole without being damaged, under a cylindrical glass bell (12 cm × 45 cm) in which only the inflorescence and the SPME probe are inserted (Figure 3).

Figure 3. The apparatus used to collect the scent of the plants used in this study.

To avoid contamination, the interior of the bell was isolated from the external environment with appropriate closing and sealing systems during the 24 h of the test (from eight in the morning to 8 the following day).

In order to be sure that the internal environment of the bell was isolated from the external environment, various blank tests were carried out.

The plants were successively used for further studies on pollination, fertility, and germination of the plants. After these studies, the plants were not in condition to be collected in an herbarium. However, these species can be recognized without ambiguities on the basis of their properties, well documented by the Figures 2 and 3. In view of the fact that the investigated taxa are rare wild plants, in order to preserve the species, we have chosen to use a single plant for our analysis.

2.2. Analysis of Volatile Organic Compounds

The SPME [4] analysis of five different samples of *Himantoglossum* has been performed. This way, the identified plants were collected and inserted in glass jar for 24 h where was present also the fiber (DVB/CAR/PDMS) of and SPME syringe. After this time the fiber was desorbed in a gas chromatographic apparatus equipped with a quadrupole mass spectrometer detector. A 50/30 µm DVB/CAR/PDMS module with 1 cm fiber (57328-U, Supelco, Milan, Italy) was employed to determine VOCs. SPME fiber was maintained in the bell jar for 24 h. The analytes were desorbed in the splitless injector at 250 °C for 2 min. Analyses were accomplished with an HP 6890 Plus gas chromatograph equipped with a Phenomenex Zebron ZB-5 MS capillary column (30-m × 0.25-mm i.d. × 0.25 µm FT) (Agilent, Milan, Italy). An HP 5973 mass selective detector in the range 1 to 800 m/z (Agilent) was utilized with helium at 0.8 mL/min as the carrier gas. The EI source was used at 70 eV. The analyses were performed by using a splitless injector. The splitless injector was maintained at 250 °C and the detector at 230 °C. The oven was held at 40 °C for 2 min, then gradually warmed, 8 °C/min, up to 250 °C and held for 10 min. Tentatively identification of aroma components was based on mass spectra and Wiley 11 and NIST 14 library comparison. Single VOC peak was considered as identified when its experimental spectrum matched with a score over 90% that present in the library. All the analyses were performed in triplicate.

3. Results

The SPME analysis of scent of *H. hircinum* showed the presence of elemicin (spicy, floral scent) as the main component (61.71%) in the presence of a relevant amount of eugenol (4.50%) (Table 2). Minor products observed in the scent were 3,4,5-trimethoxybenzaldehyde, benzyl benzoate, 3-(4,8,12-trimethyltridecyl)furan, and 9-tricosene. All the values are based on per cent of the TIC area.

Table 2. SPME-GC-MS analysis of *Himantoglossum* species.

Compound	r.t. [min.]	KI	Area [%] ± 0.03				
			Species				
			H. hircinum	*H. adriaticum*			
				Sample 1	Sample 2	Sample 3	Sample 4
β-Myrcene	8.99	991		0.31			
2,2,4,6,6-pentamethyl-3-heptene	9.31	1020		0.31			
1-Methoxy-4-methylbenzene	9.63	1030		1.63			
Limonene	9.92	1039			1.62		0.81
β-Ocimene	10.18	1050	0.17	8.23	2.85	10.28	
Methyl benzoate	11.16	1091	0.10				
2-Ethylhexyl acetate	12.18	1159	0.22				
Dodecane	13.09	1200		0.29			
Verbenone	13.53	1209			1.95		0.75
3,4-Dimethoxytoluene	13.86	1230		0.33			
Carvone	14.02	1240			0.33		
Geraniol	14.45	1260		1.49			
Tridecane	14.88	1300			0.93		
2,3-Dimethylhydroquinone	15.70	1348		1.01			0.99
Eugenol	16.05	1374	4.50				
Decanoic acid	16.08	1380			1.23		
Geranyl acetate	16.31	1388	0.35				
Tetradecane	16.64	1400		1.43	3.93	4.49	0.86
Methyleugenol	16.80	1406		2.34	5.93	18.06	4.73
Caryophyllene	17.09	1420	0.72				
2,6-Di-*t*-butylbenzoquinone	17.88	1472		0.74			1.23
Pentadecane	18.22	1500	0.31	2.98	3.41	4.19	1.16
α-Farnesene	18.33	1511	0.10				
2,5-bis(1,1-Dimethylethyl)phenol	18.41	1514		0.53			
Elemicin	19.07	1550	61.71	2.95	3.80		
Octyl hexanoate	19.42	1570	0.14		0.93		
Decyl butanoate	19.48	1590			1.59		
Hexadecane	19.72	1600		2.70	1.95		1.08
3,4,5-Trimethoxybenzaldehyde	19.84	1608	1.59				
Tetradecanal	19.86	1611		0.78			
1-Methylethyl dodecanoate	20.06	1618	0.31				
Isoelemicin	20.47	1644	0.13	0.74	0.87		

Table 2. *Cont.*

Compound	r.t. [min.]	KI	Area [%] ± 0.03				
				Species			
					H. adriaticum		
			H. hircinum	Sample 1	Sample 2	Sample 3	Sample 4
Tridecan-1-ol	20.88	1665					1.02
Heptadecane	21.14	1700	0.38	2.11	2.42	3.68	2.12
Pristane	21.18	1705		0.98	1.22		
Benzyl benzoate	22.16	1751	1.59	2.17			
Pentadecyl hexanoate	22.20	1770			22.73		
Pentadecyl isobutanoate	22.27	1776			0.86		
Octadecane	22.49	1800	0.12	0.90	1.22		1.32
Hexadecanal	22.68	1822		0.67	1.02		
Isopropyl myristate	22.84	1830	0.43		2.81		0.74
6,10,14-trimethyl-2-pentadecanone	23.11	1836		1.83	1.29		1.56
3-Amino-5-(4-morpholinylmethyl)-2-oxazolidinone	23.56	1844		17.37			
Nonadecane	23.78	1900		0.92	1.43		3.47
β-Springene	24.31	1918	0.36				
3-(4,8,12-Trimethyltridecyl)furan	24.60	1941	1.05				
Decyl decanoate	24.66	1950		5.36			
Octadecyl hexanoate	24.73	1965			2.48		
11-Hexadecen-1-ol acetate	24.94	1973		2.42			
Eicosane	25.00	2000			0.89		0.87
Octadecanal	25.23	2021		0.58			
Heneicosane	26.18	2100		0.28	2.39		4.26
Nonadecanol	26.40	2151	0.88	0.74	2.31	5.48	
Docosane	28.31	2200		0.28	0.56		
9-Tricosene	28.43	2270	1.31	6.41	5.06	31.21	40.22
Unidentified components			23.52	28.19	19.99	22.61	33.83

The scent of the sample of *H. adriaticum* collected in Abruzzo (Sample 1) showed the presence 4-amino-5-(4-morpholinylmethyl)-2-oxazolidinone, β-ocimene (green tropical woody floral vegetable scent), decyl decanoate, and 9-tricosene as main components (17.37, 8.23, 5.36, and 6.41%, respectively) (Table 2). Furthermore, several compounds are presents in relevant amounts: 1-methoxy-4-methylbenzene (1.63%), geraniol (1.49%), geranyl acetate (1.43%), methyleugenol (2.34%), pentadecane (2.98%), elemicin (2.95%), hexadecane (2.70%), heptadecane (2.11%), benzyl benzoate (2.17%), 6,10,14-trimethyl-2-pentadecanone (1.83%), and 11-hexadecen-1-ol acetate (2.42%). The sample of *H. adriaticum* collected at Marsico Nuovo (Sample 2) has an aroma where the main component was pentadecyl hexanoate (22.73%), 9-tricosene (5.06%), methyleugenol (5.93%), tetradecane (3.93%), pentadecane (3.41%), and elemicin (3.80%) (Table 2). The samples of *H. adriaticum* collected at Viggianello (samples 3 and 4) showed some similarities in the scent: the main components were 9-tricosene (31.21% in sample 3 and 40.22% in sample 4) and methyleugenol (18.06% in sample 3 and 4.73% in sample 4) (Table 2). Furthermore, β-ocimene was found in relevant amount (10.28%) only in sample 3.

4. Discussion

It is interesting to note the large differences between our reported results and those reported in the introduction section for *H. hircinum*. We did not find a correlation with the results reported by Schiestl and Cozzolino [27]. While they determined the presence of high molecular weight alkanes, we did not find these compounds in the scent. We think that the observed different results depend on the different procedures used in the determination of the scent. Schiestl and Cozzolino used an extraction of labella. Probably they determined the presence of waxy compounds present in the surface of labella but not involved in the composition of the scent. Furthermore, the components determined in that article [27] are not volatile compounds. Furthermore, in a previous work, (*E*)-ocimene, elemicin, (*E*)-3-methyl-4-decenoic acid, (*Z*)-4-decenoic acid, and lauric acid were determined as the main components of the aroma [24]. While elemicin was the main component of the scent in the sample we analyzed, we did not find the other compounds determined in the work of Kaiser [24]. Finally, we did not find (*E,Z*)- and (*E,E*)-2,6-dimethyl-3,5,7-octatrien-2-ol whose presence has been claimed in [25,26].

The scent of *H. adriaticum* has not been studied until now. It is interesting to note that all the samples we analyzed have common components, although in different amounts. Thus, methyleugenol and 9-tricosene were found in all the samples. Furthermore, some linear hydrocarbons (tetradecane, pentadecane, and hexadecane) were present. Nevertheless, sample 1 of *H. adriaticum* showed the presence of an oxazolidinone as an important component, while in sample 2 we found pentadecyl hexanoate, and in sample 3 β-ocimene was one the main components.

Based on this work, we can assume that the scent of *H. hircinum* found in Basilicata has a different composition from those described elsewhere. We can assume also that *H. hircinum* has a scent showing a different composition in comparison with the compounds found in *H. adriaticum*. Finally, we can assume that the samples of *H. adriaticum* we analyzed have some common characters, with some differences depending on the place where the samples have been collected.

5. Conclusions

This work shows the analysis of samples from Basilicata and Abruzzo of *H. hircinum* and *H. adriaticum*. The analyses have been performed by using the same procedure and the same fiber in SPME-GC-MS, allowing to have a homogenous data set. The analysis of *H. hircinum* showed a peculiar composition that differs from those observed in *H. adriaticum*. In fact, the analysis of the scent of *H. hircinum* showed the presence of elemicin in the presence of a relevant amount of eugenol. The scent of the sample of *H. adriaticum* showed the presence 4-amino-5-(4-morpholinylmethyl)-2-oxazolidinone, β-ocimene, decyl decanoate, 9-tricosene, pentadecyl hexanoate, methyleugenol, tetradecane, pentadecane, and elemicin.

The observed differences, when other head-space techniques are used, can depend both on different absorption rates of the analytes on the fiber and on variation of the scent due to natural adaptation of the plant to different environmental conditions. Furthermore, the observed differences can be due to different pollination insects. The analysis of the scent of *H. adriaticum* showed the presence of some common components (i.e., 9-tricosene and methyleugenol) that are not depending on the origin of the flowers.

Author Contributions: Conceptualization, M.D. and V.A.R.; methodology, R.R.; investigation, R.R., L.V. and M.M.; data curation, R.L.; writing—original draft preparation, review and editing, M.D., V.A.R. and R.L.; All authors have read and agreed to the published version of the manuscript.

Funding: This research received no external funding.

Institutional Review Board Statement: Not applicable.

Informed Consent Statement: Not applicable.

Data Availability Statement: Not applicable.

Conflicts of Interest: The authors declare no conflict of interest.

References

1. Delforge, P. Contribution taxonomique et nomenclaturale au genre Himantoglossum (Orchidaceae). *Natural. Belges* **1999**, *80*, 387–408.
2. Bateman, R.M.; Molnár, V.A.; Sramkó, G. In situ morphometric survey elucidates the evolutionary systematics of the Eurasian Himantoglossum clade (Orchidaceae: Orchidinae). *Peer J.* **2017**, *5*, e2893. [CrossRef] [PubMed]
3. Vöth, W. Effektive und potentielle Bestäuber von Himantoglossum Spr. Mitteilungsblatt. *Arbeitskreis Heimische Orchideen Baden-Wurttemburg* **1990**, *22*, 337–351.
4. Claessens, J.; Kleynen, J. *The Flower of the European Orchid: Form and Function*; Jean Claessens & Jacques Kleynen: Voerendaal, The Netherlands, 2011.
5. Teschner, W. Sippen Differenzierung und Bestäubung bei Himantoglossum Koch. In *Probleme der Evolution bei europäischen und mediterranen Orchideen*; Orchidee Sonderheft (Special, Issue); Senghas, K., Sundermann, H., Eds.; Brücke Verlag: Hildesheim, Germany, 1980; pp. 104–115.
6. Kropf, M.; Renner, S. Pollinator-mediated selfing in two deceptive orchids and a review of pollinium tracking studies addressing geitonogamy. *Oecologia* **2008**, *155*, 497–508. [CrossRef] [PubMed]
7. D'Auria, M.; Fascetti, S.; Racioppi, R.; Romano, V.A.; Rosati, L. Orchids from Basilicata: The Scent. In *Orchids Phytochemistry, Biology and Horticulture*; Merillon, J.-M., Kodja, H., Eds.; Springer: New York, NY, USA, 2020.
8. Künkele, S.; Lorenz, R. Zum Stand der Orchideenkartierung in Sizilien. Ein Beitrag zum OPTIMA-Projekt. "Karierung der mediterranen Orchideen". *Jahresber. Naturwiss. Ver. Wuppertal.* **1995**, *48*, 21–115.
9. GIROS. *Orchidee d'Italia. Guida alle Orchidee Spontanee*; Castello SRL: Cornaredo, MI, Italy, 2016.
10. Sramkó, G.; Molnár, A.V.; Hawkins, J.A.; Bateman, R.M. Molecular phylogenetics and evolution of the Eurasiatic orchid genus Himantoglossum s.l. *Ann. Bot.* **2014**, *114*, 1609–1626. [CrossRef] [PubMed]
11. Gölz, P.; Reinhard, H.R. Orchideen in Süditalien.—Mitt. Bl. *Arbeitskr. Heim. Orch. Baden-Württ.* **1982**, *14*, 1–124.
12. Conti, F.; Abbate, G.; Alessandrini, A.; Blasi, C. (Eds.) *An Annotated Checklist of the Italian Vascular Flora*; Palombi Editori: Roma, Italy, 2005.
13. Fascetti, S.; Soca, R.; Romolini, R.; Romano, V.A. Contributo alla conoscenza delle Orchidacee della Basilicata (Italia meridionale): Resoconto dell'escursione GIROS 2006. *GIROS Not.* **2006**, *37*, 1–10.
14. Romano, V.A.; Navazio, G.; Zampino, A. Nuove stazioni di specie rare di Orchidaceae per la Basilicata. *GIROS Not.* **2013**, *52*, 81–88.
15. Lorenz, R.; Künkele, S. Die Orchideenflora von Kalabrien und ihre Stellung innerhalb Italiens—*Jahresber. Naturwiss. Ver. Wuppertal* **1990**, *43*, 15–35.
16. Gaskett, A.C. Orchid pollination by sexual deception: Pollinator perspectives. *Biol. Rev.* **2011**, *86*, 33–75. [CrossRef] [PubMed]
17. D'Auria, M.; Lorenz, R.; Racioppi, R.; Romano, V.A. Fragrance components of *Platanthera bifolia* subsp. osca. *Nat. Prod. Res.* **2017**, *31*, 1612–1619. [CrossRef] [PubMed]
18. D'Auria, M.; Lorenz, R.; Mecca, M.; Racioppi, R.; Romano, V.A.; Viggiani, L. Fragrance components of *Platanthera bifolia* subsp. osca and Platanthera chlorantha collected in several sites in Italy. *Nat. Prod. Res.* **2020**, *34*, 2857–3861. [CrossRef] [PubMed]
19. D'Auria, M.; Lorenz, R.; Mecca, M.; Racioppi, R.; Romano, V.A. Aroma components of *Cephalanthera orchids. Nat. Prod. Res.* **2021**, *35*, 174–177. [CrossRef] [PubMed]
20. D'Auria, M.; Lorenz, R.; Mecca, M.; Racioppi, R.; Romano, V.A. The composition of the aroma of *Serapias* orchids in Basilicata (southern Italy). *Nat. Prod. Res.* **2021**, *35*, 4068–4072. [CrossRef] [PubMed]

21. D'Auria, M.; Lorenz, R.; Mecca, M.; Racioppi, R.; Romano, V.A.; Viggiani, L. Fragrance components of *Gymnadenia conopsea* and *Gymnadenia odoratissima* collected at several sites in Italy and Germany. *Nat. Prod. Res.* 2021. [CrossRef] [PubMed]

22. D'Auria, M.; Lorenz, R.; Mecca, M.; Racioppi, R.; Romano, V.A.; Viggiani, L. The scent of *Neotinea* orchids from Basilicata (Southern Italy). *Nat. Prod. Res.* **2021**. [CrossRef] [PubMed]

23. Mecca, M.; Racioppi, R.; Romano, V.A.; Viggiani, L.; Lorenz, R.; D'Auria, M. Volatile organic compounds from *Orchis* species found in Basilicata (Southern Italy). *Compounds* **2021**, *1*, 83–93. [CrossRef]

24. Kaiser, R. *Vom Duft der Orchideen*; Editiones Roche: Basel, Switzerland, 1993; p. 202.

25. Kaiser, R. Trapping, investigation and reaconstitution of flower scents. In *Perfumes, Art, Science and Technology*; Müller, P.M., Lamparsky, D., Eds.; Chaapman & Hall; Springer: Dordrecht, The Netherlands, 1994; pp. 213–250.

26. Knudsen, J.T.; Tollsten, L.; Bergström, L. G Floral scents—A checklist of volatile compounds isolated by head-space techniques. *Phytochemistry* **1993**, *33*, 253–280. [CrossRef]

27. Schiestl, F.P.; Cozzolino, S. Evolution of sexual mimicry in the orchid subtribe orchidinae: The role of preadaptations in the attraction of male bees as pollinators. *BMS Evol. Biol.* **2008**, *8*, 27. [CrossRef] [PubMed]

Article

Spatial and Temporal Variability of the Floral Scent Emitted by *Barlia robertiana* (Loisel.) Greuter, a Mediterranean Food-Deceptive Orchid

Vito Antonio Romano [1], Leonardo Rosati [2,*], Simonetta Fascetti [2], Anna Maria Roberta Cittadini [1], Rocco Racioppi [1], Richard Lorenz [3] and Maurizio D'Auria [1,*]

[1] Dipartimento di Scienze, Università della Basilicata, V.le dell'Ateneo Lucano 10, 85100 Potenza, Italy; vitoantonio_romano@libero.it (V.A.R.); anna.cittadini@hotmail.it (A.M.R.C.); rocco.racioppi@unibas.it (R.R.)

[2] Scuola di Scienze Agrarie, Forestali, Alimentari e Ambientali, Università della Basilicata, Via Ateneo Lucano 10, 85100 Potenza, Italy; simonetta.fascetti@unibas.it

[3] Arbeitskreis Heimische Orchideen Baden-Württemberg, D-69469 Weinheim, Germany; lorenz@orchids.de

* Correspondence: leonardo.rosati@unibas.it (L.R.); maurizio.dauria@unibas.it (M.D.); Tel.: +39-0971-205-480 (M.D.)

Abstract: This study on *Barlia robertiana* aims to: (1) assess whether scent is variable between populations; (2) evaluate whether scent composition may be related to geographical variables; (3) assess whether there are VOC differences during the flowering phase; and (4) assess whether there are yearly VOC variabilities. SPME sampling was used. Fourteen plants, collected along an ecological gradient, were analyzed. A multivariate analysis was performed through ordination and hierarchical cluster analysis. Compositions versus geographic distances were also analyzed using Mantel test. Seventy compounds were identified. Multivariate analyses and Mantel tests detected no correlations between VOC composition and both geographic and ecological variables. These results may suggest that there is no adaptation of floral scent to local environments. VOC compositions during the flowering phase showed a slight change but a strong variability between individuals. A huge difference was found in the pairwise comparison of the plants analyzed in different years. The high scent variability can be interpreted as a strategy of a non-rewarding but allogamous species to not allow the learning by pollinators. In fact, disrupting the association among floral scent signals with the lack of nectar may enhance the fruit set via a higher probability of being visited by insects.

Keywords: Basilicata; *Barlia robertiana*; *Himantoglossum robertianum*; mantel test; Orchidaceae; pollination syndrome; Italy; volatile compounds

Citation: Romano, V.A.; Rosati, L.; Fascetti, S.; Cittadini, A.M.R.; Racioppi, R.; Lorenz, R.; D'Auria, M. Spatial and Temporal Variability of the Floral Scent Emitted by *Barlia robertiana* (Loisel.) Greuter, a Mediterranean Food-Deceptive Orchid. *Compounds* **2022**, *2*, 37–53. https://doi.org/10.3390/compounds2010004

Academic Editor: Juan Mejuto

Received: 13 November 2021
Accepted: 4 January 2022
Published: 27 January 2022

Publisher's Note: MDPI stays neutral with regard to jurisdictional claims in published maps and institutional affiliations.

1. Introduction

Orchidaceae are one of the largest families of vascular plants [1]; since Darwin [2], they have attracted the interests of a plethora of naturalists for the amazing floral variations and the complex pollination mechanisms which they evolved [3]. Approximately one-third of Orchidaceae are believed to deceive insect pollinators [4,5]; among the mechanisms of deception, generalized food deception is one of the most common mechanisms developed by orchids for efficient pollen exportation [5,6]. These species can exploit the existing plant–pollinator relationships and achieve pollination through deception in the absence of floral rewards for pollinators. The similarity with rewarding plants determines their reproductive success; therefore, this pollination syndrome can be considered a generalized form of Batesian mimicry [7]. In order to deceive pollinators, these orchids exploit general floral signals typical for rewarding plant species, including flower color and scent [8]. However, they generally do not resemble any specific rewarding flower and they are visited by casual pollinators or exploratory pollinators [9]. For an example of floral mimicry, see [10] and references therein. It has been shown that orchids related to generalized food deception exploit bees and bumblebees that have just emerged after the winter season, blooming in early spring [11]. Moreover, reward-less species undoubtedly benefit from the simultaneous flowering of nectariferous species present in the same habitat that increase the possibility of being visited by local pollinators [12].

Floral scent, together with size, shape and color, act as signals attracting pollinators [13], and adaptations to specific pollinators are considered an important driver of evolution in angiosperms.

Barlia robertiana (Loisel.) Greuter (Orchidaceae) has non-rewarding flowers, and it is obligatorily insect-pollinated. It is a Mediterranean species [14] typical of several habitats such as clearings in scrublands and thermophilus woods, dry grasslands, roads edge, usually on bases-rich soils [15]. Recent extensions north due to climate change have been observed in western Switzerland [16] and southwestern Germany [17]. Its geographical distribution in Italy encompasses all the regions, but is lacking in some of the Alpine sectors [18]. It is particularly widespread in the Basilicata region (Southern Italy), from the coast to the mountains of the Apennines up to approximately 1000 m a.s.l. (above see level).

According to a recent molecular genetic study [19], the genus *Barlia* should be transferred to the genus *Himantoglossum*; however, a general consensus about this taxonomical rearrangement has not been reached by taxonomists; in this article, we refer to the nomenclature of the most recent checklist of the Italian vascular flora [18].

The plants are robust (up to 80–110 cm high) and early flowering, from December to April. The inflorescence is sub-cylindrical, dense and multi-flowered; it is up to 40 cm high and can develop up to 70 flowers; the color of flowers varies from red-violet to olive-green or brown-red; inside, it is covered with purplish spots [14]. The lip borders are crenate, with a papillose epidermis; the spur is conical, shorter than the ovary, turned downwards, and it does not produce nectar. The flowers give off a delicate, persistent, and easily perceptible smell. Structural particularities of epigeous and hypogeous plant organs of *B. robertiana*, in comparison with *H. hircinum*, have been interpreted as morphological adaptations to different edaphic and environmental conditions [20]. Hydroalcoholic flower extracts of *B. robertiana* revealed the presence of phenols, flavonoids, and proanthocyanidins [21]. Chromosomes ($2n = 32$), karyotypes, and the localization of ribosomal genes have been studied [22]. Unlike other Mediterranean orchids, such as the genus *Ophrys*, which are pollinated by specialized bees [3], several different groups of insects have been identified to pollinate *B. robertiana*, such as Apoidea (Hymenoptera) and Cetoniidae (Coleoptera) [23,24]. A recent study on the fruit set of *B. robertiana*, performed on the island of Mallorca (Spain), confirmed the importance of allogamy for its reproduction success [25].

Although the scent of orchids has frequently been analyzed in evolutionary studies or to identify potential chemical fragrances, knowledge about most of the species are still incomplete; moreover, studies of floral scent variations at population level including more than just a few individuals are particularly scarce. In particular, knowledge about the spatial and temporal variability of the floral scents emitted by flowers is almost completely absent.

In this study, we aim to describe the spatial and temporal variability of the spectrum of volatile compounds emitted by flowers of *B. robertiana*, testing the ability of this species to adapt to different environmental conditions to be attractive to different potential pollinators.

B. robertiana, with respect to other groups of orchids, is not plagued by taxonomic problems at species level that could determine some confounding effects due to identification discrepancies by the botanists. Moreover, it has large populations in southern Italy, and it has a broad range, spreading along a large climatic gradient, at regional level. Thus, it represents an ideal case study to explore the temporal, ecological, and spatial variability in emitted floral volatile compounds (VOCs). In particular, in this study, we aim to: (1) assess whether the chemical composition is variable between plants collected at different sites; (2) evaluate whether chemical composition of the scent is related to geographical or environmental variables; (3) assess whether there are temporal differences in VOC compositions during the flowering phase, comparing floral volatile compounds emitted at the beginning and at the end of the flowering; and (4) assess whether there is a between-year variability in VOC compositions emitted by the same plant.

2. Experimental Section

To assess spatial variability, fourteen *B. robertiana* plants were collected from different populations in the Basilicata region (Southern Italy) from the coast to the inner mountain area (Figure 1) within altitude ranges from 10 m a.s.l. to 727 m a.s.l. Plants were collected in 2017 and cultivated into pots at the campus of the University of Basilicata (Potenza, Italy). In the following year (2018), inflorescences with fully opened flowers were incapsulated in a of 6.5 glass bell (Figure 2). Sampling was performed under light conditions, in an air-conditioned room (21 ± 1 °C) to guarantee a stable temperature. VOC sampling followed the protocol we used in a recent study on the floral volatiles of the genus *Gymnospermium* [26].

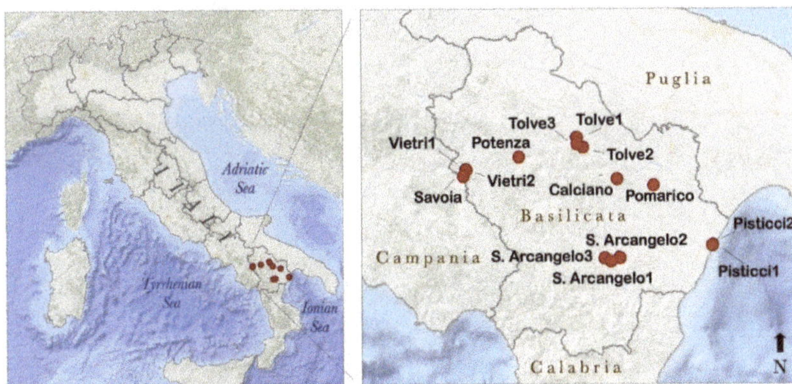

Figure 1. Study area and sampling point locations in the Basilicata region (southern Italy). Labels of the samples correspond with the municipality names where plants were collected.

Analysis of VOCs was performed using HS-SPME with a DVB/CARB/PDMS fiber. A preliminary set of analysis was conducted to optimize the sampling time: these analyses were performed at three different adsorption times of the fiber (5–24–72 h). The highest number of identified compounds (20) was detected when the SPME fiber was exposed for 24 h (Appendix A, Table A1); therefore, we used this interval of fiber exposition in this study. The fiber was exposed to the headspace and then withdrawn into the needle and transferred to a GC/MS system. A 50/30 µm DVB/CAR/PDMS module (57328-U, Supelco, Milan, Italy) was employed to determine VOCs. Analyses were accomplished with an HP 6890 Plus gas chromatograph (Agilent) equipped with a Phenomenex Zebron ZB-5 MS capillary column (30 m × 0.25 mm i.d. (inner diameter) × 0.25 µm FT) (Agilent, Milan, Italy). An HP 5973 mass selective detector (Agilent) was utilized with helium at 0.8 mL/min as the carrier gas. A splitless injector was maintained at 250 °C and the detector at 230 °C. The oven was maintained at 40 °C for 2 min, then gradually warmed, 8 °C/min, up to 250 °C and held for 10 min (Figure 3). Tentatively identification of aroma components was based on mass spectra and NIST 11 library comparison. A single VOC peak was considered as identified when its experimental spectrum matched with a score over 90% with ones present in the library and if the retention time was in agreement with the reported retention index. Retention indices were calculated using standard *n*-alkane solution (49452-U, C7-C40 saturated alkanes standard, Sigma-Aldrich, Milan, Italy).

Figure 2. A *Barlia robertiana* plant from the site of S. Arcangelo (Basilicata, Italy) in the experimental conditions for floral VOC sampling using a 50/30 μm DVB/CAR/PDMS fiber (photograph: V.A. Romano).

A multivariate analysis of VOC compositions was performed through non-metric multidimensional scaling (NMDS) as an ordination technique and hierarchical cluster analysis (HCA). Bray–Curtis dissimilarity was used for both NMDS and HCA using the relative abundance of each compounds as sample variable. For HCA, we used the unweighted pair group method with arithmetic mean (UPGMA) technique as an agglomerative method. Variables with Spearman's rho correlation coefficient > 0.65 with ordination axes were superimposed in the scatter diagram.

Concerning spatial analyses, we tested whether geographic distance and environmental variables influence the VOC compositions of samples. Site-specific soil data were not available; therefore, we approximated differences in the environmental niche using slope, aspect, altitude, and bioclimatic data. For each sample point, the most relevant phytoclimatic indices [27,28] were extracted from the high-resolution raster dataset developed to realize the bioclimate map of Italy [29]. These indices were: yearly positive temperatures (Tp = sum of the monthly mean temperatures of months with average temperatures > 0 °C); annual positive precipitation (Pp = total average precipitation of months with average temperature > 0 °C); thermicity index (T + m + M); annual ombrothermic index = (Pp/Tp); continentality index ($T_{max} - T_{min}$); ombrothermic index of the warmest summer bimester ($Ios_2 = (Pp_2/Tp_2)$). According to the approach we used for population genetics of *Centaurea filiformis* [30], the correlations between VOC compositions and both geographic and environmental distances were assessed using Mantel tests implemented in the software PAST. Geographic distances were log-transformed and environmental distances were obtained by means of analyzing the Euclidean distance after data standardization.

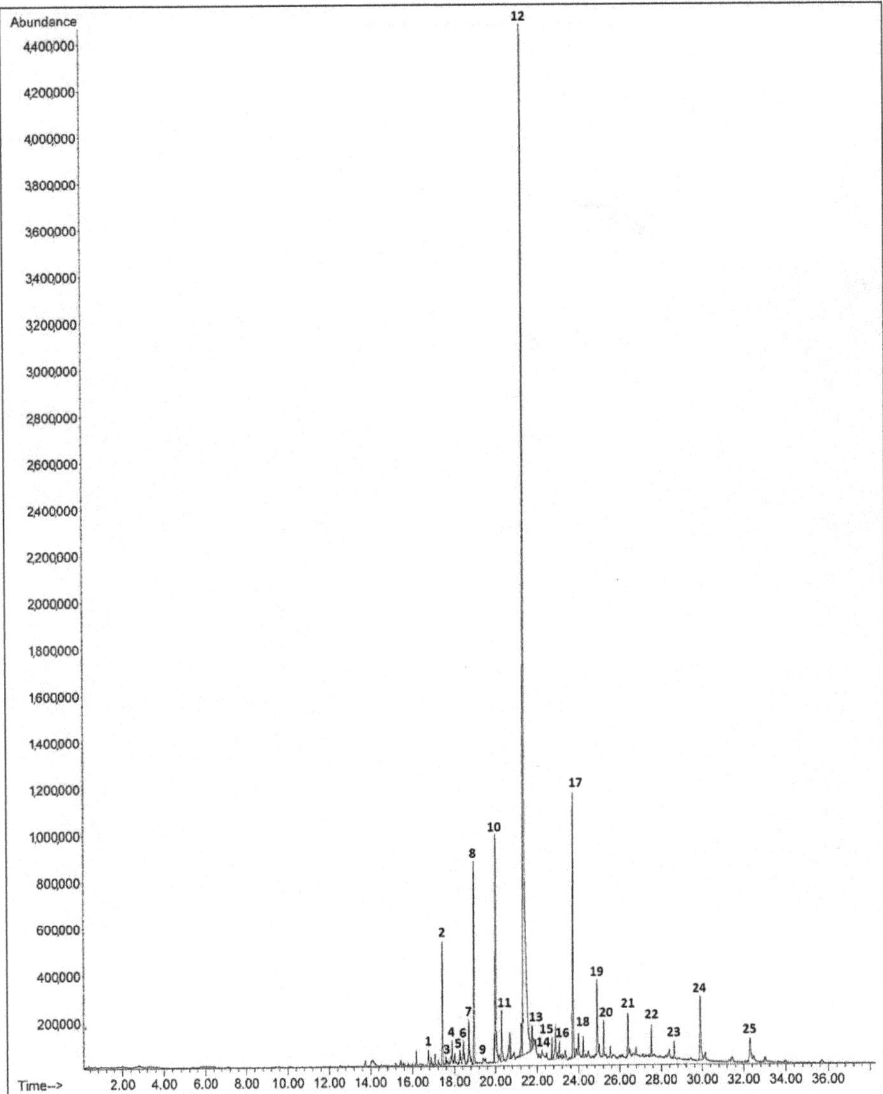

Figure 3. Gas chromatographic results of an SPME analysis of the VOCs of *Barlia robertiana*. 1: tetradecane; 2: caryophyllene; 3: sesquiphellandrene; 4: β-himalachene; 5: 1-(1,5-dimethyl-4-hexenyl)-4-methylbenzene; 6: pentadecane; 7: β-bisabolene; 8: β-sesquiphellandrene; 9: hexadecane; 10: phthalate; 11: *trans*-farnesol; 12: 2,3-Dihydrofarnesol; 13: 2,6-diidopropylnaphthalene; 14: tetradecanoic acid; 15: octadecane; 16: pentadecanal; 17: phthalate; 18: nonadecane; 19: phthalate; 20: eicosane; 21: heneicosane; 22: docosane; 23: phthalate; 24: phthalate; 25: phthalate.

To assess temporal variability of flower scent in *B. robertiana*, two individuals collected from the S. Arcangelo population were sampled twice during 2019, following the same protocol described above. The first sampling was performed in the first part of the flowering phase, approximately when 50% of the flowers were open; the second sampling at the end of the flowering, when all the flowers were open. In addition, three individuals from Potenza, Calciano and S. Arcangelo were sampled in two consecutive years (2018 and 2019), in the same experimental conditions described above, to assess the between-year stability of floral emissions. Similarity between samples has been measured through the Bray–Curtis index.

3. Results

3.1. Spatial Variability

The 14 plants analyzed to assess the spatial variability of floral scent showed the presence of 70 compounds with a high variability between samples (min 14; max 28; mean 19.4). Remarkably, no compound was found to be present in all the samples. This variability has been observed considering both the composition and abundance of VOCs. Ethyl dodecanoate was the most frequent compound; it was determined in 11 samples. Hexadecane and β-bisabolene (Figure 4a) were detected in 10 samples, δ-selinene and β-sesquiphellandrene (Figure 4e) were detected in 9 samples, whereas caryophyllene (Figure 4b), *cis*-α-bergamotene (Figure 4f), and heptadecane were found in 8 samples. Considering the most abundant compounds, each sample afforded a different result. Verbenone was the main component in Pisticci 1 and β-sesquiphellandrene in Pisticci 2; Calciano gave pristane as prevalent; in Sant'Arcangelo, the main components were alternatively α-zingiberene, verbenone, and pristane. Caryophyllene was the principal component in Tolve 1, *i*-propyl 14-methyl-pentadecanoate was the main component in Tolve 2, whereas farnesol has the same role in Tolve 3. The Pomarico sample gave *p*-menth-8-en-1-ol as a main component (Figure 4c); in the two samples of Vietri, we found β-sesquiphellandrene and citronellol (Figure 4d) as dominant, whereas the Potenza plant gave pristane as the most abundant compound. β-Sesquiphellandrene was also the main component in the sample of Savoia. Only two compounds were found as dominant in more than two samples: pristane in S. Arcangelo 3, Calciano and Potenza and β-sesquiphellandrene in Pisticci 2, Vietri 1 and Savoia. Notably, except for Vietri 1 and Savoia, these groups comprise samples located rather distantly geographically. Coherently, the overall similarities between samples were rather low, with a mean value of 18.97 ± 1.6 SE (min 0.96, max 60.94). The full VOC compositions for each sample are reported in Table 1.

The NMDS ordination resulted in a two-dimensional solution with a final stress of 0.14 (Figure 5a). The VOCs most strongly correlated with the first axis were β-sesquiphellandrene (positively) and longipinene (negatively), whereas *p*-menth-8-en-1-ol (positively) and *i*-propyl 14-methyl-pentadecanoate (negatively) were the most correlated compounds with the second axis.

We did not identify a clear geographical structure in the dataset, except for the samples from Savoia and Vietri, which seemed to group together; these are also very close to each other geographically (Figure 1), and there was a weak correlation between axis 1 and altitude. However, in the NMDS, most of the samples coming from nearby locations (e.g., Tolve 1 vs. Tolve 3 and Pisticci 1 vs. Pisticci 2) were strongly separated along the two axes, and the position of the Potenza sample, located at the higher altitude, is not coherent.

Furthermore, in the hierarchical clustering (Figure 5b) within the subcluster including samples from Vietri and Savoia a sample from Pisticci was also unexpectedly included, the locality placed at the maximum geographical distance from Vietri and Savoia (Figure 1). The lack of a significative geographical driver underlying our dataset was also confirmed by the results of the Mantel test, which showed that there is no correlation between composition in VOCs and geographical distances between samples (Mantel test: r = 0.17; *p* = 0.19).

The same result was obtained from analyzing the correlation between VOC composition and the environmental distances (i.e., ecological niche) of sample sites (Mantel test: r = 0.10; *p* = 0.28).

3.2. Temporal Variability

As for temporal differences in VOCs composition during the flowering phase, the comparison between floral volatiles emitted at the beginning and at the end of the flowering exhibited the presence of a similar number of compounds (min 15; max 22; mean 18.3). The mean value is in agreement with findings from the sampling performed to assess the spatial variability, where a mean value of 19.4 compounds was found. The VOC compositions for each sample are reported in Table 2. In particular, comparing the two flowering phases, we found the same number of compounds in sample 2 and a slight decrease in sample 1 at the end of the flowering. Additionally, in this case, despite the fact they come from the same population (S. Arcangelo), composition quantitative analysis confirmed the presence of a strong association between individuals' variability. In sample 1, the two most abundant compounds were caryophyllene and 2,3-dihydrofarnesol, whereas in sample 2, they were citronellol and β-sesquiphellandrene. Caryophyllene remained the dominant compound in the sample 1 even at the end of flowering, whereas the abundance of 2,3-dihydrofarnesol strongly decreased; then, the second most abundant compound became 4-methyltetradecane. In sample 2, instead, the dominance between the first two compounds was reversed at the end of the flowering, with β-sesquiphellandrene becoming the most abundant compound followed by citronellol. The decrease in the number of compounds in sample 1 (from 22 to 15) is linked to the non-detection at the end of flowering of various compounds which, previously, had low abundance (area% < 2), whereas only nonadecane, with an area value of 0.37%, was found in addition at the end of flowering. On the other hand, in sample 2, five compounds were no longer identified at the end of flowering, replaced by an equal number of compounds not present in the first flowering phase. Among these, α-terpineol acetate stands out for its abundance, characterized with a value of 13.83% in sample 2 at the end of flowering.

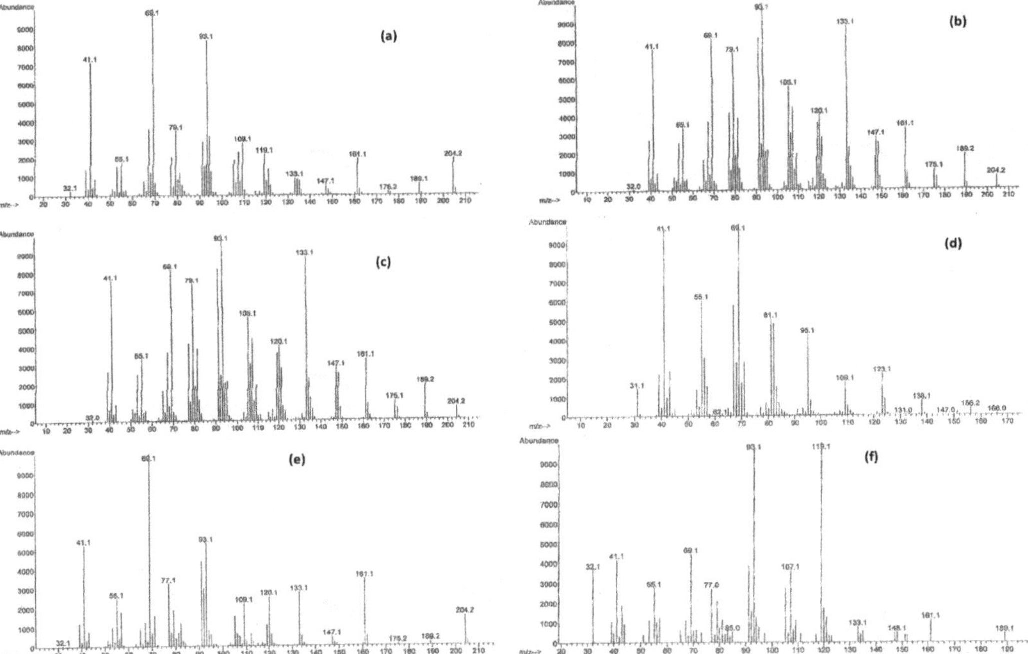

Figure 4. Mass spectra of (**a**) β-bisabolene; (**b**) caryophyllene (**c**) *p*-menth-8-en-1-ol; (**d**) citronellol; (**e**) β-sesquiphellandrene; (**f**) *cis*-α-bergamottene.

Table 1. VOCs detected from plant samples of *Barlia robertiana* sampled in the Basilicata region (S-Italy). Compounds are ordered by frequency in the table.

Sample		Pisticci 1	Pisticci 2	Calciano	S. Arcangelo 1	S. Arcangelo 2	S. Arcangelo 3	Tolve 1	Tolve 2	Tolve 3	Pomarico	Vietri 1	Vietri 2	Potenza	Savoia
Compound	r.t. [min]						Area [%] ± 0.03								
Ethyl dodecanoate	19.93	0.33		0.67	1.75	0.33	0.82	1.17	8.29			3.76	2.63	1.10	2.60
β-Bisabolene	18.43	2.13					0.96	4.06	1.05	2.13	4.61	8.98	3.17	2.00	5.32
Hexadecane	20.16	0.14	0.53		0.96			0.88	3.09		0.60	4.01	1.80	1.01	3.36
δ-Selinene	18.65	0.98		7.04	9.06	17.59	6.49		1.65			2.49	1.16		1.54
β-Sesquiphellandrene	19.35		29.06	0.91					2.94	8.39	17.59	43.08	15.14	8.97	25.05
Caryophyllene	17.46		8.51					24.63	10.11		3.68	3.07	5.21	6.75	17.96
cis-α-Bergamotene	17.58	0.42	0.53	1.73		0.77					0.77	14.04	4.15		1.14
Heptadecane	21.29		0.74	7.41	1.43			0.73	2.94			2.24	1.49		2.11
α-Pinene	7.79	0.49		3.02		0.46		0.48	0.55		0.92			7.87	
D-limonene	9.80	1.34	0.34	0.10		1.65		1.49		0.46	3.21			0.11	
Citronellol	14.36	0.57				1.34		2.53		0.5		1.77	17.96	1.53	
Methyl citronellate	15.03			0.30				0.75		0.9	0.31	3.01	3.61		1.85
Octadecane	22.85		0.23		1.30					0.21		1.24	0.66	0.31	1.25
Verbenone	13.74	45.22				31.48	1.12	3.81	0.57		1.32				
Tetradecane	17.02		0.53			0.96			1.73			3.09	1.39		2.44
α-Zingiberene	17.09	6.88	0.98	2.63	17.14	0.94	3.59								
Z-β-Farnesene	17.54	0.22				0.42	0.40		1.49				0.41	0.49	
Pristane	21.68			35.57		9.61	58.90		1.56			1.13		22.57	
i-propyl 14-methyl-pentadecanoate	26.35				3.74		1.44	1.44	12.26				1.54	0.57	
Longipinene	18.40	0.72		3.86	1.77	4.40	1.32								
Ethyl tetradecanoate	22.58	15.96		12.40					2.57					1.00	0.92
Dihydrofarnesol	22.7	0.09		0.52	1.01		0.82			22.23					
Nonadecane	24.22		0.39		2.06			0.65	0.90	0.24					
Methyl hexadecanoate	25.08			0.22				1.15	0.57		0.78		0.31		
β-Myrcene	8.99	0.36				0.37		0.23			3.26				
Tridecane	15.12								1.25			2.35	1.29		1.63
E-β-Farnesene	17.96	0.14	0.87							1.97					7.10
Humulene	18.08	0.11			0.46			0.44		0.22					
2,6-Bis(1,1-dimethylethyl)-2,5-cyclohexadiene-1,4-dione	18.11			0.19		0.64							0.37	0.63	
β-Curcumene	18.13		0.45	1.53	2.43	1.56									
α-Farnesene	18.60	0.25	0.47					16.97		0.89					
Nerolidol	19.73			0.33	1.69	0.57				0.86					
Fitone	23.98	0.16		0.28	1.32		1.12								
β-Pinene	8.72									0.18	1.13			0.39	
p-Menth-8-en-1-ol	12.26	1.35				0.83					21.68				
Decanal	13.72	0.88				2.12	0.25								
2,6-dimethyl-2,6-Octadiene	16.72			0.42	1.16		0.32								

Table 1. *Cont.*

Sample		Pisticci 1	Pisticci 2	Calciano	S. Arcangelo 1	S. Arcangelo 2	S. Arcangelo 3	Tolve 1	Tolve 2	Tolve 3	Pomarico	Vietri 1	Vietri 2	Potenza	Savoia
Compound	**r.t. [min]**	Area [%] ± 0.03													
Pentadecane	18.49	0.12	0.45	0.79											
Farnesol	21.85						1.70			36.63				19.36	
Tetradecanoic acid	22.5			0.30	0.73					0.16					
Eicosane	25.39			1.36				0.61	0.64						
cis-p-menthan-1-ol	10.93	0.46						0.77							
Linalool	11.43												0.46	0.24	
Citronellal	13.07	0.31				0.42									
α-Terpineol	13.16							0.54			5.39				
Citronellyl formate	14.52	2.42		0.33											
Bornyl acetate	14.98								1.71				1.00		
Geranyl acetone	17.52			0.21		0.37									
α-Curcumene	18.28		0.63							0.23					
Tetradecanal	21.23		0.42		2.24										
Farnesal	22.04		5.81							0.49					
2,6-diisopropylnaphthalene	22.22		0.48				1.00								
2,3-Dihydrofarnesyl acetate	22.76									0.5				0.86	
Pentadecanal	23.02		0.21			0.27									
4-methyl-3-penten-2-one	5.33	0.19													
β-Phellandrene	9.65										4.51				
γ-Terpinene	10.66										5.24				
Geraniol	14.04									0.53					
D-carvone	14.49	0.35													
Citral	14.74		1.1												
Citronellyl acid	15.56												1.53		
Citronellyl propionate	16.1	1.02													
Phytan	22.98						0.33								
Hexahydrofarnesyl acetone	23.21													0.29	
Farnesyl acetate	23.26									0.46					
Isopropyl myristate	23.35						0.47								
Methyl 9-octadecenoate	26.4													0.28	
Heneicosane	27.05								0.37						
Isopropyl linoleate	27.43									0.09					
Number of compounds		28	20	24	17	21	17	19	20	21	16	14	21	19	14

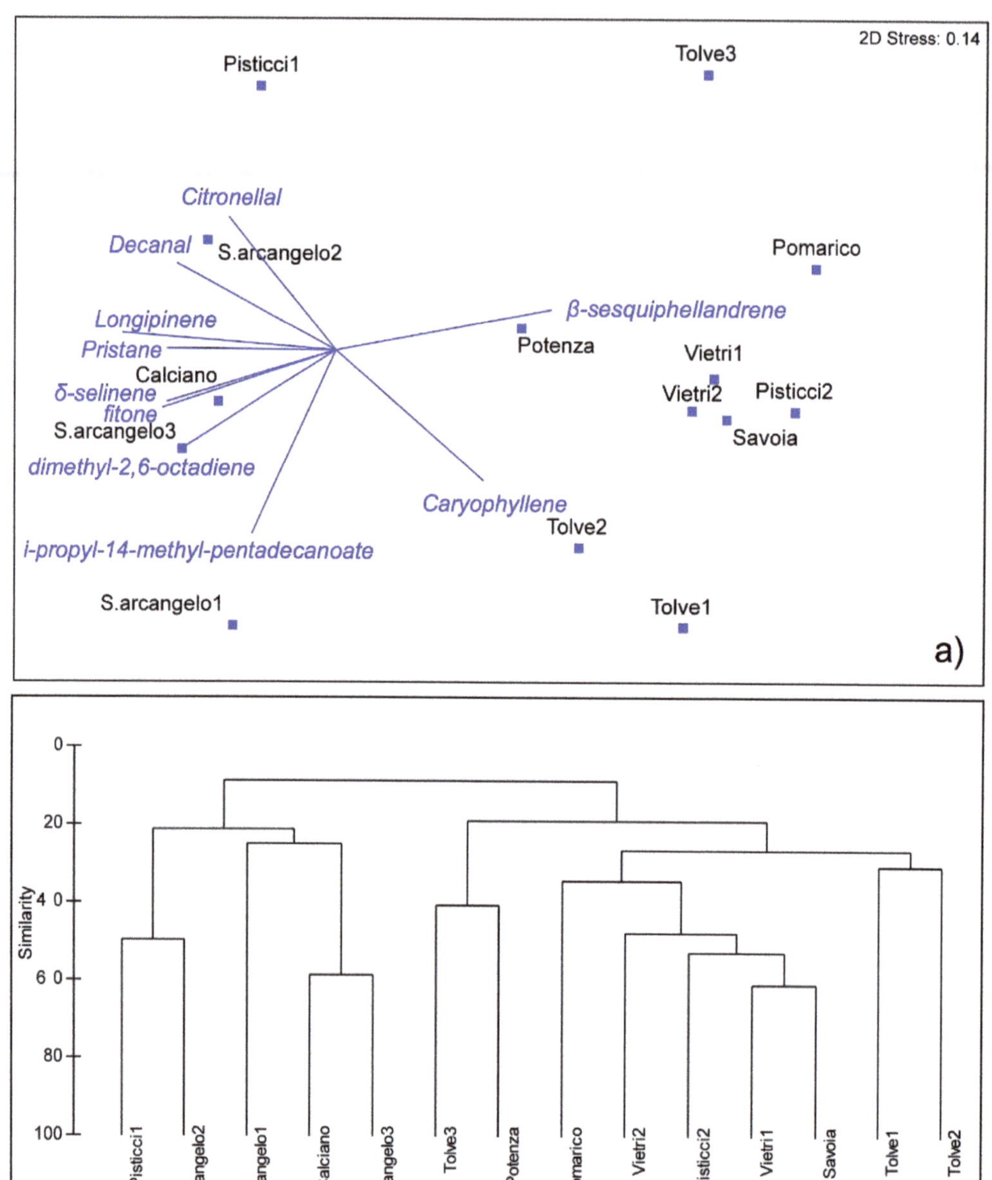

Figure 5. Multivariate analysis of VOCs emitted by flowers of *Barlia robertiana* based on the between-samples Bray–Curtis similarity: (**a**) non-metric dimensional scaling; (**b**) UPGMA hierarchical clustering. Labels of samples are accordance with Figure 1.

Table 2. Comparison of VOCs emitted by two plant samples of *Barlia robertiana* from the S. Arcangelo site (Basilicata region, S-Italy) detected during the first part of the flowering phase with respect to the last part of the flowering.

Flowering Phase		Sample 1	Sample 2	Sample 1	Sample 2
		Early		Late	
Compound	r.t. [min]	Area %			
α-Pinene	7.83		0.24		0.61
β-Myrcene	9.03		0.29		0.52
D-Limonene	9.84	0.54	1.00	0.66	2.01
Sylvestrene	11.25	0.54			
Citronellal	12.29		1.50		0.61
α-Terpineol	13.17		0.78		
Verbenone	13.45	0.70	0.81	0.54	0.23
Carvone	14.07		0.42		0.35
Citronellol	14.14		42.54		15.66
Methyl (S)-citronellate	14.24	1.28	1.06	1.23	
4-Carene	15.84				2.63
α-Terpineol acetate	15.85				13.83
(*E,Z*)-2,6-Dimethyl-2,6-octadiene	15.87		0.38		
Dihydro-b-ionone	16.20	0.42			
2-Methylene-4,8,8-trimethyl-4-vinylbicyclo[5.2.0]nonane	16.70	2.52			
Caryophyllene	17.12	34.56	3.34	32.35	
α-Bergamotene	17.29	1.10	0.50		2.74
(*Z*)-β-Farnesene	17.47		1.50	0.57	1.99
trans-Geranylacetone	17.50				0.73
(*E*)-β-Farnesene	17.60		5.29		
4-Methyltetradecane	17.65			30.68	
α-longipinene	17.97				0.36
Pentadecane	18.17	4.27		3.78	
δ-Selinene	18.30				0.32
β-Bisabolene	18.42	4.85	3.82	6.53	6.98
β-Sesquiphellandrene	19.35		24.95		44.13
Elemene	19.54	1.14			
Hexadecane	19.65	0.98		0.37	
8-Heptadecene	20.80	1.47		1.60	
2,3-Dihydrofarnesol	21.08	34.80	11.28	17.15	1.02
2,6-Diisopropylnaphthalene	21.23	0.98			
Octadecane	22.41	0.98		1.12	
(E)-5-Octadecene	22.55	1.47			
Isopropyl miristate	22.78	0.89	0.90	0.83	0.23
Phytone	22.90			1.09	
4-Octadecylmorfoline	23.60	1.40			
Nonadecane	23.70			0.37	
Methyl hexadecanoate	24.05	0.42			
(*E,Z*)-5,9,13-Trimethyl-4,8,12-tetradecatrienal	24.32	1.61			
trans-Geranylgeraniol	26.50	1.61			
Number of compounds		**22**	**18**	**15**	**18**

As for inter-annual variations in VOC compositions, unexpectedly, a huge difference was found in the pairwise comparison of the three plants consecutively analyzed in 2018 and 2019 (Table 3). These plants came from the populations of Calciano, Potenza and S. Arcangelo 2; the 2018 samples were also used for spatial variability analyses. The plant from Calciano, as described above, had pristane and ethyl tetradecanoate as main components; surprisingly, these two compounds were completely absent from the sample of the following year, substituted by α-terpinolene and *(E)*-β-farnesene. Additionally, the other two individuals analyzed gave similar results for the main components: in the plant from Potenza, verbenone and D-carvone were not detected in the second year of sampling, which gave pristane and farnesol as the two dominant VOCs. Regarding the plant from S. Arcangelo, verbenone, the main component in 2018, was almost absent in 2019 (0.81%), whereas δ-selinene, which was the second most abundant, was no more detected. Instead, the main component in 2019 resulted citronellol and β-sesquiphellandrene. Consequently, the inter-annual between-sample similarity, measured through the Bray–Curtis index, reached only very low values, lower than 9% (Calciano 4.6%, Arcangelo 5.1%, Potenza 8.9).

Table 3. Inter-annual variations in VOCs emitted by three individuals of *Barlia robertiana* sampled in two consecutive years.

		Calciano	Calciano	Potenza	Potenza	S. Arcangelo 2	S. Arcangelo 2
Year of Sampling		2018	2019	2018	2019	2018	2019
Compound	**r.t. [min]**				Area (%)		
α-Pinene	7.83	3.02	10.63	6.43	7.87	0.46	0.24
β-Pinene	8.74		0.42	0.35	0.39		
Sulcatone	8.91			0.19			
β-Myrcene	9.03		0.65	0.33		0.37	0.29
β-Phellandrene	9.65			0.06			
2-Ethyl-1-hexanol	9.83			0.18			
D-limonene	9.84	0.10	1.31	0.95	0.11	1.65	1.00
Eucaliptol	9.88		0.35	0.35			
α-Terpinolene	11.05		32.45				
trans-Sabinene idrate	11.13			1.46			
6-Methyl-2-pyridinecarboxyaldehyde	11.19			0.36			
Undecane	11.3			1.03			
Linalool	11.43				0.24		
Nonanal	11.45			1.35			
Rose oxide	11.64			0.59			
2-Phenylethanol	11.67			0.42			
p-Menth-8-en-1-ol	12.26			1.66		0.83	
Citronellal	12.29		1,12	0.4		0.42	1.50
α-Terpineol	13.17		0.45	0.17			0.78
1-Butoxy-2-ethylhexane	13.23			0.18			
Verbenone	13.45		4.72	55.18		31.48	0.81
Decanal	13.72					2.12	
Carvone	14.07						0.42
Citronellol	14.14				1.53	1.34	42.54
cis-Octahydro-3a-methyl-2H-inden-2-one	14.23		1.80				
Methyl (S)-citronellate	14.24		1.80				1.06
D-carvone	14.49			7.92			
Citronellyl formate	14.52	0.33					
(E)-cinnamaldehyde	14.59		8.25				
Methyl citronellate	15.03	0.30					
Tridecane	15.12			1.55			
Citronellyl acid	15.56			0.72			
Cinnamyl alcohol	15.80			0.15			
4-Carene	15.84		1.53				
(E,Z)-2,6-Dimethyl-2,6-octadiene	15.87						0.38
2,6-dimethyl-2,6-octadiene	16.72	0.42					

Table 3. *Cont.*

		Calciano	Calciano	Potenza	Potenza	S. Arcangelo 2	S. Arcangelo 2
Year of Sampling		2018	2019	2018	2019	2018	2019
Compound	r.t. [min]				Area (%)		
Tetradecane	17.02					0.96	
α-zingiberene	17.09	2.63		0.97		0.94	
Caryophyllene	17.12		1.74	0.32	6.75		3.34
α-Bergamotene	17.29		0.64				0.50
(Z)-β-Farnesene	17.47		1.77		0.49	0.42	1.50
Trans-Geranyl acetone	17.50		0.37				
Geranyl acetone	17.52	0.21		0.09		0.37	
cis-α-bergamotene	17.58	1.73		0.67		0.77	
(E)-β-Farnesene	17.60		14.22				5.29
2,6-bis(1,1-dimethylethyl)-2,5-cyclohexadiene-1,4-dione	18.11	0.19			0.63	0.64	
β-curcumene	18.13	1.53		0.31		1.56	
Longipinene	18.40	3.86		0.34		4.40	
β-Bisabolene	18.42		5.22		2.00		3.82
Pentadecane	18.49	0.79					
δ-selinene	18.65	7.04		1.37		17.59	
β-sesquiphellandrene	19.35	0.91	1.98		8.97		24.95
Hexadecane	19.65		0.35		1.01		
Nerolidol	19.73	0.33				0.57	
Ethyl dodecanoate	19.93	0.67			1.10	0.33	
2,3-Dihydrofarnesol	21.08						11.28
Heptadecane	21.29	7.41					
Pristane	21.68	35.57			22.57	9.61	
Farnesol	21.85				19.36		
Farnesal	22.04		0.54				
Tetradecanoic acid	22.5	0.30					
Ethyl tetradecanoate	22.58	12.40		1.23			
Dihydrofarnesol	22.7	0.52					
2,3-Dihydrofarnesyl acetate	22.76				0.86		
Isopropyl miristate	22.78						0.90
Octadecane	22.85				0.31		
pentadecanal	23.02					0.27	
Hexahydrofarnesyl acetone	23.21				0.29		
Fitone	23.98	0.28					
methyl hexadecanoate	25.08	0.22					
Eicosane	25.39	1.36					
i-propyl 14-methyl-pentadecanoate	26.35				0.57		
methyl 9-octadecenoate	26.4				0.28		
Number of compounds		**24**	**22**	**31**	**19**	**21**	**18**

4. Discussion

In this study, we identified a very high number of VOCs emitted by *Barlia robertiana*: considering all the analyzed samples, more than 100 compounds were identified. These results largely encompass the findings of our preliminary study [31].

As for VOC compositions, notably, in a study performed in Spain [32], Gallego et al. found α-pinene, β-pinene, and limonene as the main components of the floral scent of *B. robertiana*. In our samples, instead, α-pinene and limonene were detected, and only at low percentages, and they were not always present (Table 1), whereas β-pinene was found with even less frequency and lower percentages. On the other hand, the compounds that most characterize the plants from Italian populations have been not detected in the samples carried out in Spain or, therein, they were only present with very low values; this was the case with verbenone, for example. Considering the influence that analytical tools and sampling procedures may have on the results obtained from VOC analysis, we argue that most of detected differences in the floral scents of *B. robertiana* are due to an intrinsic extreme capability of this species to vary its floral emissions, both qualitatively and quantitatively. At present, the data collected do not allow hypothesizing how much of this variability is under genetic control and how much depends on contingent environmental conditions. What it was possible to ascertain in this study, confirming the preliminary data shown by [31] based on a minor number of samples, is that the variability of floral scents is not related to the geographical distance between populations, nor to the main environmental characteristics of the growth sites. However, even if based on only three samples, the observed variation of the VOCs emitted by the same individual in two consecutive years would seem to indicate a poor genetic determinism for this phenomenon. The mechanisms behind this variability could be related to an intrinsic plasticity of metabolic pathways that lead to the synthesis of VOCs in *B. robertiana*, but no studies have specifically investigated this aspect thus far.

Several of the detected compounds, such as verbenone and α-zingiberene, are known to act as pheromones [33,34]; however, this specific function probably is not specifically used by *B. robertiana*. These results may suggest that there is no adaptation of floral scent to local environments or specific communities of pollinators. In fact, the wide spectrum of VOCs emitted can allow *B. robertiana* to attract different species of insects also belonging to very distant taxonomic groups, as evidenced by some studies on its pollinators [24,25] relying on a large plethora of possible pollinators. The strategy of *B. robertiana* to attract pollinators manly involves early flowering, showiness and long-lasting inflorescence, traits that can be advantageous for exploiting the first insects that emerged from winter hibernation. Floral scent, a key trait for interaction between plants and insects [35,36], plays an important role for floral mimicry in deceptive species. In this context, a huge variation in flora scent, such as that highlighted by *B. robertiana*, can be considered an effective strategy for a rewardless, but allogamous, species to avoid that visiting insect learn to avoid such flowers. In fact, some studies [37] has been highlighted as rewardlessness can be a dangerous strategy [38]. Despite the causes of rewardlessness are still little known, just a study on *B. robertiana* showed for the first time the reproductive advantage of the lack of nectar [24]. However, it must be stressed again that this reproductive advantage can only occur if pollinating insects do not learn to associate the floral signals of a species with the lack of nectar inside the flowers.

5. Conclusions

We believe that the high variability highlighted in our study about the floral scent emitted within the same population or individual, and the lack of evidence of correlations between flora scents and environmental features (both geographical and ecological) has to be interpreted as an effective strategy carried out by *Barlia robertiana* to not allow the learning by pollinators, associating floral signals such as scent with a lack of nectar. Similar strategies have been highlighted for other rewardless orchids, such as *Ophrys sphegodes* Mill. [36]. However, further in-depth studies are needed to investigate several aspects that could not be addressed here. As recently pointed out [35], to investigate specific intra-species variations of floral scent, exploring the less investigated factors that could explain floral scent variations and their mechanisms is a promising research field in light of the evolutionary ecology to which more attention should be given in the future.

Author Contributions: Conceptualization, V.A.R. and L.R.; methodology, L.R., V.A.R. and M.D.; formal analysis, L.R., A.M.R.C., R.R. and M.D.; investigation, V.A.R.; data curation, V.A.R. and M.D.; writing—original draft preparation, L.R., S.F., R.L. and M.D. All authors have read and agreed to the published version of the manuscript.

Funding: This research received no external funding.

Institutional Review Board Statement: Not applicable.

Informed Consent Statement: Not applicable.

Data Availability Statement: Not applicable.

Conflicts of Interest: The authors declare no conflict of interest.

Appendix A

Table A1. VOC analysis of *Barlia robertiana* at different adsorption times.

Compound	r.t. [min]	KI	Adsorption Time (h)		
			5	12	24
			Area [%] $\pm$ 0.03		
D-limonene	9.86	1022	1.35	1.04	0.93
nonanal	11.36	1103		2.87	
decanal	13.29	1203		3.35	
4,6,6-trimethylbicyclo[3.1.1]ept-3-en-2-one	13.46	1212	6.28	2.01	1.54
methyl (S)-(-)-citronellate	14.24	1258	3.44	2.64	2.68
α/β-caryophyllene	17.12	1474	60.36	35.16	35.32
1,3-di-isopropylnaphthalene	17.13	1668		1.71	
trans-α-bergamotene	17.28	1433		1.04	0.41
6,10-dimethyl-5,9-undecadien-2-one	17.51	1453	10.92		
(E),(Z),α/β-3-metylene-7,11-dimethyl-1,6,10-dodecatriene	17.54	1459		2.20	
β-farnesene	17.60	1463		25.60	20.71
humulene	17.66	1467		1.08	
pentadecane	18.17	1499	3.66	5.17	3.98
β-bisabolene	18.42	1515	7.03	7.56	5.60
diethyltoluamide	19.55	1571			0.97
hexadecane	19.65	1600		1.30	4.71
(E),(Z)-8-heptadecene	20.80	1676			1.42
2,6-diisopropylnaphthalene	21.23	1695		1.71	
heptadecane	21.75	1700		2.01	0.41
2,3-dihydrofarnesyl acetate	22.60	1805			2.92
isopropyl myristate	22.78	1827	5.16	25.83	7.15
(E),(Z)-5,9,13-trimethyl-4,8,12-tetradecatrienal	23.03	1840		1.19	1.58
7-acetyl-6-ethyl-1,1,4,4-tetramethyltetraline	23.33	1843		0.74	0.73
galaxolide	23.35	1850		0.74	1.46
4-octadecyl morfoline	23.60	1880	1.80		0.97
methyl hexadecanoate	24.05	1927			0.77

References

1. Christenhusz, M.J.M.; Byng, J.W. The number of known plants species in the world and its annual increase. *Phytotaxa* **2016**, *261*, 201–207. [CrossRef]
2. Darwin, C. *On the Various Contrivances by Which British and Foreign Orchids Are Fertilised by Insects, and on the Good Effects of Intercrossing*; Murray, J., Ed.; University of Chicago Press: London, UK, 1862.
3. Vereecken, N.J.; Dafni, A.; Cozzolino, S. Pollination Syndromes in Mediterranean Orchids—Implications for Speciation, Taxonomy and Conservation. *Bot. Rev.* **2010**, *76*, 220–240. [CrossRef]
4. Ackerman, J.D. Mechanisms and evolution of food-deceptive pollination systems in orchids. *Lindleyana* **1986**, *1*, 108–113.
5. Jersáková, J.; Johnson, S.D.; Kindlmann, P. Mechanisms and evolution of deceptive pollination in orchids. *Biol. Rev.* **2006**, *81*, 219. [CrossRef]
6. Johnson, S.D.; Schiestl, F.P. *Floral Mimicry*, 1st ed.; Oxford University Press: Oxford, NY, USA, 2016.

7. Kunze, J.; Gumbert, A. The combined effect of color and odor on flower choice behavior of bumble bees in flower mimicry systems. *Behav. Ecol.* **2001**, *12*, 447–456. [CrossRef]
8. Galizia, C.G.; Kunze, J.; Gumbert, A.; Borg-Karlson, A.-K.; Sachse, S.; Markl, C.; Menzel, R. Relationship of visual and olfactory signal parameters in a food-deceptive flower mimicry system. *Behav. Ecol.* **2005**, *16*, 159–168. [CrossRef]
9. Johnson, S. Batesian mimicry in the non-rewarding orchid *Disa* pulchra, and its consequences for pollinator behaviour. *Biol. J. Linn. Soc.* **2000**, *71*, 119–132. [CrossRef]
10. D'Auria, M.; Lorenz, R.; Mecca, M.; Racioppi, R.; Antonio Romano, V. Aroma components of *Cephalanthera* orchids. *Nat. Prod. Res.* **2021**, *35*, 174–177. [CrossRef]
11. Heinrich, B. Bee flowers: A hypothesis on flower variety and blooming times. *Evolution* **1975**, *29*, 325–334. [CrossRef]
12. Johnson, S.D.; Peter, C.I.; Nilsson, L.A.; Ågren, J. Pollination success in a deceptive orchid is enhanced by co-occuring rewarding magnet plants. *Ecology* **2003**, *84*, 2919–2927. [CrossRef]
13. Dötterl, S.; Vereecken, N.J. The chemical ecology and evolution of bee–flower interactions: A review and perspectives. *Can. J. Zool.* **2010**, *88*, 668–697. [CrossRef]
14. Delforge, P. *Orchidées d'Europe, d'Afrique du Nord et du Proche-Orient: La Bible des Orchidophiles, Plus de 600 Espèces et de Nombreuses Variétés et Illustrées*; Guide Delachaux; 4e éd.; Revue et augmentée; Delachaux et Niestlé: Paris, France, 2016; ISBN 978-2-603-02407-2.
15. Baumann, H.; Künkele, S.; Lorenz, R. *Orchideen Europas: Mit angrenzenden Gebieten*; Ulmer-Naturführer; Eugen Ulmer KG: Stuttgart, Germany, 2006; ISBN 978-3-8001-4162-3.
16. Wartmann, B. Orchideen als "Neophyten" in der Schweiz? *AGEO Orchis* **2020**, *1*, 9–13.
17. Vögtlin, J. *Himantoglossum robertianum* (Loisel.) Delforge am Isteiner Klotz. *Berichte Bot. Arbeitsgemeinschaft Südwestdtsch.* **2008**, *5*, 128.
18. Bartolucci, F.; Peruzzi, L.; Galasso, G.; Albano, A.; Alessandrini, A.; Ardenghi, N.M.G.; Astuti, G.; Bacchetta, G.; Ballelli, S.; Banfi, E.; et al. An updated checklist of the vascular flora native to Italy. *Plant Biosyst.* **2018**, *152*, 179–303. [CrossRef]
19. Sramkó, G.; Molnár, A.V.; Hawkins, J.A.; Bateman, R.M. Molecular phylogenetics and evolution of the Eurasiatic orchid genus Himantoglossum sl. *Ann. Bot.* **2014**, *114*, 1609–1626. [CrossRef]
20. Colombo, P.; Giardina, S.; Perrone, A. Studio morfoanatomico di *Himantoglossum robertianum* ed *H. hircinum* (Orchidaceae) della Sicilia. *J. Eur. Orch.* **2009**, *41*, 359–388.
21. Bazzicalupo, M.; Burlando, B.; Denaro, M.; Barreca, D.; Trombetta, D.; Smeriglio, A.; Cornara, L. Polyphenol Characterization and Skin-Preserving Properties of Hydroalcoholic Flower Extract from *Himantoglossum robertianum* (Orchidaceae). *Plants* **2019**, *8*, 502. [CrossRef] [PubMed]
22. D'Emerico, S.; Galasso, I.; Pignone, D.; Scrugli, A. Localization of rDNA loci by Fluorescent In Situ Hybridization in some wild orchids from Italy (Orchidaceae). *Caryologia* **2001**, *54*, 31–36. [CrossRef]
23. Doneddu, M. Osservazioni sull'impollinazione di *Barlia robertiana* ad opera dei coleotteri *Tropinota squalida* e *Oxythrea funesta* (Cetoniidae) in Sardegna. *GIROS Orch. Spont. Eur.* **2015**, *58*, 262–265.
24. Smithson, A.; Gigord, L.D.B. Are there fitness advantages in being a rewardless orchid? Reward supplementation experiments with *Barlia robertiana*. *Proc. R. Soc. Lond. Ser. B Biol. Sci.* **2001**, *268*, 1435–1441. [CrossRef]
25. Sánchez Rosa, E.J. Estudi de la Biologia Reproductiva de L'orquídia Gegant (*Barlia robertiana*) a L'illa de Mallorca. Ph.D. Thesis, Universitat del les Illes Baleares, Palma, Spain, 2014.
26. Rosati, L.; Romano, V.A.; Cerone, L.; Fascetti, S.; Potenza, G.; Bazzato, E.; Cillo, D.; Mecca, M.; Racioppi, R.; D'Auria, M.; et al. Pollination features and floral volatiles of *Gymnospermium scipetarum* (Berberidaceae). *J. Plant Res.* **2019**, *132*, 49–56. [CrossRef]
27. Rivas-Martınez, S.; Rivas-Saenz, S.; Penas-Merino, A. Worldwide Bioclimatic classification system. *Glob. Geobot.* **2011**, *1*, 1–638.
28. Canu, S.; Rosati, L.; Fiori, M.; Motroni, A.; Filigheddu, R.; Farris, E. Bioclimate map of Sardinia (Italy). *J. Maps* **2015**, *11*, 711–718. [CrossRef]
29. Pesaresi, S.; Galdenzi, D.; Biondi, E.; Casavecchia, S. Bioclimate of Italy: Application of the worldwide bioclimatic classification system. *J. Maps* **2014**, *10*, 538–553. [CrossRef]
30. Farris, E.; Filigheddu, R.; Mameli, G.; Falanga, V.; Vanetti, I.; Rosati, L.; Binelli, G. Is population genetic structure of vascular plants shaped more by ecological or geographic factors? A study case on the Mediterranean endemic *Centaurea filiformis* (Asteraceae). *Plant Biol.* **2018**, *20*, 936–947. [CrossRef] [PubMed]
31. D'Auria, M.; Fascetti, S.; Racioppi, R.; Romano, V.A.; Rosati, L. Orchids from Basilicata: The Scent. In *Orchids Phytochemistry, Biology and Horticulture*; Merillon, J.-M., Kodja, H., Eds.; Springer International Publishing: Cham, Switzerland, 2020; pp. 1–22. ISBN 978-3-030-11257-8.
32. Gallego, E.; Gelabert, A.; Roca, F.J.; Perales, J.F.; Guardino, X. Identification of volatile organic compounds (VOC) emitted from three European orchid species with different pollination strategies: Two deceptive orchids (*Himantoglossum robertianum* and *Ophrys apifera*) and a rewarding (*Gymnadenia conopsea*). *J. Biodivers. Environ. Sci.* **2012**, *2*, 18–29.
33. McBrien, H.L.; Millar, J.G.; Rice, R.E.; McElfresh, J.S.; Cullen, E.; Zalom, F.G. Sex Attractant Pheromone of the Red-Shouldered Stink Bug *Thyanta pallidovirens*: A Pheromone Blend with Multiple Redundant Components. *J. Chem. Ecol.* **2002**, *28*, 1797–1818. [CrossRef] [PubMed]
34. Bakthavatsalam, N. Semiochemicals. In *Ecofriendly Pest Management for Food Security*; Academic Press: Cambridge, MA, USA, 2016; pp. 563–611.

35. Raguso, R.A. Start making scents: The challenge of integrating chemistry into pollination ecology. *Entomol. Exp. Appl.* **2008**, *128*, 196–207. [CrossRef]
36. Schiestl, F.P.; Johnson, S.D. Pollinator-mediated evolution of floral signals. *Trends Ecol. Evol.* **2013**, *28*, 307–315. [CrossRef] [PubMed]
37. Smithson, A.; Macnair, M.R. Negative frequency-dependent selection by pollinators on artificial flowera without rewards. *Evolution* **1997**, *51*, 715–723. [CrossRef]
38. Moya, S.; Ackerman, J.D. Variation in the floral fragrance of *Epidendrum ciliare* (Orchidaceae). *Nord. J. Bot.* **1993**, *13*, 41–47. [CrossRef]

Article

Volatile Organic Compounds from *Orchis* Species Found in Basilicata (Southern Italy)

Marisabel Mecca [1], Rocco Racioppi [1], Vito A. Romano [1], Licia Viggiani [1], Richard Lorenz [2] and Maurizio D'Auria [1,*]

[1] Dipartimento di Scienze, Università della Basilicata, V.le dell'Ateneo Lucano 10, 85100 Potenza, Italy; marisabelmecca@libero.it (M.M.); rocco.racioppi@unibas.it (R.R.); vitoantonio_romano@libero.it (V.A.R.); licia.viggiani@unibas.it (L.V.)

[2] Arbeitskreis Heimische Orchideen Baden-Württemberg, D-69469 Weinheim, Germany; lorenz@orchids.de

[*] Correspondence: maurizio.dauria@unibas.it; Tel.: +39-0971-205-480

Abstract: This study is part of a project devoted to determining the scent of all the orchid species present in Basilicata. All the analyses were performed by using the solid-phase microextraction technique coupled with gas chromatography-mass spectrometry. The scent of eight species belonging to the *Orchis* genus was investigated. In the case of *O. anthropophora*, caryophyllene, tetradecanal and hexadecanal were the main components of the aroma; in *O. purpurea*, 3,5-dimethoxytoluene and elemicin were found; in *O. italica*, caryophyllene and 4-(3-hydroxy-2-methoxyphenyl)butan-2-one were found; in *O. pauciflora*, linalool and 1,4-dimethoxybenzene were found; in *O. mascula*, linalool was found; in *O. quadripunctata*, penta- and heptadecane were found; in *O. provincialis*, β-farnesene and farnesal were found; and in *O. pallens*, curcumene was the main product.

Keywords: *Orchis*; scent; gas chromatography; mass spectrometry; solid-phase microextraction

Citation: Mecca, M.; Racioppi, R.; Romano, V.A.; Viggiani, L.; Lorenz, R.; D'Auria, M. Volatile Organic Compounds from *Orchis* Species Found in Basilicata (Southern Italy). *Compounds* **2021**, *1*, 83–93. https://doi.org/10.3390/compounds1020008

Academic Editor: Juan Mejuto

Received: 23 June 2021
Accepted: 18 August 2021
Published: 2 September 2021

Publisher's Note: MDPI stays neutral with regard to jurisdictional claims in published maps and institutional affiliations.

1. Introduction

Within the families of flowering plants, the Orchidaceae family is one of largest, with more than 28,000 species. The scent of some orchids has a relevant importance in perfume industries. However, the emission of volatile organic compounds from an orchid can have a relevant role in the life of the plant considering the possible effect of these compounds in attracting pollinators, or in defense against pathogens. One third of all the orchid species are food-deceptive species because the flowers do not contain nectar and the volatile organic compounds emitted mimic the floral signal of rewarding plants to attract pollinators. Furthermore, many compounds show antimicrobial and antifungal activities [1]. Volatile organic compounds are mainly terpenes, phenylpropanoid derivatives and fatty acid derivatives.

Some years ago, we started a project devoted to determining the floral scent of all the orchid species found in Basilicata (Southern Italy). The main feature of this study is the use of the same chemical method in order to determine the scent. Several methods can be used to determine VOCs. For example, FT-IR has been used to determine the composition of volatile mixtures [2,3], and, probably, GC-MS and FT-IR can be considered complementary methods in the analysis of complex mixtures [4]. We decided to use the solid-phase microextraction (SPME) procedure [5]. SPME analysis needs the exposure of a fiber, contained in the needle of a syringe, to a scent. The adsorbed components are then thermally desorbed into the injection sector of the gas chromatographic apparatus. The use of a single procedure allows obtaining a homogeneous dataset, also considering that SPME can suffer from the different absorption rates of the single components in the fiber [5]. This way, significant results were obtained in the characterization of the scent of *Platanthera bifolia* subsp. *osca* [6–8], *Platanthera chlorantha* [7,8], *Cephalanthera* orchids [9], *Serapias* orchids [8,10], *Gymnadenia* orchids [11], *Barlia robertiana* [8] and *Neotinea* orchids [12].

In this article, we want to continue with the realization of our project, showing the chemical composition of *Orchis* species found in Basilicata. This orchid genus is a very common one, where it is diffused in all of Europe, and in Basilicata, there are nine species, *Orchis anthropophora, O. italica, O. mascula, O. pallens, O. pauciflora, O. provincialis, O. purpurea, O. quadripunctata* and *O. simia*. In this study, we will report the results obtained in the determination of the scent in *O. anthropophora, O. purpurea, O. italica, O. pauciflora, O. mascula, O. quadripunctata, O. provincialis* and *O. pallens*.

In some of these species, some previous results have been reported in the literature. Nilsson reported a data headspace analysis of *O. mascula*, showing the presence of tricyclene (23.6%), α-pinene (15.6%) and *E*-ocimene (30.5%) as the main components, and linalool in a low quantity (2.4%) [13,14]. The same species, in a headspace analysis of the scent, showed the presence of limonene (8.37%), 1,8-cineole (11.74%), *E*-ocimene (23.25%) and linalool (11.89%) [15]. In a work where hexane extracts were considered, pentacosane (12.07%), heptacosane (46.00%) and nonacosane (27.97%) were determined by the same research group [16]. The same species has been analyzed using SPME (with PDMS-DVB fiber), showing the presence of limonene (11.67%), *E*-ocimene (26.68%) and linalool (13.15%) [17,18]. In an analysis of *O. mascula* through SPME (with PDMS-DVB fiber) where white and purple flowers were analyzed, the authors found in purple flowers (*Z*)-3-hexenyl acetate (6.46%), limonene (10.67%), *E*-ocimene (22.68%) and linalool (13.15%), while in the white flowers, the same compounds were found in a different ratio (12.20%, 12.88%, 16.30%, 3.46%) [19]. In *O. italica*, only one article reported the composition of the scent, obtained through hexane extraction, where tricosane (37.05%), pentacosane (16.88%), heptacosane (20.65%), nonacosane (8.14%) and 5-pentacosene (5.44%) were observed [16]. The same article also examined the scent of *O. provincialis*, showing the presence of pentacosane (12.07%), heptacosane (46.00%) and nonacosane (27.97%) [16]. Schiestl and Cozzolino also examined the scent of *O. quadripunctata* and found tricosane (14.86%), pentacosane (29.62%), heptacosane (32.51%) and nonacosane (9.21%) [16]. Headspace analysis of *O. pauciflora* showed in its scent nonanal (4.88%), 2-methyl-6-methylene-3,7-octadien-2-ol (30.49%), myrcene (25.87%) and *E*-ocimene (8.26%) [15]. Two articles were related to the analysis of the aroma components of *O. simia*: In the first one, where dynamic headspace analysis was performed, ethyl acetophenone (6.79%), α-pinene (32.68%), β-pinene (6.10%), sabinene (5.23%), myrcene (5.45%), eucalyptol (7.89%) and linalool (7.41%) were found [20]. In the second article, where SPME was used, nonanal (5.47%), (*Z*)-3-hexenyl acetate (3.21%), decanal (2.15%), α-pinene (11.49%), myrcene (7.12%), limonene (4.26%) and β-phellandrene (9.03%) were found [19]. The scent of *O. pallens* has been determined in a work where SPME (with Carbowax-PDMS fiber) was used; in this case, phenethyl alcohol, β-farnesene, α-farnesene and farnesol were determined [21]. Finally, in a headspace analysis of *O. anthropophora*, nonanal (9.46%), undecane (8.72%), benzeneacetaldehyde (4.87%), α-pinene (5.77%), limonene (5.22%), 1,8-cineole (7.49%), β-caryophyllene (22.33%) and caryophyllodienol (9.91%) were found as components of its aroma [22,23], while in a study where hexane extracts were examined, tricosane (10.82%), pentacosane (24.47%), heptacosane (26.92%), nonacosane (8.00%), 9-pentacosene (8.50%) and 9-heptacosene (7.68%) were found [16].

The above-reported data show that very different results can be obtained by using different GC-MS analytical methods able to characterize the aroma components, showing that the use of a homogenous method can provide valuable information on the scent of these species. In this work, the same HS-SPME-GC-MS method was used in order to characterize the scent of eight species of the *Orchis* genus.

2. Experimental Section

2.1. Plant Material

The sample of *Orchis anthropophora* was collected at Tolve (PZ) (359 m a.s.l.) on 18 April 2018. The sample of *Orchis italica* was collected at Tolve (PZ) (343 m a.s.l.) on 23 April 2018. The sample of *Orchis mascula* was collected at Sasso di Castalda (PZ)

(1090 m a.s.l.) on 22 May 2018. The sample of *Orchis pallens* was collected at Serra di Crispo at Terranova del Pollino (PZ) (1882 m a.s.l.) on 18 June 2018. The sample of *Orchis pauciflora* was collected at Madonna di Sasso at Sasso di Castalda (PZ) (1882 m a.s.l.) (1330 m a.s.l.) on 9 May 2018. The sample of *Orchis provincialis* was collected at Sasso di Castalda (PZ) (1069 m a.s.l.) on 30 April 2018. The sample of *Orchis purpurea* was collected at the campus of the University of Basilicata at Macchia Roma in Potenza (714 m a.s.l.) on 5 May 2018. The sample of *Orchis quadripunctata* was collected at Bosco Ralle at Satriano di Lucania (PZ) (1034 m a.s.l.) on 11 June 2018. The plants were collected by Vito Antonio Romano.

The plants were successively used for further studies on the impollination, fertility and germination of the plants. After these studies, the plants were not in condition to be collected in a herbarium. However, these species can be recognized without ambiguities on the basis of their properties, well documented in Figures 1 and 2.

Figure 1. (**a**) Orchis anthropophora; (**b**) Orchis purpurea; (**c**) Orchis italica; (**d**) Orchis pauciflora. Photos of V. A. Romano.

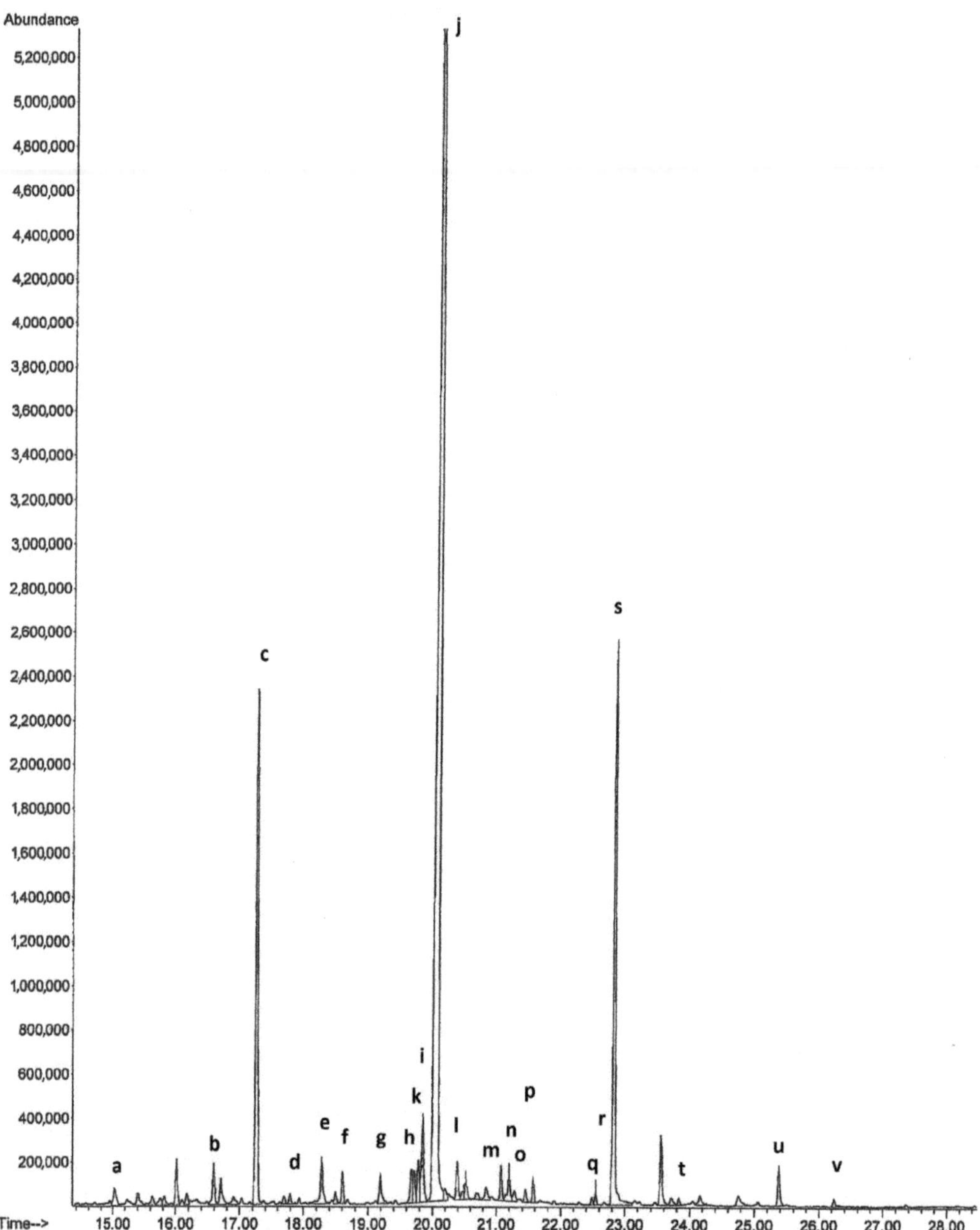

Figure 2. Chromatogram of volatile organic compounds from *Orchis anthropophora*: (a) tridecane; (b) tetradecane; (c) caryophyllene; (d) humulene; (e) pentadecane; (f) tridecanal; (g) 1-(1-methylethyl)-5-methyl-1,2,3,4-tetrahydronaphthalene; (h) 2-allyl-4-methylphenol; (k) ethyl dodecanoate; (i) hexadecane; (j) tetradecanal; (l) megastigmatrienone; (m) 4-(1-methylethyl)-1,6-dimethyl-1,2,3,4-tetrahydronaphthalene; (n) heptadecane; (o) pristane; (p) pentadecanal; (q) ethyl tetradecanoate; (r) octadecane; (s) hexadecanal; (t) nonadecane; (u) isopropyl palmitate; (v) heneicosane.

To prevent plant damage to the whole plant from a population in Basilicata, a large portion of soil all around the plant was removed from its habitat and placed in a greenhouse for a few days of acclimatization.

Following this period, for three days, the plant was placed under a bell jar. In view of the fact that the investigated taxa are rare wild plants, in order to preserve the species, we chose to use a single plant for our analysis.

2.2. Analysis of Volatile Organic Compounds

SPME [4] analysis of eight different samples of Orchis was performed. This way, the identified plants were collected and inserted in a glass jar for 24 h where a fiber (DVB/CAR/PDMS) and SPME syringe were also present. After this time, the fiber was desorbed in a gas chromatographic apparatus equipped with a quadrupole mass spectrometer detector. A 50/30 μm DVB/CAR/PDMS module with a 1 cm fiber (57328-U, Supelco, Milan, Italy) was employed to determine VOCs. The SPME fiber was maintained in the bell jar for 24 h. The analytes were desorbed in the splitless injector at 250 °C for 2 min. Analyses were accomplished with an HP 6890 Plus gas chromatograph equipped with a Phenomenex Zebron ZB-5 MS capillary column (30 m × 0.25 mm i.d. × 0.25 μm FT) (Agilent, Milan, Italy). An HP 5973 mass selective detector (Agilent) was utilized with helium at 0.8 mL/min as the carrier gas. The analyses were performed by using a splitless injector. The splitless injector was maintained at 250 °C, and the detector at 230 °C. The oven was held at 40 °C for 2 min, then gradually warmed, 8 °C/min, up to 250 °C and held for 10 min. Tentative identification of aroma components was based on mass spectra and Wiley 11 and NIST 14 library comparison. A single VOC peak was considered as identified when its experimental spectrum matched with a score over 90% present in the library. All the analyses were performed in triplicate.

To avoid contamination on the sample due, for example, to volatile organic compounds emitted from the soil, analysis of *Orchis anthropophora* was conducted on a single flower without the presence of soil, showing that this type of contamination does not exist. Otherwise, all the analyses were carried out by inserting the flowering plant in a glass bell jar and isolating the plant from the soil.

3. Results

Orchis anthropophora (Figure 1a) returned the results reported in Table 1 and Figure 2. The main volatile organic compounds detected were caryophyllene (11.32%), tetradecanal (57.17%) and hexadecanal (12.10%). Caryophyllene has a scent described as sweet, woody and terpenic, while the aroma of tetradecanal is described as fatty, waxy, amber, incense, citrus peel and musk.

The SPME analysis of *Orchis purpurea* (Figure 1b) showed that the main components of the aroma were aromatic compounds such as 3,5-dimethoxytoluene (35.29%) and elemicin (4.76%) (Table 1). The scent of elemicin is described as spicy and floral. When *Orchis italica* (Figure 1c) was examined, eucalyptol (3.93%), caryophyllene (47.29%) and 4-(3-hydroxy-2-methoxyphenyl)butan-2-one (4.96%) were the most abundant compounds found in the scent (Table 1). The scent of eucalyptol is recognized as eucalyptus, herbal, camphoreous and medicinal. *Orchis pauciflora* (Figure 1d) showed the presence of linalool (26.12%), 1,4-dimethoxybenzene (15.01%), germacrene D (9.34%) and 6,10,14-trimethyl-4,8,12-tetradecadrienal (3.60%). The scent of linalool is described as citrus, floral, sweet and woody, while that of 1,4-dimethoxybenzene is sweet, green and hay.

The SPME analysis of the scent of *Orchis mascula* (Figure 3a) showed the presence of eucalyptol (7.80%), linalool (21.28%), tetradecane (3.70%), pentadecane (4.41%) and 6,10,14-trimethyl-2-pentadecanone (Table 1 and Figure 4). *Orchis quadripunctata* (Figure 3b) had a scent composition where only hydrocarbons were found. Thus, tridecane (9.70%), tetradecane (9.38%), pentadecane (26.85%) and heptadecane were the main components of the scent (Table 1). In the case of *Orchis provincialis* (Figure 3c), the volatile organic compounds found in the SPME analysis were β-farnesene (44.16%), 3,7,11-trimethyl-

2,6,10-dodecatrienal (29.25%) and 6,10,14-trimethyl-2-pentadecanone (6.32%) (Table 1). β-Farnesene's scent is described as woody, citrus, herbal and sweet. Finally, the aroma components of *Orchis pallens* (Figure 3d) were α-zingiberene (14.67%), di-*epi*-α-cedrene (10.64%), β-curcumene (33.29%) and diethyltoluamide (13.95%) (Table 1). α-Zingiberene has a scent described as spicy, fresh and sharp, while that of di-*epi*-α-cedrene is described as woody, cedar, sweet and fresh.

Figure 3. (**a**) Orchis mascula; (**b**) Orchis quadripunctata; (**c**) Orchis provincialis; (**d**) Orchis pallens. Photos of V. A. Romano.

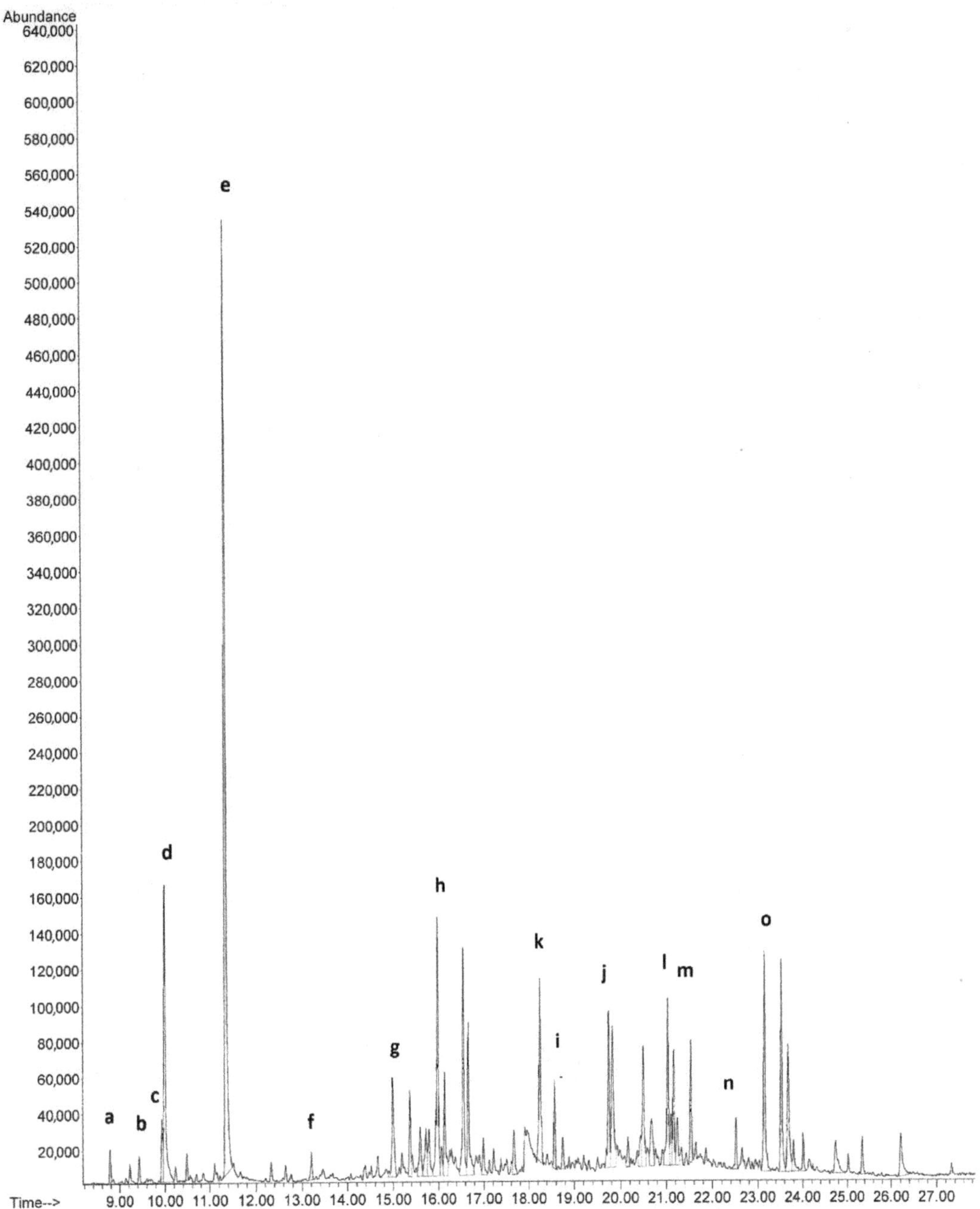

Figure 4. Chromatogram of volatile organic compounds from *Orchis mascula*; (a) β-terpinene; (b) 2,2,4,6,6-pentamethyl-3-heptene; (c) limonene; (d) eucalyptol; (e) linalool; (f) dodecane; (g) tridecane; (h) tetradecane; (k) germacrene D; (i) β-sesquiphellandrene; (j) hexadecane; (l) heptadecane; (m) pristane; (n) octadecane; (o) 6,10,14-trimethyl-2-pentadecanone.

Table 1. SPME-GC-MS analysis of *Orchis* species.

Compound	r.t. (min.)	KI	Area (%) ± 0.03							
			Species							
			O. anthropophora	*O. purpurea*	*O. italica*	*O. pauciflora*	*O. mascula*	*O. quadripunctata*	*O. provincialis*	*O. pallens*
Limonene	9.65	1039				0.72	1.26		1.52	
Eucalyptol	10.02	1042			3.93	1.95	7.80	7.80	0.95	
β-terpinene	10.20	1071					0.63			
Linalool	11.35	1110				26.12	21.28			
1,4-Dimethoxybenzene	12.62	1163				15.01				
α-Terpineol	13.17	1198							1.42	
Dodecane	13.20	1200					0.59			
3,5-Dimethoxytoluene	14.04	1264		35.29						
Tridecane	15.04	1300	0.54			1.26	2.88	9.70		
Caproleic acid	15.90	1358		0.91						
2,4,4,6,6,8,8-Heptamethyl-2-nonene	16.09	1363								1.34
1,2,4-Trimethoxybenzene	16.30	1378				2.53				
α-Cubebene	16.45	1386				0.74				
Tetradecane	16.71	1400	0.76			2.46	3.70	9.38	0.45	1.55
Dodecanal	16.75	1409		0.45						
β-Ylangene	17.17	1422				0.81				
cis-α-Bergamotene	17.22	1441							0.29	
Caryophyllene	17.27	1446	11.32	0.53	47.29					1.85
α-Zingiberene	17.32	1448								14.67
β-Farnesene	17.62	1445				1.69			44.16	2.76
Humulene	17.80	1452	0.20							
Epi-Bicyclosesquiphllandrene	17.89	1460				1.31				
2,6-Di-*t*-butylbenzoquinone	17.92	1469						3.28		
Germacrene D	18.18	1482				9.34				
Pentadecane	18.29	1500	1.27	0.71	2.21	2.80	4.41	26.85		3.02
Di-*epi*-α-Cedrene	18.62	1503								10.64
Tridecanal	18.50	1510	0.26							
β-Curcumene	18.57	1517							0.73	33.29
Methyl dodecanoate	18.65	1521							0.35	
β-Sesquiphellandrene	18.75	1543					0.71			
Elemicin	19.08	1550		4.76						
1-(1-Methylethyl)-5-methyl-1,2,3,4-tetrahydronaphthalene	19.19	1565	0.82							
Nerolidol	19.32	1567								
Diethyltoluamide	19.54	1571							0.98	
Methyl 3,5-dimethoxybenzoate	19.60	1574			2.63					13.95
1(10),5-Germacradien-4-ol	19.62	1579				0.49				
Ethyl dodecanoate	19.72	1592	0.69		2.29	1.63			0.98	
Hexadecane	19.78	1600	1.01		2.27	1.64	3.47	7.78	0.83	2.91

Table 1. *Cont.*

Compound	r.t. (min.)	KI	Area (%) ± 0.03							
			Species							
			O. anthropophora	*O. purpurea*	*O. italica*	*O. pauciflora*	*O. mascula*	*O. quadripunctata*	*O. provincialis*	*O. pallens*
Tetradecanal	20.07	1611	57.17	0.89						
Benzophenone	20.28	1621		0.75						
Megastigmatrienone	20.38	1631	0.82							
4-(3-Hydroxy-2-methoxyphenyl)butan-2-one	20.69	1640			4.96					
Heptadecane	21.20	1700	0.91	1.44	3.48	1.24	2.44	10.63	1.55	2.19
Pristane	21.28	1705	0.31				1.26			1.55
Pentadecanal	21.45	1710	0.29							
Farnesal	21.53	1730							29.25	
Tetradecanoic acid	21.93	1758		0.89						
Benzyl benzoate	22.15	1765		2.23						
Ethyl tetradecanoate	22.49	1790	0.13			0.36			0.50	
Octadecane	22.55	1800	0.22	1.12	1.38	0.41	1.06	3.48	0.69	1.23
Isopropyl myristate	22.77	1816		0.73						
Hexadecanal	22.83	1820	12.10							
5,9,13-Trimethyl-4,8,12-tetradecatrienal	23.01	1855		1.22						
6,10,14-Trimethyl-2-pentadecanone	23.14	1864				3.60	5.78		6.32	1.71
Nonadecane	23.83	1900	0.15	1.72	1.97				0.34	
Hexadecanoic acid	24.49	1958		0.65						
Eicosane	24.94	2000		1.22					0.38	
Isopropyl palmitate	25.38	2023	0.89		1.97				0.94	
Heinecosane	26.24	2100	0.19		2.88				0.67	
Nonadecanol	26.88	2172		0.41						
Docosane	27.23	2200		0.69					0.32	
Tricosane	28.42	2300			1.44					

4. Discussion

It is interesting to note the large differences between our reported results and those reported in the Introduction section. For *O. anthropophora*, two different analyses are available [16,20]. While in the work of Cozzolino [16], only hydrocarbons with an extremely high molecular weight were found, the other article [20] reported that β-caryophyllene was the main component. In our study, β-caryophyllene was present, but the main component was tetradecanal. In the case of *O. purpurea*, no other results on the composition of the scent are available. For *O. italica*, only an article published by Cozzolino is available [16], and, also in this case, only high-molecular weight hydrocarbons were found. In our experiment, on the contrary, β-caryophyllene was the main component of the scent.

The scent of *O. pauciflora* has been determined through headspace analysis, showing the presence of 2-methyl-6-methylene-3,7-octadiene-2-ol as the main component [15]. However, in our analysis, linalool and 1,4-dimethoxybenzene were found as the main components. *O. mascula* was the object of an intense study where several different analytical methods were used. This way, headspace analysis returned *E*-ocimene as the main component of the scent [13,14]. This result was confirmed by SPME analysis [15,17]. In our analysis, as it is evident considering Figure 4, linalool was the main component of the aroma. For *O. quadripunctata*, the work of Schiestl and Cozzolino found only hydrocarbons [16]. Only hydrocarbons were found in this work, but with a significant difference in the molecular weight of the detected compounds. In the case of *O. provincialis*, the work of Schiestl and Cozzolino determined only the presence of hydrocarbons [17], while the presence of relevant amounts of β-farnesene was determined in this study. Finally, while an SPME analysis of the scent of *O. pallens* found phenethyl alcohol, farnesene and farnesol [21], our analysis of the same species found β-curcumene as the main component.

5. Conclusions

This work shows the analysis of *Orchis* samples from Basilicata. The analyses were performed by using the same procedure and the same fiber in SPME-GC-MS, which allowed achieving a homogenous dataset. The analyses showed different scent compositions from those determined on samples deriving from different sites. These observed differences, when SPME of other headspace techniques is used, can depend both on the different absorption rates of the analytes on the fiber and on the variation in the scent due to natural adaptation of the plant to different environmental conditions, due, for example, to different pollination insects. A completely different consideration can be found in the work of Schiestl and Cozzolino, where a completely different analytical method was used (hexane extraction of labellum, and GC-MS analysis of the extracts). In their case, only hydrocarbons were determined. Probably, their analytical procedure was not the correct method for the determination of the orchid scent.

Author Contributions: Conceptualization, M.D. and V.A.R.; methodology, R.R.; investigation, R.R., L.V. and M.M.; data curation, R.L.; writing—original draft preparation, M.D. and V.A.R.; writing—review and editing, M.D., V.A.R. and R.L. All authors have read and agreed to the published version of the manuscript.

Funding: This research received no external funding.

Data Availability Statement: Not applicable.

Conflicts of Interest: The authors declare no conflict of interest.

References

1. Ramya, M.; Jang, S.; An, H.-R.; Lee, S.-Y.; Park, P.-M.; Park, P.H. Volatile Organic Compounds from Orchids: From Synthesis and Function to Gene Regulation. *Int. J. Mol. Sci.* **2020**, *21*, 1160. [CrossRef]

2. Apolonski, A.; Maiti, K.S. Towards a standard operating procedure for revealing hidden volatile organic compounds in breath: The Fourier-transform IR spectroscopy case. *Appl. Opt.* **2021**, *60*, 4217–4224. [CrossRef] [PubMed]

3.   Maiti, K.S.; Lewton, M.; Fill, E.; Apolonskiy, A. Sensitive spectroscopic breath analysis by water condensation. *J. Breath Res.* **2018**, *12*, 046003. [CrossRef] [PubMed]

4.   Maiti, K.S.; Lewton, M.; Fill, E.; Apolonski, A. Human beings as islands of stability: Monitoring body states using breath pro-files. *Sci. Rep.* **2019**, *9*, 16167. [CrossRef] [PubMed]

5.   Pawliszyn, J. *Solid-Phase Microextraction: Theory and Practice*; VCH: New York, NY, USA, 1997.

6.   D'Auria, M.; Racioppi, R. Characterization of the volatile fraction of mastic oil and mastic gum. *Nat. Prod. Res.* **2020**, 1–4. [CrossRef]

7.   D'Auria, M.; Lorenz, R.; Racioppi, R.; Romano, V.A. Fragrance components of Platanthera bifolia subsp. osca. *Nat. Prod. Res.* **2017**, *31*, 1612–1619. [CrossRef]

8.   D'Auria, M.; Lorenz, R.; Mecca, M.; Racioppi, R.; Romano, V.A.; Viggiani, L. Fragrance components of Platanthera bifolia subsp. osca and Platanthera chlorantha collected in several sites in Italy. *Nat. Prod. Res.* **2020**, *34*, 2857–3861. [CrossRef]

9.   D'Auria, M.; Fascetti, S.; Racioppi, R.; Romano, V.A.; Rosati, L. Orchids from Basilicata: The Scent. In *Reference Series in Phytochemistry*; Springer Science and Business Media LLC: Berlin, Germany, 2020; pp. 1–22.

10.   D'Auria, M.; Lorenz, R.; Mecca, M.; Racioppi, R.; Romano, V.A. Aroma components of Cephalanthera orchids. *Nat. Prod. Res.* **2021**, *35*, 174–177. [CrossRef]

11.   D'Auria, M.; Lorenz, R.; Mecca, M.; Racioppi, R.; Romano, V.A. The composition of the aroma of Serapias orchids in Basilicata (Southern Italy). *Nat. Prod. Res.* **2020**, 1–5. [CrossRef]

12.   D'Auria, M.; Lorenz, R.; Mecca, M.; Racioppi, R.; Romano, V.A.; Viggiani, L. Fragrance components of Gymnadenia conopsea and Gymnadenia odoratissima collected at several sites in Italy and Germany. *Nat. Prod. Res.* **2020**, 1–5. [CrossRef]

13.   D'Auria, M.; Lorenz, R.; Mecca, M.; Racioppi, R.; Romano, V.A.; Viggiani, L. The scent of Neotinea orchids from Basilicata (Southern Italy). *Nat. Prod. Res.* **2021**, 1–3. [CrossRef]

14.   Nilsson, L.A. Anthecology of Orchis mascula (Orchidaceae). *Nord. J. Bot.* **1983**, *3*, 157–179. [CrossRef]

15.   Jacquemyn, H.; Brys, R.; Honnay, O.; Hutchings, M.J. Biological Flora of the British Isles:Orchis mascula(L.) L. *J. Ecol.* **2009**, *97*, 360–377. [CrossRef]

16.   Salzmann, C.C.; Cozzolino, S.; Schiestl, F.P. Floral scent in food-deceptive orchids: Species specificity and sources of variabil-ity. *Plant Biol.* **2007**, *9*, 720–729. [CrossRef]

17.   Schiestl, F.P.; Cozzolino, S. Evolution of sexual mimicry in the orchid subtribe orchidinae: The role of preadaptations in the attraction of male bees as pollinators. *BMC Evol. Biol.* **2008**, *8*, 27. [CrossRef] [PubMed]

18.   Dormont, L.; Delle-Vedove, R.; Bessière, J.-M.; Hossaert-Mc Key, M.; Schatz, B. Rare white-flowered morphs increase the re-productive success of common purple morphs in a food-deceptive orchid. *New Phytol.* **2010**, *185*, 300–310. [CrossRef] [PubMed]

19.   Dormont, L.; Fort, T.; Bessière, J.-M.; Proffit, M.; Hidalgo, E.G.; Buatois, B.; Schatz, B. Sources of floral scent variation in the food-deceptive orchid Orchis mascula. *Acta Oecol.* **2020**, *107*, 103600. [CrossRef]

20.   Dormont, L.; Delle-Vedove, R.; Bessière, J.-M.; Schatz, B. Floral scent emitted by white and coloured morphs in orchids. *Phytochemistry* **2014**, *100*, 51–59. [CrossRef]

21.   Schatz, B.; Geoffroy, A.; Dainat, B.; Bessière, J.-M.; Buatois, B.; Hossaert-Mckey, M.; Selosse, M.-A. A case study of modified interactions with symbionts in a hybrid mediterranean orchid. *Am. J. Bot.* **2010**, *97*, 1278–1288. [CrossRef]

22.   Barták, P.; Bednář, P.; Čáp, L.; Ondráková, L.; Stránský, Z. SPME-A valuable tool for investigation of flower scent. *J. Sep. Sci.* **2003**, *26*, 715–721. [CrossRef]

23.   Jacquemyn, H.; Brys, R.; Hutchings, M.J. Biological flora of the British Isles: Orchis anthropophora (L.) All. (Aceras anthropopho-rum (L.) W.T. Aiton). *J. Ecol.* **2011**, *99*, 1551–1565. [CrossRef]

Article

Estimation of Avocado Oil (*Persea americana* Mill., Greek "Zutano" Variety) Volatile Fraction over Ripening by Classical and Ultrasound Extraction Using HS-SPME–GC–MS

Marinos Xagoraris [1], Eleni Galani [1], Lydia Valasi [1], Eleftheria H. Kaparakou [1] Panagiota-Kyriaki Revelou [1,2], Petros A. Tarantilis [1] and Christos S. Pappas [1,*]

[1] Laboratory of Chemistry, Department of Food Science and Human Nutrition, Agricultural University of Athens, 75 Iera Odos, 11855 Athens, Greece; mxagor@aua.gr (M.X.); stud516036@aua.gr (E.G.); lydia.valasi@aua.gr (L.V.); elek@aua.gr (E.H.K.); p.revelou@aua.gr (P.-K.R.); ptara@aua.gr (P.A.T.)

[2] Department of Food Science and Technology, University of West Attica, Ag. Spyridonos Str., Egaleo, 12243 Athens, Greece

* Correspondence: chrispap@aua.gr; Tel.: +30-2105294262

Abstract: The study of flavors and fragrances is a topic of rising interest from both marketing and scientific perspectives. Over the last few years, the cultivation of avocados has accelerated in Greece, with production levels elevated by 300%. There has been increasing attention from a number of growers and consumers on avocado oil, the volatiles of which form a key part of consumers' purchase decisions. A previously unevaluated Zutano cultivar was chosen for this study. Extraction of the pulp oil was performed during three phases of ripening using Soxhlet and ultrasound techniques. Headspace-solid-phase microextraction (HS-SPME) and gas chromatography–mass spectrometry (GC–MS) were utilized in order to analyze the isolated volatile fraction. At least 44 compounds, including mainly terpenoids (61.7%) and non-terpenoid hydrocarbons (35.9%), presented in the Zutano variety, while (1*S*,6*S*,7*S*,8*S*)-1,3-dimethyl-8-propan-2-yltricyclo[4.4.0.02,7]dec-3-ene (a-copaene) and (1*R*,9*S*,Z*)-4,11,11-trimethyl-8-methylenebicyclo[7.2.0]undec-4-ene (β-caryophyllene) were in higher abundance. The composition of the volatiles was unaffected by the extraction techniques but was influenced by the ripening stage. Thus, during maturation, the volatile fraction fluctuates, with a significantly higher abundance of terpenoids during the fourth day of the ripe stage, whilst it decreases during over-ripening. These findings demonstrate that the Zutano variety can be used to produce an aromatic oil and hence could be used, among others, as an ingredient in cosmetic products.

Keywords: Zutano variety; avocado oil; Soxhlet extraction; ultrasound-assisted extraction; volatiles; ripening; over-ripe; HS-SPME–GC–MS

Citation: Xagoraris, M.; Galani, E.; Valasi, L.; Kaparakou, E.H.; Revelou, P.-K.; Tarantilis, P.A.; Pappas, C.S. Estimation of Avocado Oil (*Persea americana* Mill., Greek "Zutano" Variety) Volatile Fraction over Ripening by Classical and Ultrasound Extraction Using HS-SPME–GC–MS. *Compounds* 2022, 2, 25–36. https://doi.org/10.3390/compounds2010003

Academic Editor: Juan Mejuto

Received: 6 December 2021
Accepted: 31 December 2021
Published: 17 January 2022

Publisher's Note: MDPI stays neutral with regard to jurisdictional claims in published maps and institutional affiliations.

1. Introduction

The avocado (*Persea americana* Mill.) is a subtropical/tropical tree which is traditionally cultivated in Central America [1]. The growth of the avocado fruit depends on the cultivar and/or environmental conditions, and it is harvested when the fruit is horticulturally mature [1]. The production of avocados has become more widespread worldwide in the last few decades, with seven million tons produced in 2019 [2]. In Greece, avocados are mainly produced in Crete, Peloponnese, and Rhodes, with an average of 9.3 tons produced in 2019 [2]. It is an emblematic crop for Crete, with 90% of the total Greek production concentrated in the prefecture of Chania. Furthermore, it is notable that production increased by about 300% between 2014 and 2019 [3].

Avocado oil is gaining considerable attention from business and research sectors, demonstrated by the expanding literature. It is renowned for its uses in cosmetics, the food processing industries, and edible oil [4,5]. Avocado oil is obtained from the fleshy mesocarp (pulp) of the fruit and has a high nutrition value, biological effects, and healing properties [4,6,7].

The process of avocado oil extraction is quite similar to olive oil extraction. Generally, the extra virgin oil is extracted from high-quality fruit with minimal levels of rot and physiological disorders, while some rots or physiological disorders are permitted in virgin oil. In each case, extraction is carried out using only mechanical methods at low temperatures (<50 °C). For pure avocado oil, quality is not important, with low acidity, color, and a bland flavor, and for mixed avocado oil, blends with other oils are allowed. According to Woolf, [8] avocado oil might be classified as "extra virgin", "virgin", "pure", or "mixed". However, no global standardized physicochemical measurements have been implemented for such a categorization. The composition, quality, and yield of avocado oil are dependent on several factors, including fruit variety [9–11], harvesting time, and ripening stage [10,12,13]. The majority of scientists have focused their endeavors on assessing the "Hass" and "Fuerte" varieties or others with commercial value, while the "Zutano" variety has been studied by few researchers [14], i.e., there is a dearth of data regarding the cultivation of this variety within the Mediterranean. A further feature of the avocado is that if the fruit were to remain on the tree, ripening would not occur. Ripening takes place over a period of time, i.e., between 3–4 and 18–21 days following harvest [8]. The time duration is impacted by storage parameters, amongst additional extrinsic factors. Over-ripening may also occur. Different extraction techniques, conditions, and solvents are factors determining the avocado oil quality and yield [4,5]. An overview of extraction methods used in the last 20 years includes mainly liquid extraction using Soxhlet apparatus [15–27], homogenization [26,28], and microwave-assisted extraction (MAE) [19,25,27]. Several studies have focused on supercritical fluids [15,16,19–22,29] and mechanical extraction by cold pressure [17,18,30,31]. Lesser-used methods include extraction by enzymes [32] and ultrasound-assisted extraction (UAE) [15,16,19,30].

Volatile compounds of avocado oil have a well-established role in aroma profiles and are one of the most important characteristics of the quality of the product. The combination of different volatile compounds forms the aroma character of avocado oil [33,34]. Some studies demonstrate that volatile compounds may differ depending on the variety [35,36], oil extraction solvents, or treatments [25,37]. One possible influence on the aromatic profile of avocado oil is the ripening phase, which, to date, has been poorly investigated.

In the current work, the volatiles from Greek Zutano avocado oil were acquired during three ripening phases, i.e., breaking, ripe, and overripe. Two diverse extraction methods were employed, i.e., Soxhlet extraction (SE) and ultrasound-assisted extraction (UAE). Headspace-solid-phase microextraction (HS-SPME) and gas chromatography–mass spectrometry (GC–MS) techniques were utilized to characterize the respective volatile fractions obtained.

2. Materials and Methods

2.1. Avocado Fruit Samples

Avocado fruits (*Persea americana* Mill., "Zutano" variety) were provided directly from producers at commercial maturity (firm) during the 2020 harvest year. The fruits were located in the Greek island of Crete (35°28′28.1″ N, 23°56′53.3″ E). The samples were stored in the dark at ambient temperature (24 ± 1 °C) for one day (breaking), four days (ripe), and eight days (overripe) (Figure 1). Then, samples were cut and lyophilized by freeze drying on a VirTis Freezemobile 25EL (SP Industries, 935 Mearns Rd, Warminster, PA, USA) to remove the water, and the solid residue was stored for 24 h at −20 ± 1 °C until oil extraction. Moreover, the percentage (% w/w) of dry matter (>19% w/w) was calculated according to Greek legislation for avocado commercial standards [38].

Figure 1. Ripening stages of avocado fruit (*Persea americana* Mill., Greek "Zutano" variety).

2.2. Avocado Oil Extraction

Avocado oils were extracted by classical SE and UAE techniques. The AOAC Official Method 948.22 was applied for the SE with some modifications. Approximately 10 g of avocado pulp powder was mixed with 625 mL petroleum ether (purity 99.0%) in a Soxhlet apparatus for 6 h at 50 °C. UAE was performed in a Grant ultrasonic water bath (Grant Instruments Ltd., Cambridge, UK) ($300 \times 140 \times 150$ mm^3 internal dimensions) at the fixed frequency of 35 kHz. Approximately 10 g of avocado pulp powder was mixed with 80 mL of petroleum ether in an Erlenmeyer flask for 30 min at 25 °C. The organic solvent of each extract was totally evaporated under reduced pressure at 35 °C using a Laborota 4000 efficient rotary evaporator (Heidolph Instruments GmbH & Co. KG, Schwabach, Germany). The previous procedure was performed in triplicate and the received oily extracts were refrigerated at -20 ± 1 °C in a totally filled storing flask until GC analysis.

2.3. Isolation and Analysis of Avocado Oil Volatile Fraction

The isolation and analysis of the volatile compounds were performed using HS-SPME–GC–MS according to Xagoraris with few modifications [39]. An amount of 4 g of avocado oil alongside 1 μL of β-ionone (Alfa Aesar, Ward Hill, MA, USA) were placed into a 15 mL screw-top glass vial with PFTE/silicone septa. The vials were equilibrated for 30 min in a water bath at 60 °C under stirring at 700 rpm. Subsequently, the HS-SPME procedure was carried out using a triple-phase divinylbenzene/carboxen/polydimethylsiloxane (DVB/CAR/PDMS) fiber 50/30 μm needle (Supelco, Bellefonte, PA, USA) with a length of 1 cm. The needle was inserted into the vial and exposed to the headspace for 30 min.

The analysis of volatile compounds was performed using a Thermo GC-Trace ultra, coupled with a Thermo mass spectrometer DSQ II (Thermo Fisher Scientific Inc., Waltham, MA, USA). The GC inlet temperature was 260 °C in splitless mode for 3 min with a 0.8 mm injector liner (SGE International Pty Ltd., Ringwood, Australia). The column used was a Restek Rtx-5MS (30 m $\times$ 2.25 mm i.d., 0.25 μm film thickness) (Restek, Bellefonte, PA, USA). The carrier gas was helium at a 1 mL·min^{-1} flow rate. The column was maintained at 40 °C, held for 6 min, then heated to 120 °C at a rate of 5 °C·min^{-1}, then heated to 160 °C at a rate of 3 °C·min^{-1}, then heated to 250 °C at a rate of 15 °C·min^{-1} and held at

250 °C for 1 min [39]. The temperature conditions of the mass spectrometer were: transfer line (290 °C), source (240 °C), and quadrupole (150 °C). Electron impact was 70 eV, and mass spectra were recorded at the 35–650 mass range. Retention index (RI) values were calculated using n-alkane (C8-C20) standards (Supelco, Bellefonte, PA, USA). The peak identification was carried out with the Wiley 275 mass spectra library and masses spectral data and arithmetic index provided by Adams [40]. Quantification of volatile compounds was accomplished by dividing the peak areas of the compounds by the peak area of the internal standard (β-ionone) and multiplying this ratio by the initial concentration of the internal standard.

2.4. Statistical Analysis

All chromatographic data were acquired by analysis of variance (ANOVA) and multivariate analysis of variance (MANOVA) using the SPSS v.25 (IBM, SPSS, Statistics) software. The mean values were calculated in Microsoft Excel 2013.

3. Results and Discussion

3.1. Estimation of Avocado Oil Yield

SE and UAE are classical methods that are commonly used for avocado oil extraction [7]. These processes have total solvent penetration into the oil membranes of avocado fruit [7]. Furthermore, these have been widely used to determine the theoretical maximum oil yield of avocado [41]. However, some factors, including variety, drying method, organic solvent, and temperature, should be taken into account in oil recovery. For each instance, the current data concur with previous publications [7]. Avocado oils were weighed to measure the oily mass, and all yields were calculated from 46.76 to 66.22% (*w/w*) for SE and 31.97 to 54.54% (*w/w*) for the UAE method. Similarly to our results, in a previous study, SE produced a 64.76% (*w/w*) oil yield and UAE produced a 54.63% (*w/w*) oil yield [19].

3.2. Volatile Compounds Analysis

The volatile compounds and their semi-quantification in avocado oil are expressed as average values and are summarized in Table 1. The identified fraction was characterized by at least 44 components, including terpenoids, hydrocarbons, aldehydes, and ketones. Terpenoids were the dominant fraction of volatiles, with an average relative abundance of 61.7%, whilst hydrocarbons (non-terpenoids) were 35.9%. Thus, the avocado oil fragrance from the Zutano cultivar was characterized by many terpenoids with high abundance. The results show that the volatile fraction of this variety of oil is rich in terpenoids.

As reported previously by Tan [5], the quality and quantity of volatile compounds detected in avocado oil are affected by several factors such as variety, extraction conditions (e.g., organic solvent, temperature, time), and analytical technique (isolation or analysis parameters). Avocado oil obtained from *P. americana* Mill. sourced from a Mexico City regional market was studied by Moreno [25], who identified 36 volatile substances using four diverse extraction methods. It should be noted that the largest number of compounds (15 volatiles) were identified using microwaves and Soxhlet and hexane as extractors. Furthermore, in a recent study by Liu [37], 40 volatile compounds were detected using different extraction methods (squeezing, supercritical carbon dioxide, and aqueous). Nineteen volatile materials were identified by Bukykkurt [33] in cold-pressed avocado oil grown in regions of Turkey.

A typical chromatogram of avocado oil's ultrasound technique is presented in Figure 2. However, no qualitative changes were observed among the chromatograms that emerged from different extraction techniques or ripening stages. The total ion chromatograms reveal the capture of seven peaks that characterize the dominant volatile profile of avocado oil from Zutano cultivar. These peaks include 2,6,6-trimethylbicyclo[3.1.1]hept-2-ene (α-pinene); 1-methyl-4-(prop-1-en-2-yl)cyclohex-1-ene (D-limonene); 4,10-dimethyl-7-propan-2-yltricyclo[4.4.0.0^{1,5}]dec-3-ene (α-cubebene); (1*S*,6*S*,7*S*,8*S*)-1,3-dimethyl-8-propan-2-yltricyclo[4.4.0.0^{2,7}]dec-3-ene (a-copaene); (1*R*,9*S*,*Z*)-4,11,11-trimethyl-8-methylenebicyclo[7.2.0]undec-4-ene (β-caryophyllene); 2,6-dimethyl-6-(4-methylpent-3-enyl)bicyclo[3.1.1]hept-2-ene (α-bergamotene); and (1*E*,4*E*,8*E*)-2,6,6,9-tetramethylcycloundeca-1,4,8-triene (humulene). Similarly, Buyukurt [33] identified D-limonene, α-cubebene, β-caryophyllene, and β-curcumene as the most abundant compounds in avocado oil, while Moreno [25] confirmed the above results.

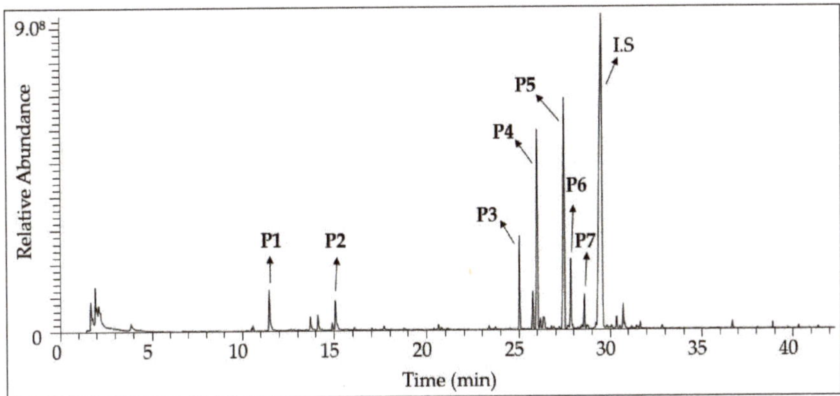

Figure 2. A characteristic gas chromatogram of avocado oil (*Persea americana* Mill., Greek "Zutano" variety) from UAE. (**P1**) 2,6,6-trimethylbicyclo[3.1.1]hept-2-ene (α-pinene); (**P2**) 1-methyl-4-(prop-1-en-2-yl)cyclohex-1-ene (D-limonene); (**P3**) 4,10-dimethyl-7-propan-2-yltricyclo[4.4.0.0^{1,5}]dec-3-ene (α-cubebene); (**P4**) (1*S*,6*S*,7*S*,8*S*)-1,3-dimethyl-8-propan-2-yltricyclo[4.4.0.0^{2,7}]dec-3-ene (a-copaene); (**P5**) (1*R*,9*S*,*Z*)-4,11,11-trimethyl-8-methylenebicyclo[7.2.0]undec-4-ene (β-caryophyllene); (**P6**) 2,6-dimethyl-6-(4-methylpent-3-enyl)bicyclo[3.1.1]hept-2-ene (α-bergamotene); (**P7**) (1*E*,4*E*,8*E*)-2,6,6,9-tetramethylcycloundeca-1,4,8-triene (humulene); (**I.S**) Internal Standard.

3.3. Estimation of Volatiles over Extraction Method

Although the SE and UAE techniques provided different oil yields, the analysis of the volatile compounds by HS-SPME–GC–MS showed similar results. In both techniques, a solvent is used for extraction, which permeates the oily cells to engage with the lipid compounds [41]. The main difference between these techniques was the time and temperature of extraction. In SE, the avocado fruit is repeatedly brought into contact with an organic solvent in a relatively mid–high temperature of 50 °C for 6 h, whereas in UAE, the avocado fruit has been used for accelerated extraction [30] in a relatively controlled mid–low temperature of 25 °C for 15 min.

The extraction techniques did not have statistically significant differences between their qualification and quantification results. Only three components including octan-2-one, (6*Z*)-7,11-dimethyl-3-methylidenedodeca-1,6,10-triene (β-farnesene), and (1*S*,2*S*,4*R*)-1-ethenyl-1-methyl-2,4-bis(prop-1-en-2-yl)cyclohexane (β-elemene) could potentially vary between the SE and UAE methods. However, an analysis of the variance indicated that the *p*-values were higher than 0.05 for all the volatile compounds.

It is evident that the mass transfer and molecular affinity amongst petroleum ether and the targeted compounds exerted a higher influence on acquiring and retaining the volatiles. In this context, comparing the advantages and drawbacks between SE and UAE, the latter appeared more cost-effective and environmentally advantageous.

Table 1. Volatile compounds isolated from headspace of avocado oil (mg·kg^{-1}).

Volatile Compounds	CAS Number	RT [a]	RI [b]	Soxhlet			UAE		
				Breaking	Ripe	Overripe	Breaking	Ripe	Overripe
Hydrocarbons (Non Terpenoids)									
1-ethyl-2-methylcyclohexane	3728-54-9	9.7	883	0.20	0.15	0.21	0.20	0.00	0.20
(1*R*,3*S*)-1-ethyl-3-methylcyclohexane	3728-55-0	9.8	887	0.33	0.15	0.19	0.33	0.01	0.22
nonane	111-84-2	10.2	897	1.34	0.74	1.02	1.34	0.04	1.02
propylcyclohexane	1678-92-8	11.2	922	0.20	0.06	0.10	0.20	0.02	0.30
2,6-dimethyloctane	2051-30-1	11.6	932	0.93	0.15	0.85	0.93	0.00	0.54
3-ethyl-2-methylheptane	14676-29-0	11.8	937	0.78	0.43	0.82	0.78	0.00	0.70
1,1,2,3-tetramethylcyclohexane	6783-92-2	12.3	951	0.19	0.04	0.24	0.46	0.00	0.25
4-ethyloctane	15869-86-0	12.4	953	0.40	0.16	0.48	0.40	0.01	0.36
4-methylnonane	17301-94-9	12.7	961	0.74	0.30	0.95	0.74	0.07	0.69
2-methylnonane	871-83-0	12.8	964	0.85	0.28	1.00	0.85	0.06	0.70
3-methylnonane	5911-04-6	13.0	970	1.19	0.47	1.62	1.19	0.04	1.07
1-methyl-2-propylcyclohexane	4291-79-6	13.6	986	0.86	0.32	1.64	0.86	0.06	1.32
decane	124-18-5	14.2	1001	4.45	1.80	6.32	4.47	0.51	5.08
butylcyclohexane	1678-93-9	15.2	1032	0.57	0.00	0.50	0.57	0.00	0.40
dodecane	112-40-3	20.6	1200	0.19	0.06	0.07	0.19	0.15	0.10
(1*S*,2*S*,3*R*,4*S*,6*R*,7*R*,8*S*)-1,2-dimethyl-8-propan-2-yltetracyclo[4.4.0.0^{2,4}.0^{3,7}]decane (cyclosativene)	22469-52-9	25.8	1369	0.64	0.81	0.54	0.64	1.01	0.62
(1*S*,2*S*,4*R*)-1-ethenyl-1-methyl-2,4-bis(prop-1-en-2-yl)cyclohexane (β-elemene)	515-13-9	26.3	1387	0.36	0.44	0.25	0.44	0.67	0.35
tetradecane	629-59-4	26.8	1401	0.10	0.01	0.10	0.10	0.08	0.10
10,10-dimethyl-2,6-dimethylenebicyclo[7.2.0]undecane	136296-38-3	27.7	1427	0.10	0.20	0.10	0.10	0.17	0.13
Terpenoids									
2,6,6-trimethylbicyclo[3.1.1]hept-2-ene (α-pinene)	7785-70-8	11.5	929	1.79	2.57	0.74	1.80	1.95	0.36
7-methyl-3-methyleneocta-1,6-diene (β-myrcene)	123-35-3	13.7	988	0.96	0.94	0.33	0.96	0.64	0.39
1-methyl-4-propan-2-ylbenzene (p-cymene)	99-87-6	14.9	1022	0.82	0.44	0.96	0.82	0.23	0.75
1-methyl-4-(prop-1-en-2-yl)cyclohex-1-ene (D-limonene)	138-86-3	15.1	1028	0.99	1.50	0.27	0.99	1.20	0.31
1-isopropyl-4-methylcyclohexa-1,4-diene (γ-terpinene)	99-85-4	16.1	1057	0.00	0.10	0.00	0.00	0.11	0.00
1-methyl-4-(propan-2-ylidene)cyclohex-1-ene	586-62-9	17.0	1085	0.15	0.07	0.04	0.15	0.07	0.03
1-methyl-4-(prop-1-en-2-yl)benzene (p-cymenene)	1195-32-0	17.2	1090	0.09	0.14	0.01	0.09	0.14	0.02
4,10-dimethyl-7-propan-2-yltricyclo[4.4.0.0^{1,5}]dec-3-ene (α-cubebene)	17699-14-8	25.0	1346	1.48	1.70	1.19	1.70	2.28	1.42
(1*S*,6*S*,7*S*,8*S*)-1,3-dimethyl-8-propan-2-yltricyclo[4.4.0.0^{2,7}]dec-3-ene (a-copaene)	3856-25-5	26.0	1377	3.80	5.27	3.66	3.81	7.15	4.24
(1*R*,9*S*,*Z*)-4,11,11-trimethyl-8-methylenebicyclo[7.2.0]undec-4-ene (β-caryophyllene)	87-44-5	27.5	1422	5.78	8.07	4.32	5.78	9.84	4.73
2,6-dimethyl-6-(4-methylpent-3-enyl)bicyclo[3.1.1]hept-2-ene (α-bergamotene)	13474-59-4	27.8	1432	1.05	1.55	1.09	1.05	2.24	1.21
(6*Z*)-7,11-dimethyl-3-methylidenedodeca-1,6,10-triene (β-farnesene)	28973-97-9	28.4	1450	0.05	0.07	0.03	0.05	0.13	0.08
(1*E*,4*E*,8*E*)-2,6,6,9-tetramethylcycloundeca-1,4,8-triene (humulene)	6753-98-6	28.6	1454	0.54	0.76	0.45	0.54	0.93	0.52
(1a*R*,4a*S*,7*R*,7a*S*,7b*S*)-1,1,7-trimethyl-4-methylidene-2,3,4a,5,6,7,7a,7b-octahydro-1a*H*-cyclopropa[e]azulene (alloaromadendrene)	25246-27-9	28.7	1458	0.13	0.16	0.11	0.13	0.20	0.14
(1*S*,4a*S*,8a*R*)-7-methyl-4-methylidene-1-propan-2-yl-2,3,4a,5,6,8a-hexahydro-1*H*-naphthalene (γ-muurolene)	30021-74-0	29.2	1473	0.08	0.15	0.08	0.08	0.21	0.12
(1*S*,4a*S*,8a*R*)-4,7-dimethyl-1-propan-2-yl-1,2,4a,5,6,8a-hexahydronaphthalene (α-muurolene)	31983-22-9	30.0	1497	0.08	0.10	0.05	0.05	0.14	0.09
(4*S*)-1-methyl-4-(6-methylhepta-1,5-dien-2-yl)cyclohexene (β-bisabolene)	495-61-4	30.3	1506	0.14	0.23	0.24	0.14	0.36	0.30
(1*R*,4a*S*,8a*S*)-7-methyl-4-methylidene-1-propan-2-yl-2,3,4a,5,6,8a-hexahydro-1*H*-naphthalene (γ-cadinene)	39029-41-9	30.5	1511	0.04	0.08	0.04	0.04	0.09	0.06
(1*S*,8a*R*)-4,7-dimethyl-1-propan-2-yl-1,2,3,5,6,8a-hexahydronaphthalene (δ-cadinene)	483-76-1	30.7	1517	0.40	0.65	0.51	0.40	0.66	0.57
4-isopropyl-1,6-dimethyl-1,2,3,4,4a,7-hexahydronaphthalene	16728-99-7	31.1	1530	0.04	0.07	0.05	0.04	0.08	0.06
(1*S*)-4,7-dimethyl-1-propan-2-yl-1,2-dihydronaphthalene (α-calacorene)	21391-99-1	31.4	1538	0.04	0.08	0.09	0.05	0.08	0.09

Table 1. *Cont.*

Volatile Compounds	CAS Number	RT [a]	RI [b]	Soxhlet			UAE		
				Breaking	Ripe	Overripe	Breaking	Ripe	Overripe
Aldehydes									
nonanal	124-19-6	17.7	1104	0.08	0.11	0.07	0.08	0.16	0.02
Ketones									
octan-2-one	111-13-7	10.6	907	0.56	0.28	0.48	0.27	0.12	0.19
Others									
trans-decahydronaphthalene	493-02-7	16.1	1057	0.34	0.10	0.25	0.27	0.02	0.27
(1S,4S)-1,6-dimethyl-4-propan-2-yl-1,2,3,4-tetrahydronaphthalene (calamenene)	72937-55-4	30.8	1519	0.13	0.24	0.19	0.13	0.23	0.22

[a] RT: Retention time (min); [b] RI: Experimental retention index.

3.4. Estimation of Volatiles over Ripening

The volatile fraction resulting from avocado oil extraction with petroleum ether and isolated by the HS-SPME technique gave mainly terpenoids and hydrocarbons (non-terpenoids). In contrast to previous studies, trace amounts of aldehydes, ketones, and other compounds were detected [25,33,34,36,37], which may be attributed to the sampling method or to the differences among avocado varieties.

The ripening stages (breaking, ripe, and overripe) indicated major differences in the semi-quantification of volatile compounds. The hydrocarbon (non-terpenoids) contents were found to be 14.41, 6.57, and 17.62 mg·kg^{-1} for breaking, ripe, and overripe samples over SE and 14.80, 2.89, and 14.14 mg·kg^{-1} over UAE, respectively. The corresponding terpenoids contents were found to be 18.44, 24.68, and 14.27 mg·kg^{-1} for breaking, ripe, and overripe samples over SE and 18.67, 28.74, and 15.50 mg·kg^{-1} over UAE. The above results show that, during maturation, the volatile fraction fluctuates. On the fourth day of maturation, the abundance of hydrocarbons was significantly lower ($p < 0.05$) compared with the first day, while it increased on the eighth day. In contrast, the abundance of terpenoids was significantly higher ($p < 0.05$) on the fourth day of maturation. This change in volatility was observed in both cases of the extraction techniques (Figure 3).

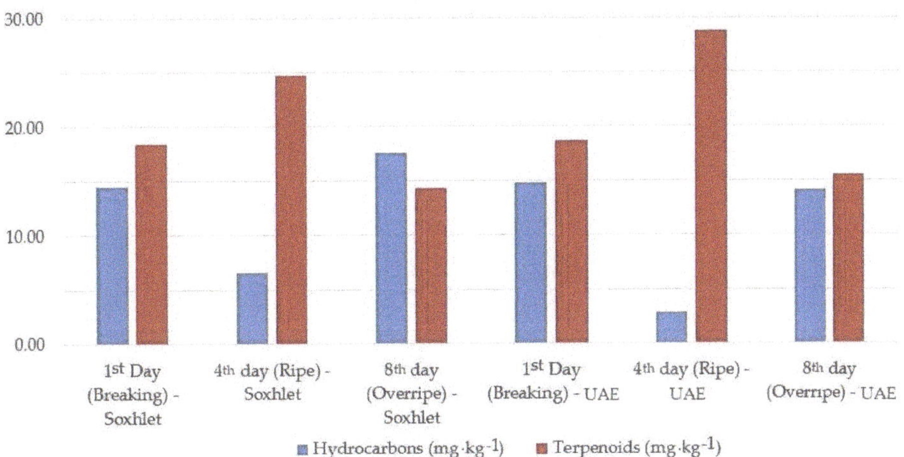

Figure 3. The variation of volatility during three ripening stages.

The abundance of hydrocarbons and terpenoids was examined by multivariate analysis of variance in comparison with the ripening stages. The Scheffe post hoc test was used to investigate which pairs of means were significant, and detailed results are presented in Table 2. It is evident that 21 of 44 total volatiles were statistically significant ($p < 0.05$).

Three volatiles significantly differed between the ripening stages, including 1-methyl-4-(propan-2-ylidene)cyclohex-1-ene; 1-methyl-4-(prop-1-en-2-yl)benzene (p-cymenene); and (1*S*,8a*R*)-4,7-dimethyl-1-propan-2-yl-1,2,3,5,6,8a-hexahydronaphthalene (δ-cadinene).

Table 2. The statistically significant volatile compounds between ripening stages.

	Multiple Comparisons [a]			
No.	Volatile Compounds	Ripening Stages in Pairs		p-Value [b]
1	2,6-dimethyloctane	Breaking Ripe Overripe	Ripe Overripe Breaking	0.021 0.049 0.373
2	4-methylnonane	Breaking Ripe Overripe	Ripe Overripe Breaking	0.066 0.047 0.859
3	2-methylnonane	Breaking Ripe Overripe	Ripe Overripe Breaking	0.047 0.047 1.000
4	1-methyl-2-propylcyclohexane	Breaking Ripe Overripe	Ripe Overripe Breaking	0.064 0.011 0.077
5	decane	Breaking Ripe Overripe	Ripe Overripe Breaking	0.046 0.019 0.364
6	butylcyclohexane	Breaking Ripe Overripe	Ripe Overripe Breaking	0.002 0.004 0.131
7	10,10-dimethyl-2,6-dimethylenebicyclo[7.2.0]undecane	Breaking Ripe Overripe	Ripe Overripe Breaking	0.037 0.061 0.716
8	2,6,6-trimethylbicyclo[3.1.1]hept-2-ene (α-pinene)	Breaking Ripe Overripe	Ripe Overripe Breaking	0.408 0.024 0.056
9	7-methyl-3-methyleneocta-1,6-diene (β-myrcene)	Breaking Ripe Overripe	Ripe Overripe Breaking	0.486 0.091 0.039
10	1-methyl-4-(prop-1-en-2-yl)cyclohex-1-ene (D-limonene)	Breaking Ripe Overripe	Ripe Overripe Breaking	0.133 0.008 0.025
11	1-isopropyl-4-methylcyclohexa-1,4-diene (γ-terpinene)	Breaking Ripe Overripe	Ripe Overripe Breaking	0.000 0.000 1.000
12	1-methyl-4-(propan-2-ylidene)cyclohex-1-ene	Breaking Ripe Overripe	Ripe Overripe Breaking	0.001 0.008 0.000
13	1-methyl-4-(prop-1-en-2-yl)benzene (p-cymenene)	Breaking Ripe Overripe	Ripe Overripe Breaking	0.003 0.000 0.001
14	(1*R*,9*S*,*Z*)-4,11,11-trimethyl-8-methylenebicyclo[7.2.0]undec-4-ene (β-caryophyllene)	Breaking Ripe Overripe	Ripe Overripe Breaking	0.053 0.022 0.366
15	(1*E*,4*E*,8*E*)-2,6,6,9-tetramethylcycloundeca-1,4,8-triene (humulene)	Breaking Ripe Overripe	Ripe Overripe Breaking	0.060 0.039 0.781
16	(1*R*,4a*S*,8a*S*)-7-methyl-4-methylidene-1-propan-2-yl-2,3,4a,5,6,8a-hexahydro-1*H*-naphthalene (γ-cadinene)	Breaking Ripe Overripe	Ripe Overripe Breaking	0.036 0.070 0.604
17	(1*S*,8a*R*)-4,7-dimethyl-1-propan-2-yl-1,2,3,5,6,8a-hexahydronaphthalene (δ-cadinene)	Breaking Ripe Overripe	Ripe Overripe Breaking	0.005 0.043 0.025
18	4-isopropyl-1,6-dimethyl-1,2,3,4,4a,7-hexahydronaphthalene	Breaking Ripe Overripe	Ripe Overripe Breaking	0.021 0.089 0.171
19	(1*S*)-4,7-dimethyl-1-propan-2-yl-1,2-dihydronaphthalene (α-calacorene)	Breaking Ripe Overripe	Ripe Overripe Breaking	0.008 0.192 0.004
20	trans-decahydronaphthalene	Breaking Ripe Overripe	Ripe Overripe Breaking	0.026 0.046 0.640
21	(1*S*,4*S*)-1,6-dimethyl-4-propan-2-yl-1,2,3,4-tetrahydronaphthalene (calamenene)	Breaking Ripe Overripe	Ripe Overripe Breaking	0.009 0.213 0.023

[a] Based on observed means. [b] The mean difference is significant at the 0.05 level.

There is a lack of published data relating to the differences between the variations of volatile components of avocado oil and the ripening stages of avocado fruit. The majority of literature reports have focused on the aroma of avocado fruit; however, a limited number of studies have investigated the aroma of avocado oil [25,33–37]. Nonetheless, similar results can be drawn from the study of avocado fruit volatiles. Pereira [42] reported that the sesquiterpenes of avocado fruit decreased during ripening. Moreover, similar results have been reported in other climacteric fruits, in which series of changes in metabolic biosynthesis occur during storage ripening [43]. In particular, the monoterpenes in mango fruits (*Mangifera indica* L. "Kensington Pride") have increased on the fourth day of ripening and decreased afterwards [43].

Furthermore, in the same fruit, the monoterpenes, sesquiterpenes, and aromatics were determined at a higher total amount in ripened mango compared with the unripe and overripe stages [44]. In another study by Zidi on figs (*Ficus carica* L.) [45], β-caryoophyllene and D-limonene increased significantly from the unripe to the ripe stage and were suppressed in the fully ripe stage.

Ethylene is well-known to control the storage duration and rate of ripening of climacteric fruits. A potential hypothesis is that the rise could be linearly correlated with ethylene synthesis. Several studies reported that climacteric fruits including apple (*Malus domestica* Borkh.) [46], tomato (*Solanum lucopersicum* L.) [47], and mango [48] undergo a rapid production of terpenes which depends on the response of ethylene. Thus, ethylene plays a key role in the metabolic events of volatiles during ripening [49]. Nevertheless, numerous underlying processes are yet to be delineated and merit additional study [50]. Another possible interpretation of the suppression of terpenoids over the later ripening stages (over-ripe) could be correlated with the presence of terpenoid hydroperoxides. The mesocarp of avocado fruit contains idioblastic cells that contain oil sacs and sesquiterpene hydroperoxides [7,51,52]. During maturation or enzymatic reaction, a degradation of a primary wall of the parenchyma cells occurs, which releases the oil from the idioblastic cells, and then the released hydroperoxides act on the terpenoids.

4. Conclusions

In summary, this work shows the analysis of avocado oil extracted from the Zutano variety by two (SE and UAE) techniques. In this context, petroleum ether volatile fractions were estimated over three ripening stages (breaking, ripe, and overripe) using HS-SPME–GC–MS. The Zutano variety, which is cultivated in the Crete region, gave a fragrant oil which has not been previously studied. This cultivar is characterized from seven main volatile compounds, including 2,6,6-trimethylbicyclo[3.1.1]hept-2-ene (α-pinene), 1-methyl-4-(prop-1-en-2-yl)cyclohex-1-ene (D-limonene), 4,10-dimethyl-7-propan-2-yltricyclo[4.4.0.0^{1,5}]dec-3-ene (α-cubebene), (1S,6S,7S,8S)-1,3-dimethyl-8-propan-2-yltricyclo[4.4.0.0^{2,7}]dec-3-ene (a-copaene), (1R,9S,Z)-4,11,11-trimethyl-8-methylenebicyclo[7.2.0]undec-4-ene (β-caryophyllene), 2,6-dimethyl-6-(4-methylpent-3-enyl)bicyclo[3.1.1]hept-2-ene (α-bergamotene), and (1E,4E,8E)-2,6,6,9-tetramethylcycloundeca-1,4,8-triene (humulene). The analyzed fractions consisted of a high content of terpenoids with an average relative abundance of over 61.7%. ANOVA revealed that the extraction methods did not have statistically significant differences between their qualification and semi-quantification results. In contrast, the application of MANOVA between the ripening stages and some volatiles indicated that the p-values were lower than 0.05. Even though ripening is one significant factor that affects volatiles, additional research is required to approve the above results. This study could form the foundation for additional research on the impact of ethylene and the metabolism of avocado oil volatiles.

Author Contributions: Conceptualization, M.X. and C.S.P.; methodology, M.X., L.V., E.H.K., P.-K.R. and C.S.P.; software, M.X., L.V., E.H.K. and P.-K.R.; validation, M.X. and E.G.; formal analysis, E.G.; investigation, M.X., E.G., L.V. and E.H.K.; resources, E.G.; data curation, M.X. and E.G.; writing—original draft preparation, M.X.; writing—review and editing, L.V., E.H.K., P.-K.R., C.S.P. and P.A.T.; visualization, C.S.P. and P.A.T.; supervision, C.S.P.; project administration, C.S.P. and

P.A.T.; funding acquisition, E.G. and P.-K.R. All authors have read and agreed to the published version of the manuscript.

Funding: This research received no external funding.

Institutional Review Board Statement: Not applicable.

Informed Consent Statement: Not applicable.

Data Availability Statement: Not applicable.

Acknowledgments: The authors would like to thank Ioannis Lathourakis and Antonios Galanis for the sponsorship of *Persea americana* Mill. "Zutano" variety avocados from the Greek island of Crete, Chania.

Conflicts of Interest: The authors declare no conflict of interest.

References

1. Caballero, B.; Finglas, P.M.; Toldrá, F. (Eds.) *Encyclopedia of Food and Health*; Academic Press: Cambridge, MA, USA, 2016; ISBN 978-0-12-384947-2.
2. FAOSTAT. Available online: https://www.fao.org/faostat/en/#data/QCL/visualize (accessed on 24 November 2021).
3. Kourgialas, N.N.; Dokou, Z. Water management and salinity adaptation approaches of Avocado trees: A review for hot-summer Mediterranean climate. *Agric. Water Manag.* **2021**, *252*, 106923. [CrossRef]
4. Flores, M.; Saravia, C.; Vergara, C.; Avila, F.; Valdés, H.; Ortiz-Viedma, J. Avocado Oil: Characteristics, Properties, and Applications. *Molecules* **2019**, *24*, 2172. [CrossRef] [PubMed]
5. Tan, C.X. Virgin avocado oil: An emerging source of functional fruit oil. *J. Funct. Foods* **2019**, *54*, 381–392. [CrossRef]
6. Bhuyan, D.J.; Alsherbiny, M.A.; Perera, S.; Low, M.; Basu, A.; Devi, O.A.; Barooah, M.S.; Li, C.G.; Papoutsis, K. The Odyssey of Bioactive Compounds in Avocado (*Persea americana*) and their Health Benefits. *Antioxidants* **2019**, *8*, 426. [CrossRef]
7. Cervantes-Paz, B.; Yahia, E.M. Avocado oil: Production and market demand, bioactive components, implications in health, and tendencies and potential uses. *Compr. Rev. Food Sci. Food Saf.* **2021**, *20*, 4120–4158. [CrossRef] [PubMed]
8. Woolf, A.; Wong, M.; Eyres, L.; McGhie, T.; Lund, C.; Olsson, S.; Wang, Y.; Bulley, C.; Wang, M.; Friel, E.; et al. Avocado oil. In *Gourmet and Health-Promoting Specialty Oils*; Elsevier: Amsterdam, The Netherlands, 2009; pp. 73–125. ISBN 978-1-893997-97-4.
9. El-Zeftawi, B. Physical and chemical changes in fruit of seven avocado cultivars at Mildura. *Aust. J. Agric. Res.* **1978**, *29*, 81–88. [CrossRef]
10. Vekiari, S.A.; Papadopoulou, P.P.; Lionakis, S.; Krystallis, A. Variation in the Composition of Cretan Avocado Cultivars during Ripening: Seasonal Variation in the Composition of Cretan Avocados. *J. Sci. Food Agric.* **2004**, *84*, 485–492. [CrossRef]
11. Takenaga, F.; Matsuyama, K.; Abe, S.; Torii, Y.; Itoh, S. Lipid and fatty acid composition of mesocarp and seed of avocado fruits harvested at northern range in Japan. *J. Oleo Sci.* **2008**, *57*, 591–597. [CrossRef]
12. Ozdemir, F.; Topuz, A. Changes in dry matter, oil content and fatty acids composition of avocado during harvesting time and post-harvesting ripening period. *Food Chem.* **2004**, *86*, 79–83. [CrossRef]
13. Mahendran, T.; Brennan, J.G.; Hariharan, G. Aroma volatiles components of 'Fuerte' Avocado (*Persea americana* Mill.) stored under different modified atmospheric conditions. *J. Essent. Oil Res.* **2019**, *31*, 34–42. [CrossRef]
14. Ali, S.; Plotto, A.; Scully, B.T.; Wood, D.; Stover, E.; Owens, N.; Pisani, C.; Ritenour, M.; Anjum, M.A.; Nawaz, A.; et al. Fatty acid and volatile organic compound profiling of avocado germplasm grown under East-Central Florida conditions. *Sci. Hortic.* **2020**, *261*, 109008. [CrossRef]
15. Tan, C.X.; Hean, C.G.; Hamzah, H.; Ghazali, H.M. Optimization of ultrasound-assisted aqueous extraction to produce virgin avocado oil with low free fatty acids. *J. Food Process. Eng.* **2018**, *41*, e12656. [CrossRef]
16. Tan, C.X.; Chong, G.H.; Hamzah, H.; Ghazali, H.M. Comparison of subcritical CO2 and ultrasound-assisted aqueous methods with the conventional solvent method in the extraction of avocado oil. *J. Supercrit. Fluids* **2018**, *135*, 45–51. [CrossRef]
17. Dos Santos, M.A.Z.; Alicieo, T.V.R.; Pereira, C.M.P.; Ramis-Ramos, G.; Mendonça, C.R.B.; Dos Santos, M.A.Z. Profile of Bioactive Compounds in Avocado Pulp Oil: Influence of the Drying Processes and Extraction Methods. *J. Am. Oil Chem. Soc.* **2013**, *91*, 19–27. [CrossRef]
18. Krumreich, F.D.; Borges, C.D.; Mendonça, C.R.B.; Jansen-Alves, C.; Zambiazi, R.C. Bioactive compounds and quality parameters of avocado oil obtained by different processes. *Food Chem.* **2018**, *257*, 376–381. [CrossRef] [PubMed]
19. Reddy, M.; Moodley, R.; Jonnalagadda, S.B.; Jonnalagadda, S.B. Fatty acid profile and elemental content of avocado (*Persea americana* Mill.) oil –effect of extraction methods. *J. Environ. Sci. Health Part B* **2012**, *47*, 529–537. [CrossRef]
20. Abaide, E.; Zabot, G.L.; Tres, M.V.; Martins, R.F.; Fagundez, J.L.; Nunes, L.F.; Druzian, S.; Soares, J.F.; Prá, V.D.; Silva, J.R.; et al. Yield, composition, and antioxidant activity of avocado pulp oil extracted by pressurized fluids. *Food Bioprod. Process.* **2017**, *102*, 289–298. [CrossRef]
21. Mostert, M.E.; Botha, B.M.; Du Plessis, L.M.; Duodu, K.G. Effect of fruit ripeness and method of fruit drying on the extractability of avocado oil with hexane and supercritical carbon dioxide. *J. Sci. Food Agric.* **2007**, *87*, 2880–2885. [CrossRef]

22. Corzzini, S.C.; Barros, H.D.; Grimaldi, R.; Cabral, F. Extraction of edible avocado oil using supercritical CO_2 and a CO_2/ethanol mixture as solvents. *J. Food Eng.* **2017**, *194*, 40–45. [CrossRef]

23. Espinosa-Alonso, L.G.; Paredes-López, O.; Valdez-Morales, M.; Oomah, B.D. Avocado Oil Characteristics of Mexican Creole Genotypes: Mexican Creole Avocado Oil Properties. *Eur. J. Lipid Sci. Technol.* **2017**, *119*, 1600406. [CrossRef]

24. Yanty, N.A.M.; Marikkar, J.M.N.; Long, K. Effect of Varietal Differences on Composition and Thermal Characteristics of Avocado Oil. *J. Am. Oil Chem. Soc.* **2011**, *88*, 1997–2003. [CrossRef]

25. Moreno, A.O.; Dorantes, L.; Galíndez, J.; Guzmán, R.I. Effect of Different Extraction Methods on Fatty Acids, Volatile Compounds, and Physical and Chemical Properties of Avocado (*Persea americana* Mill.) Oil. *J. Agric. Food Chem.* **2003**, *51*, 2216–2221. [CrossRef]

26. Meyer, M.D.; Terry, L.A. Development of a Rapid Method for the Sequential Extraction and Subsequent Quantification of Fatty Acids and Sugars from Avocado Mesocarp Tissue. *J. Agric. Food Chem.* **2008**, *56*, 7439–7445. [CrossRef]

27. Ortiz, M.A.; Dorantes, A.L.; Gallndez, M.J.; CRdenas, S.E. Effect of a Novel Oil Extraction Method on Avocado (*Persea americana* Mill.) Pulp Microstructure. *Plant Foods Hum. Nutr.* **2004**, *59*, 11–14. [CrossRef]

28. Meyer, M.D.; Terry, L.A. Fatty Acid and Sugar Composition of Avocado, Cv. Hass, in Response to Treatment with an Ethylene Scavenger or 1-Methylcyclopropene to Extend Storage Life. *Food Chem.* **2010**, *121*, 1203–1210. [CrossRef]

29. Barros, H.D.F.Q.; Coutinho, J.P.; Grimaldi, R.; Godoy, H.T.; Cabral, F.A. Simultaneous Extraction of Edible Oil from Avocado and Capsanthin from Red Bell Pepper Using Supercritical Carbon Dioxide as Solvent. *J. Supercrit. Fluids* **2016**, *107*, 315–320. [CrossRef]

30. Martínez-Padilla, L.P.; Franke, L.; Xu, X.-Q.; Juliano, P. Improved Extraction of Avocado Oil by Application of Sono-Physical Processes. *Ultrason. Sonochem.* **2018**, *40*, 720–726. [CrossRef] [PubMed]

31. del Pilar Ramírez-Anaya, J.; Manzano-Hernández, A.J.; Tapia-Campos, E.; Alarcón-Domínguez, K.; Castañeda-Saucedo, M.C. Influence of Temperature and Time during Malaxation on Fatty Acid Profile and Oxidation of Centrifuged Avocado Oil. *Food Sci. Technol.* **2018**, *38*, 223–230. [CrossRef]

32. Schwartz, M.; Olaeta, J.A.; Undurraga, P. Mejoramiento del rendimiento de extracción del aceite de palta (aguacate). In Proceedings of the VI World Avocado Congress (Actas VI Congreso Mundial del Aguacate) 2007, Viña Del Mar, Chile, 12–16 November 2007.

33. Kilic-Buyukkurt, O. Characterization of Aroma Compounds of Cold-Pressed Avocado Oil Using Solid-Phase Microextraction Techniques with Gas Chromatography–Mass Spectrometry. *J. Raw Mater. Process. Foods* **2021**, *2*, 1–7.

34. Haiyan, Z.; Bedgood, D.R.; Bishop, A.G.; Prenzler, P.D.; Robards, K. Endogenous Biophenol, Fatty Acid and Volatile Profiles of Selected Oils. *Food Chem.* **2007**, *100*, 1544–1551. [CrossRef]

35. Pino, J.A.; Rosado, A.; Aguero, J. Volatile Components of Avocado (*Persea americana* Mill.) Fruits. *J. Essent. Oil Res.* **2000**, *12*, 377–378. [CrossRef]

36. de Sousa Galvao, M.; Nunes, M.L.; Constant, P.B.L.; Narain, N. Identification of Volatile Compounds in Cultivars Barker, Collinson, Fortuna and Geada of Avocado (*Persea americana* Mill.) Fruit. *Food Sci. Technol.* **2016**, *36*, 439–447. [CrossRef]

37. Liu, Y.-J.; Gong, X.; Jing, W.; Lin, L.-J.; Zhou, W.; He, J.-N.; Li, J.-H. Fast Discrimination of Avocado Oil for Different Extracted Methods Using Headspace-Gas Chromatography-Ion Mobility Spectroscopy with PCA Based on Volatile Organic Compounds. *Open Chem.* **2021**, *19*, 367–376. [CrossRef]

38. Commission Regulation (EC) No 387/2005 of 8 March 2005 Amending (EC) Regulation No 831/97 Laying down Marketing Standards Applicable to Avocados—Publications Office of the EU. Available online: https://op.europa.eu/en/publication-detail/-/publication/dd5847ac-2cc1-431f-b945-8c2ecb500f40 (accessed on 24 November 2021).

39. Xagoraris, M.; Revelou, P.-K.; Dedegkika, S.; Kanakis, C.D.; Papadopoulos, G.K.; Pappas, C.S.; Tarantilis, P.A. SPME-GC-MS and FTIR-ATR Spectroscopic Study as a Tool for Unifloral Common Greek Honeys' Botanical Origin Identification. *Appl. Sci.* **2021**, *11*, 3159. [CrossRef]

40. Adams, R.P. *Identification of Essential Oil Components by Gas Chromatography/Mass Spectrometry*, 4th ed.; Allured Publishing Corporation: Carol Stream, IL, USA, 2007; ISBN 978-1-932633-21-4.

41. Satriana, S.; Supardan, M.D.; Arpi, N.; Wan Mustapha, W.A. Development of Methods Used in the Extraction of Avocado Oil. *Eur. J. Lipid Sci. Technol.* **2019**, *121*, 1800210. [CrossRef]

42. Pereira, M.E.C.; Tieman, D.M.; Sargent, S.A.; Klee, H.J.; Huber, D.J. Volatile Profiles of Ripening West Indian and Guatemalan-West Indian Avocado Cultivars as Affected by Aqueous 1-Methylcyclopropene. *Postharvest Biol. Technol.* **2013**, *80*, 37–46. [CrossRef]

43. Lalel, H.J.D.; Singh, Z.; Tan, S.C.; Agustí, M. Maturity Stage at Harvest Affects Fruit Ripening, Quality and Biosynthesis of Aroma Volatile Compounds in 'Kensington Pride' Mango. *J. Hortic. Sci. Biotechnol.* **2003**, *78*, 225–233. [CrossRef]

44. Lalel, H.J.D.; Singh, Z.; Tan, S.C. Aroma Volatiles Production during Fruit Ripening of 'Kensington Pride' Mango. *Postharvest Biol. Technol.* **2003**, *27*, 323–336. [CrossRef]

45. Zidi, K.; Kati, D.E.; Bachir-bey, M.; Genva, M.; Fauconnier, M.-L. Comparative Study of Fig Volatile Compounds Using Headspace Solid-Phase Microextraction-Gas Chromatography/Mass Spectrometry: Effects of Cultivars and Ripening Stages. *Front. Plant Sci.* **2021**, *12*, 667809. [CrossRef]

46. Schaffer, R.J.; Friel, E.N.; Souleyre, E.J.F.; Bolitho, K.; Thodey, K.; Ledger, S.; Bowen, J.H.; Ma, J.-H.; Nain, B.; Cohen, D.; et al. A Genomics Approach Reveals That Aroma Production in Apple Is Controlled by Ethylene Predominantly at the Final Step in Each Biosynthetic Pathway. *Plant Physiol.* **2007**, *144*, 1899–1912. [CrossRef]

47. Kovács, K.; Fray, R.G.; Tikunov, Y.; Graham, N.; Bradley, G.; Seymour, G.B.; Bovy, A.G.; Grierson, D. Effect of Tomato Pleiotropic Ripening Mutations on Flavour Volatile Biosynthesis. *Phytochemistry* **2009**, *70*, 1003–1008. [CrossRef] [PubMed]

48. Pandit, S.S.; Kulkarni, R.S.; Chidley, H.G.; Giri, A.P.; Pujari, K.H.; Köllner, T.G.; Degenhardt, J.; Gershenzon, J.; Gupta, V.S. Changes in Volatile Composition during Fruit Development and Ripening of 'Alphonso' Mango. *J. Sci. Food Agric.* **2009**, *89*, 2071–2081. [CrossRef]
49. Defilippi, B.G.; Manríquez, D.; Luengwilai, K.; González-Agüero, M. Chapter 1 Aroma Volatiles. In *Advances in Botanical Research*; Elsevier: Amsterdam, The Netherlands, 2009; Volume 50, pp. 1–37, ISBN 978-0-12-374835-5.
50. Gapper, N.E.; McQuinn, R.P.; Giovannoni, J.J. Molecular and Genetic Regulation of Fruit Ripening. *Plant Mol. Biol.* **2013**, *82*, 575–591. [CrossRef]
51. Platt, K.A.; Thomson, W.W. Idioblast Oil Cells of Avocado: Distribution, Isolation, Ultrastructure, Histochemistry, and Biochemistry. *Int. J. Plant Sci.* **1992**, *153*, 301–310. [CrossRef]
52. Platt-Aloia, K.A.; Oross, J.W.; Thomson, W.W. Ultrastructural Study of the Development of Oil Cells in the Mesocarp of Avocado Fruit. *Bot. Gaz.* **1983**, *144*, 49–55. [CrossRef]

Article

Effects of Electrolytes on the Dediazoniation of Aryldiazonium Ions in Acidic MeOH/H$_2$O Mixtures

Sonia Losada-Barreiro and Carlos Bravo-Díaz *

Department of Química-Física, Universidad de Vigo, Rua Abelleiras s/n, 36310 Vigo, Spain; sonia@uvigo.es
* Correspondence: cbravo@uvigo.es

Abstract: Aryldiazonium, ArN_2^+, ions decompose spontaneously through the formation of highly reactive aryl cations that undergo preferential solvation by water, showing a low selectivity towards the nucleophiles present in their solvation shell. In this work, we investigate the effects of electrolytes (NaCl, LiCl, and LiClO$_4$) on the dediazoniation of 2-, 3-, and 4-methylbenzenediazonium ions in acidic MeOH/H$_2$O mixtures. In the absence of electrolytes, the rates of dediazoniation, k_{obs}, increase modestly upon increasing the MeOH content of the reaction mixture. At any solvent composition, the rate of ArN_2^+ loss is the same as that for product formation. The main dediazoniation products are cresols (ArOH) and methyl phenyl ethers (ArOMe). Only small amounts (less than 5%) of the reduction product toluene (ArH), which are detected at high percentages of MeOH. Quantitative yields of are obtained at any solvent composition. The addition of LiCl or NaCl ([MCl] = 0–1.5 M) to the reaction mixtures has a negligible effect on k_{obs} but leads to the formation, in low yields (<10%), of the ArCl derivative. The addition of LiClO$_4$ (0–1.5 M) to 20% MeOH/H$_2$O mixtures has a negligible effect on both k_{obs} and on the product distribution. However, at 99.5% MeOH, the addition of the same amounts of LiClO$_4$ leads to a modest decrease in k_{obs} but to a significant decrease in the yields of ArOMe. Results are interpreted in terms of the preferential solvation of perchlorate ions by the aryl cations, removing MeOH molecules from the solvation shell.

Keywords: solvolysis; aryldiazonium ions; perchlorate anions

Citation: Losada-Barreiro, S.; Bravo-Díaz, C. Effects of Electrolytes on the Dediazoniation of Aryldiazonium Ions in Acidic MeOH/H$_2$O Mixtures. *Compounds* **2022**, *2*, 54–67. https://doi.org/10.3390/compounds2010005

Academic Editors: Nuno Basílio and Maria Lurdes Santos Cristiano

Received: 12 December 2021
Accepted: 9 February 2022
Published: 15 February 2022

Publisher's Note: MDPI stays neutral with regard to jurisdictional claims in published maps and institutional affiliations.

1. Introduction

Aryldiazonium, ArN_2^+, salts have been profusely employed in organic chemistry for years [1–3]. The chemistry of these compounds has been originally carried out in aqueous solutions [4], but chemists went further and developed new synthetic methods that use reactions of diazonium salts in organic solvents [5–8]. In recent years, their chemistry has been exploited in the synthesis of unnatural amino acids [5], inpalladium-catalyzed cross-coupling reactions [7,9,10] and, for example, as nitrogen-based Lewis acids [6,11].

Aryldiazonium ions are inherently reactive molecules that undergo a wide variety of chemical transformations that can be carried out under milder conditions, often at ambient temperature and pH levels [1,2,5,12–14]. Their reactivity is frequently dominated by the loss of the N$_2$ moiety, leading to the formation of heterolytic products, such as phenols, haloderivatives, and ethers. Scheme 1 shows some of the most common reactions of aryldiazonium ions, including the formation of diazoethers (O-coupling), azo dye (C-coupling) reactions, and their attachments to surfaces and nanoparticles.

In the last decades, after the pioneering work by J. Pinson et al. [15], the reductive properties of aryldiazonium salts have been exploited as a very potent method for surface functionalization [16–20]. The method is relatively simple, easy to process, is fast, and can be employed to functionalize massive surfaces that are either flat, or are nanomaterials of various shapes and sizes [17]. The formed surfaces are robust and resistant to heat, chemical degradation, ultrasonication, and, most importantly, the grafting of surfaces is a useful and practical approach that can be applied to a variety of conducting and insulating substrates.

Briefly, the grafting procedure comprises the formation of highly reactive aryl radicals that attack the surface and that are generated, for example, through the electrochemical reduction of ArN_2^+. Variations of the method include the reaction of generated radicals with already-grafted aryl species, leading to the formation of multilayer films [17,20]. Scheme 1B illustrates the use of ArN_2^+ for the grafting of both spherical nanoparticles and flat surfaces.

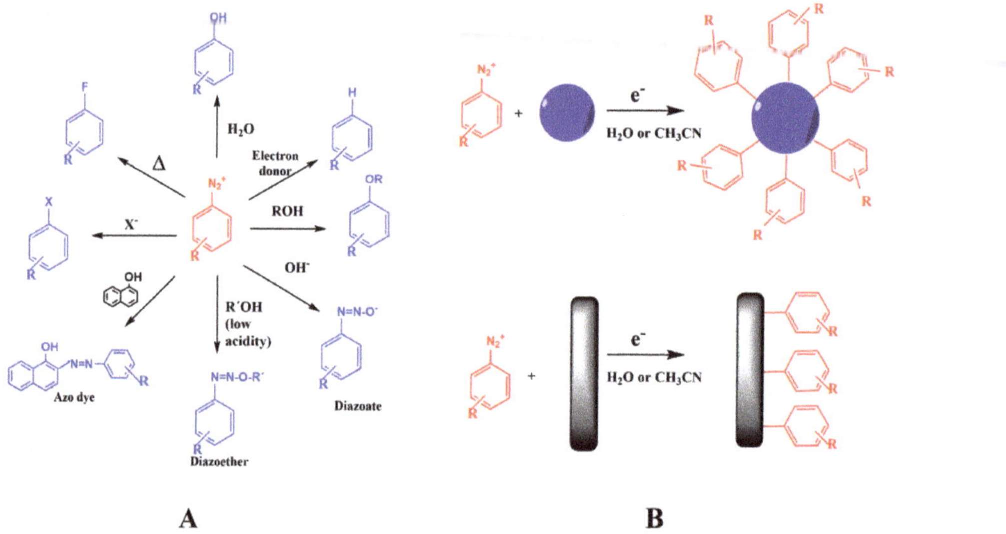

Scheme 1. (**A**) Some common reactions of aryldiazonium ions in solution. (**B**) Illustrative examples of the use of aryldiazonium ions as reagents to functionalize surfaces.

Aryldiazonium ions can be easily prepared from their anilines both in situ and in the solid state [1,21], and most of them can be easily handled under ambient conditions. Nevertheless, the stability of ArN_2^+ depends strongly on the nature of substituents R and counterions X^-. For example, aryldiazonium chlorides are highly unstable and can be explosive above 0 °C, as well as aryldiazonium perchlorates [12]. Nowadays, most synthetically prepared aryldiazonium salts are formulated as tetrafluoroborates because of the high stability and availability of the BF_4^- anion [3,12].

In an aqueous solution, in the dark, and in the absence of reducing species, ArN_2^+ undergoes a rich chemistry, comprising aromatic *ipso*-substitutions, *N*-terminal addition reactions, and *O*-coupling reactions, among others. However, the fate of the reactions strongly depends on the acidity of the solution. For instance, we showed, in previous dediazoniation studies [2,14,22–24] that, in an aqueous acid solution and in mixed alcohol–water solvents ($[H_3O^+] > 10^{-2}$ M), in the dark and in the absence of reductants, the spontaneous decomposition of aryldiazonium ArN_2^+ salts proceeds through a S_N1-type heterolytic mechanism, where a highly reactive aryl cation is formed, reacting with nucleophiles present in their solvation shell (dissociation + addition mechanism, $D_N + A_N$, Scheme 2). Recent dediazoniation research suggests, however, that ArN_2^+ may decompose heterolytically through borderline S_N1–S_N2 mechanisms. Surprisingly, upon moderately decreasing the acidity, reactions involving the formation of diazohydroxides, ArN_2OH, diazoethers, ArN_2OR, and diazoates, ArN_2O^-, become competitive and may even constitute the main decomposition pathway [2].

In this work, we analyze the effects of the solvent composition and of added electrolytes (NaCl, LiCl, and $LiClO_4$) on the esolvolytic dediazoniation of 2-, 3-, and 4-methylbenzenediazonium ions (2MBD, 3MBD, and 4MBD, respectively) in acidic $MeOH/H_2O$ mixtures (20% and 99.5%). The aim of the manuscript is two-fold: (1) to complement previous kinetic studies that were focused on the effects of electrolytes on short-lived carbocations, particularly on the effects of the solvent composition on the solvation shell of the *ipso* carbon of aryldiazonium ions, and (2) to investigate if the addition of electrolytes may modify the composition of the solvation shell of aryl carbocations.

$$(ArN_2^+)_{solv} \xrightarrow[\text{slow}]{k_{HET}} \left[Ar^+ \bullet N_2\right] \xrightarrow[\text{fast}]{Nu\ (H_2O,\ ROH,\ X^-)} ArOH + ArOR + ArX \qquad A$$

$$\xrightarrow{Nu^-,\ NuH} Ar\text{-}Nu(H) + N_2 \qquad B$$

Scheme 2. Illustrative representations of the (**A**) ionic or heterolytic S_N1 ($D_N + A_N$) dediazoniation mechanism in the presence of various nucleophiles comprising the formation of an ion–molecule pair that traps (with very low selectivity) nucleophiles present in its solvation shell. (**B**) S_N2 dediazoniation mechanism. In both cases, heterolytic products are formed.

For this purpose, we determined the dediazoniation rate constants for ArN_2^+ loss and for product formation, as well as the product distribution at various solvent compositions. Reported thermodynamic and kinetic data, obtained upon the solvolysis of ArN_2^+ in a number of alcohol–water mixtures (MeOH, EtOH, 2,2,2-trifluoroethanol, BuOH), are consistent with the S_N1 mechanism; that is, the rate-determining formation of a highly reactive aryl cation that traps the nucleophiles in its solvation shell [14,22–29]. The major dediazoniation products are the corresponding methylphenyl ethers and cresols, and the equal amounts of products that are produced at water molar fractions of 0.34–0.36, suggesting that the aryldiazonium ions undergo preferential solvation by water, and that the preferential solvation around the *ipso* carbon reflects the experimental product yield obtained [2,14,22].

The rates of the reactions of stable carbocations and aryldiazonium ions with water, alcohols, and anions have been described in terms of the Ritchie's equation $\log(k/k_0) = N_+$, based on the assumption that the relative reactivities of two nucleophiles are controlled by the differences in their N_+ values, which are considered independent of the electrophilicities of the reaction partners. Later, Mayr et al. [30–32] demonstrated that the rates of these reactions can be described in terms of their electrophilicity and nucleophilicity parameters, $\log(k) = s(E + N)$, where s is a nucleophile-specific parameter. A theoretical interpretation of the physical meaning of s has been published and a comprehensive list of N and E parameters can be found elsewhere [33]. The development of nucleophilicity scales has intrigued chemists for years, and Grunwald and Winstein proposed a relationship for the solvolyses of S_N1 reactions on the basis of the solvent-ionizing power Y and a substrate-specific parameter m whose value was one for t-butyl chloride [34–36], $\log(k/k_W) = mY$. This equation holds for S_N1 reactions, where the nucleophilic participation of the solvent in the rate-determining step is negligible [2,35].

In this work, we are particularly interested in analyzing the effects of ClO_4^- ions on solvolytic dediazoniations. Aryldiazonium perchlorates are particularly effective for the Heck palladium-catalyzed arylation of olefins when reactions are carried out in alcohol–water mixtures because of the high yields obtained and because they are cheaper reagents than those with other counterions [37]. However, cautions must be taken, because some aryl diazonium perchlorates have been reported to be explosive when prepared in the solid state [38,39]. Conversely, they can be conveniently generated in situ [37]. Perchlorate is a commercially available anion that forms salts with many cations (NH_4^+, Li^+, Na^+, K^+, etc.). Probably the most common form of perchlorate includes ammonium perchlorate (frequently used as a solid rocket oxidant and as ignition source in fireworks) and potassium perchlorate (used in road flares and in air bag inflation systems). Other commercial perchlorate counterions include H^+, Li^+, Na^+, K^+, NH_4^+, Al^{3+}, and $N_2H_5^+$. Perchlorate is also formed in laboratory waste as a byproduct of perchloric acid. It is a very poor complexing agent, similar to other weak anions, such as tetrafluoroborate or trifluoromethanesulfonates (triflate, $CF_3SO_3^-$), making it very useful in metal cation chemistry [40]. The high solubility in both aqueous and non-aqueous media, together with the highly delocalized monovalent charge over the four oxygen atoms and its large volume, allows it to be widely used to adjust ionic strength in kinetic experiments. From the thermodynamic point of view, perchlorates are expected to be powerful oxidizers ($E^0 = -1.229$ V), but the stability of ClO_4^- in solution is governed by kinetics and not thermodynamics, and, therefore, they do not easily oxidize [40,41].

2. Materials and Methods

2.1. Materials

The aryldiazonium 2-, 3-, and 4-methylbenzenediazonium (2MBD, 3MBD, and 4MBD, respectively) were prepared as tetrafluorborate salts under nonaqueous conditions, as described elsewhere [42,43]. The reagents used in their preparation were from Sigma-Aldrich (Taufkirchen, Germany). They were stored in the dark at low temperatures ($T < 5\,°C$) to minimize their decomposition and were recrystallized periodically. All chemicals were of the maximum purity available and were used without further purification. methyl phenyl ethers, ArOMe, cresols, ArOH, chlorotoluenes, ArCl, and toluene were from Sigma-Aldrich (Germany). The coupling agent, sodium 2-Naphthol-6-sulfonate ((2N6S), was from Pfaltz & Bauer (USA). The salts NaCl, LiCl, and $LiClO_4$ were from Sigma-Aldrich (Germany) and other materials employed were from Panreac (Barcelona, Spain). All solutions were prepared by using deionized water (resistivity > 18 MΩ·cm).

Hereafter, solution compositions are given by their percent of MeOH (by volume). Molar concentrations were calculated by ignoring the small excess volume of mixed solvents [44].

2.2. Instrumentation

UV-VIS spectra, and some kinetic experiments, were followed on a Beckman DU-640 UV–VIS or on an Agilent HP-6456 diode array spectrophotometer, both equipped with cell carriers thermostated with water from a refrigerated/heating circulator, Julabo F1 2-ED, and were attached to computers for data storage. A product analysis was carried out with the aid of a WATERS HPLC system, equipped with a model 2487 dual-λ absorbance detector, a model 717 automatic injector, a model 600 quaternary pump, and a computer for data storage. The dediazoniation products were separated with a reverse phase column (Microsorb-MV C-18 (Rainin, 25 cm length, 4.6 mm internal diameter, and 5 μm particle size) using an acidic 70/30 (v/v) MeOH/H_2O mobile phase, containing 10^{-4} M HCl. The injection volume was set at 25 μL in all runs and the UV detector was set at 220 nm. Details on the method can be found elsewhere [42,45].

2.3. Methods

Kinetic data were obtained both chromatographically and spectrophotometrically. Observed rate constants, k_{obs}, were obtained by fitting the percentage of yield-time or absorbance-time data to the integrated first order Equation (1), where X stands for the experimental measured UV–VIS absorbance or product yields. Runs were done at $T = 35 \pm 0.1\,°C$ (2MBD and 3MBD) and at $T = 60 \pm 0.1\,°C$ (4MBD) with aryldiazonium ions, ArN_2^+, as the limiting reagents. The reported k_{obs} values are the average of duplicate or triplicate experiments, with deviations lower than 7%.

$$\ln\left(\frac{X_t - X_\infty}{X_0 - X_\infty}\right) = -k_{obs}t \tag{1}$$

Spectrophotometric kinetic data were obtained by monitoring ArN_2^+ loss at a suitable wavelength to minimize interferences with other components of the solution (dediazoniation products) as much as possible. Linear variations (not shown) in the absorbance of ArN_2^+ in aqueous and MeOH solutions up to $[ArN_2^+] = 2.0 \times 10^{-4}$ M (cc. ≥ 0.999) were found, keeping with the predictions of Beer's law. ArN_2^+ solutions were prepared by dissolving the corresponding aryldiazonium salt in the appropriate acidic (HCl) MeOH/H_2O mixtures to diminish diazotate formation [2]. Final concentrations were approximately 1.0×10^{-4} M and [HCl] = 0.01 M. The stock ArN_2^+ solutions were kept in the dark at low temperatures ($T < 5\,°C$) to minimize their photochemical and/or spontaneous decomposition (Zollinger, 1994) and were used immediately or within a time period of less than 60 min.

Preliminary HPLC experiments showed that, in the absence of ClO_4^-, up to five decomposition products were detected to different extents. The most frequent were cresols, ArOH, and the methyl phenyl ethers (ArOMe), and, on occasion, depending on the particular experimental conditions, chlorotoluenes (when investigating Cl^- dependence), as well as the reduction product, toluene (ArH), that was detected in low yields when employing high percentages of MeOH in the solvent mixture. The calibration curves for all these products were obtained by employing authentic commercial samples. Linear (cc. > 0.999) calibration absorbance–concentration plots were obtained and employed for converting HPLC peak areas into concentrations. The following equation Yield = 100 [dediazoniation product]/[$ArN_2^+{}_T$] was employed to calculate the yields of the various dediazoniation products.

When studying the effects of the perchlorate salts, a new chromatographic peak was found that was not observed in its absence, with a retention time lower than that of ArOH but higher than that of the front peak (see Figure S1, Supplementary Materials). The area of this new peak increased linearly upon the increasing [$LiClO_4$]. In dediazoniations, where sulfate ions were employed, the aryldiazonium sulfate anion, Ar-OSO_4^- was formed and could be isolated [46,47]. As the possibility of the formation of the aryldiazonium perchlorate derivative $ArOClO_3$ exists, which is explosive in the solid state [3,38], we made no attempts in identifying nor isolating this new product. In spite of this, the failure in identifying this new derivative does not invalidate the main conclusions of the work because: (i) the peak area is proportional to the added [$LiCLO_4$] and (ii) dediazoniation products are formed competitively and their rate of formation is the same as that of other dediazoniation products (as determined from the variations in its peak areas with time), and equal to that of ArN_2^+ loss.

The rates of the formation of dediazoniation products were obtained by employing a derivatization method, as described elsewhere [48]. To minimize side reactions that may occur upon the injection of ArN_2^+ in the HPLC system (metal parts, solvent, etc.), TRIS buffer ([TRIS] = 0.05 M) solutions of the coupling agent 2N6S, that allows for the rapid formation of a stable azo dye, were employed [48]. The dediazoniations were quenched at convenient times, as described elsewhere [42,48]. The derivatization reaction was carried out under pseudo-first order conditions ([2N6S] > 20 [ArN_2^+]). The final pH was adjusted to a pH of approximately 8, because naphthoxide ions are much more reactive than their protonated forms. It is not advisable to use lower acidities, because the competing reactions of aryldiazonium ions with OH^-, to form diazotates, becomes significant [2].

The experimental conditions were chosen so that the coupling reaction was essentially over by the time the reagents were mixed, i.e., azo dye formation is much faster than dediazoniations (at least 100 times faster) and ArN_2^+ is effectively quenched at any solvent composition. Figure 1 illustrates the determination of the rates of product formation (ArOMe, Figure 1A) and of ArN_2^+ loss by monitoring the decrease in the absorbance of the azo dye formed (Figure 1B).

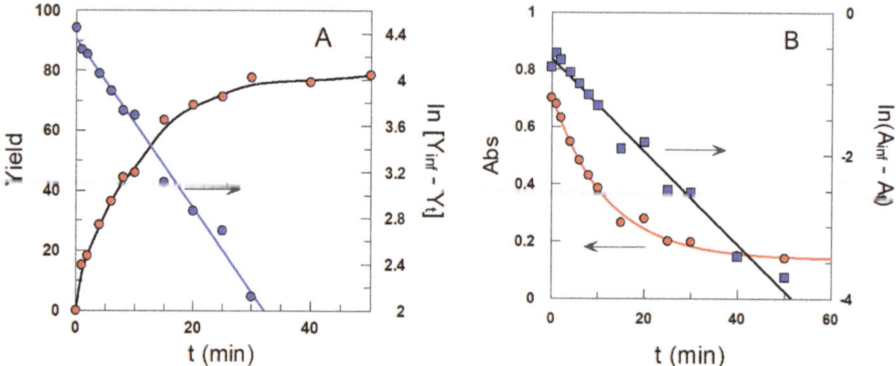

Figure 1. Illustrative examples of determination of the rate constant for product formation (ArOMe) by employing the chromatographic (**A**) and ArN_2^+ loss by monitoring the decrease in the absorbance of the azo dye formed in the derivatization method (**B**) under the same experimental conditions. The k_{obs} values obtained were similar (~12 × 10^{-4} s^{-1}) with differences lower than 10%. Experimental conditions: 99.5% MeOH/H_2O, [2MBD] = 1 × 10^{-4} M, [HCl] = 0.01 M, T = 35 °C.

3. Results

3.1. Spontaneous Dediazoniation of 2-, 3-, and 4-Methylbenzenediazonium Ions: Effects of Solvent (MeOH/H_2O) Composition on the Observed Rate Constant, k_{obs}

The effects of the solvent composition on the observed rate constant k_{obs} for the spontaneous dediazoniation of 2MBD, 3MBD, and 4MBD were explored by modifying the percentage of MeOH in the reaction mixture, as shown in Figure 2. Solvolytic rate constants k_{obs} increased smoothly at low percentages of MeOH, but more drastically at high percentages of methanol. The k_{obs} values at 0% MeOH, k_{obs} = 6 × 10^{-4} s^{-1} (2MBD), 8 × 10^{-4} s^{-1} (3MBD), and 9.5 × 10^{-4} s^{-1} (4MBD) aligned with the reported values obtained by different techniques, including N_2 evolution [49] at pH = 1.6–1.8, as well as HPLC and VIS–UV spectrophotometry [14]. The increase in k_{obs}, upon changing the methanol content, was modest, less than two-fold, and much lower than those reported in other solvolytic reactions. It is, however, aligned with literature reports that indicate that the rates of heterolytic dediazoniation of a number of arenediazonium ions vary by a factor of only 9 in 19 solvents [1].

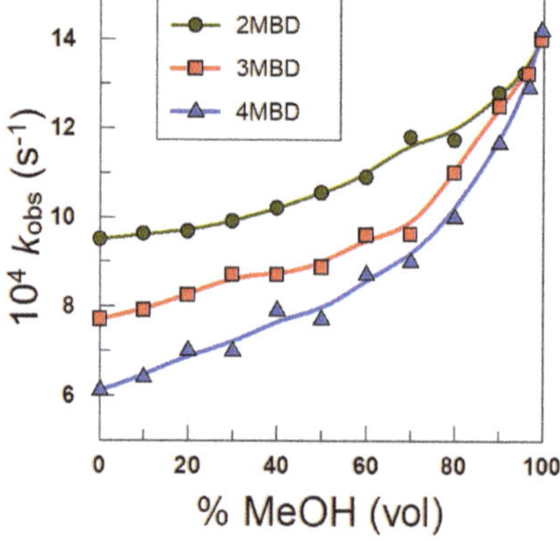

Figure 2. Effects of solvent composition on the solvolytic rate constants for the spontaneous decomposition of 2-, 3-, and 4-methylbenzenediazonium ions in acidic MeOH/H_2O mixtures. Experimental conditions: [ArN_2^+] = ~10^{-4} M, [HCl] = 10^{-2} M, T = 35 °C (2MBD, 3MBD), T = 60 °C (4MBD).

3.2. Effects of Added Electrolytes on Dediazoniation Rate Constants

The effects of electrolytes ([NaCl] = 0–1 M, [LiCl] = 0–1 M, and [LiCLO$_4^-$] = 0–1.5 M) on k_{obs} were determined at two representative solvent compositions (20% and 99.5% MeOH), as seen in Figure 3. At 20% MeOH, k_{obs} values, in the absence of salt, were similar to those in Figure 1. The addition of NaCl, LiCl, or LiClO$_4$ did not have a significant effect on k_{obs} and the values remained essentially constant. However, when the percentage of MeOH increased to 99.5%, the addition of NaCl and LiCl (up to 1.5 M) did not significantly change k_{obs}, but, upon increasing [LiClO$_4$], k_{obs} values decreased by approximately 40%. The observed decrease in k_{obs} values with the presence of ClO$_4^-$ ions, though somewhat modest, was unexpected and quite significant because no changes in k_{obs} were found in the presence of other salts and the decrease in k_{obs} was only detected at high percentages of MeOH. This is an important decrease that deserves further investigation, because we noted that k_{obs} values only had a two-fold change when going from 0 to 100% MeOH, as seen in Figure 3.

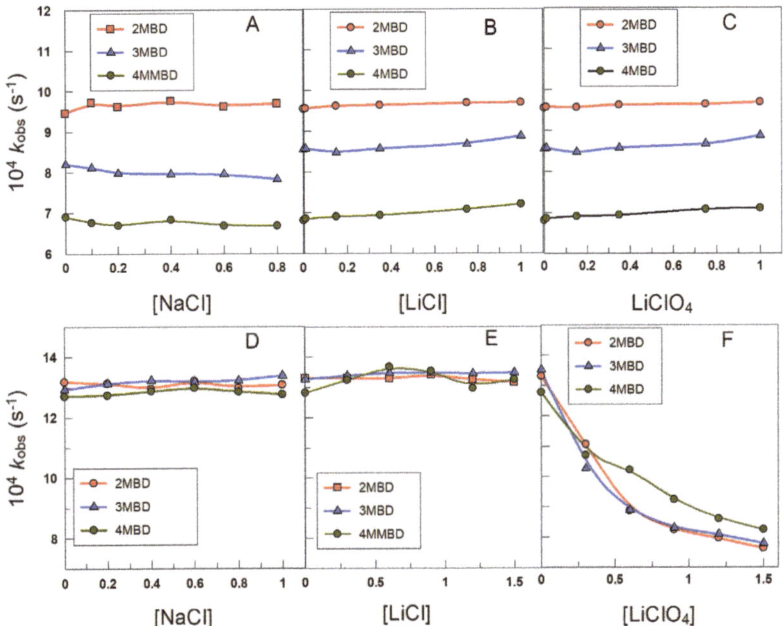

Figure 3. Effects of added electrolytes on k_{obs} for dediazoniation of 2MBD, 3MBD, and 4MBD in 20% MeOH/H$_2$O (**A–C**) and in 99.5% MeOH/H$_2$O (**D–F**) mixtures. k_{obs} values were determined spectrophotometrically by monitoring ArN$_2^+$ loss at λ = 305 nm (2MBD) and λ = 310 nm (3MBD, 4MBD). Experimental conditions: [ArN$_2^+$] = 10^{-4} M, [HCl] = 10^{-2} M, T = 35 °C (2MBD, 3MBD), T = 60 °C (4MBD).

To obtain further insights into the dediazoniation process, we employed the chromatographic technique, as illustrated in Figure 4, to determine k_{obs} for product formation, in the absence and in the presence of added electrolytes, at two selected solvent compositions (20% and 99.5% MeOH). The k_{obs} values for product formation were the same as those obtained spectrophotometrically for ArN$_2^+$ loss, confirming that products are formed competitively, in keeping with the predictions of the D$_N$ + A$_N$ mechanism shown in Scheme 2A.

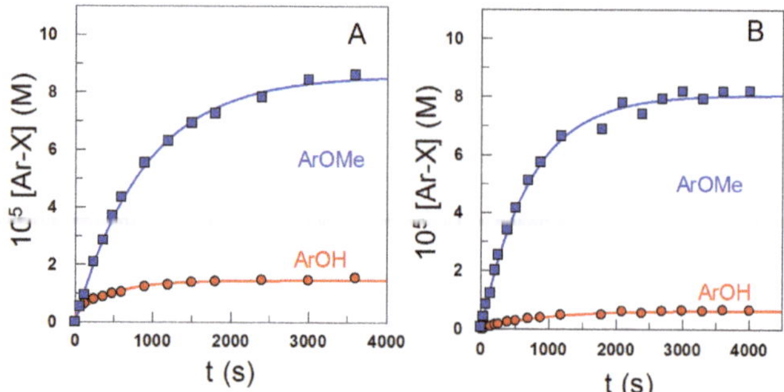

Figure 4. Illustrative determination of the rates of dediazoniation product formation in 99.5% MeOH/H$_2$O mixture as determined by HPLC. Rates are determined by fitting the variation in the concentration of a particular dediazoniation product with the time to a first order kinetic Equation (1). (**A**) 2MBD. (**B**) 4MBD. Experimental conditions: [2MBD]$_0$ = [4MBD$_0$] = 10^{-4} M, [HCl] = 10^{-2} M; T = 35 °C (2MBD), T = 60 °C (4MBD).

3.3. Effects of Added Electrolytes on Product Distribution

Figure 5 shows the effects of the solvent composition on the product distribution of 3- and 4MBD in the absence of added electrolytes. Only the heterolytic products (cresol and methyl phenyl ethers) were formed in significant yields, and the formation of the reduction product Ar-H was only detected in highly alcoholic solutions (4MBD) but its yield was very low when compared with those of heterolytic products; therefore, such a mechanism (homolytic) can be neglected. In all cases, the quantitative conversion to products was achieved in all composition ranges. The results are in agreement with published data [30,32,34].

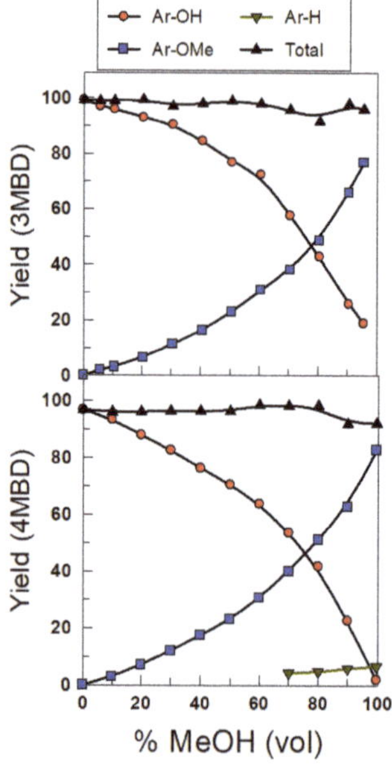

Figure 5. Variation of the percentage of solvolytic dediazoniation products in MeOH/H$_2$O binary mixtures in the absence of electrolytes. Experimental conditions were the same as those in Figure 2.

To analyze the effects of added electrolytes, we selected two methanol compositions, 20% and 99.5% MeOH. At a low methanol content, the addition of salts (NaCl and LiClO$_4$)

up to 1.5 M had a negligible effect on the production distribution, as illustrated in Figure 6. However, this was not the case at high percentages of MeOH, Figure 7.

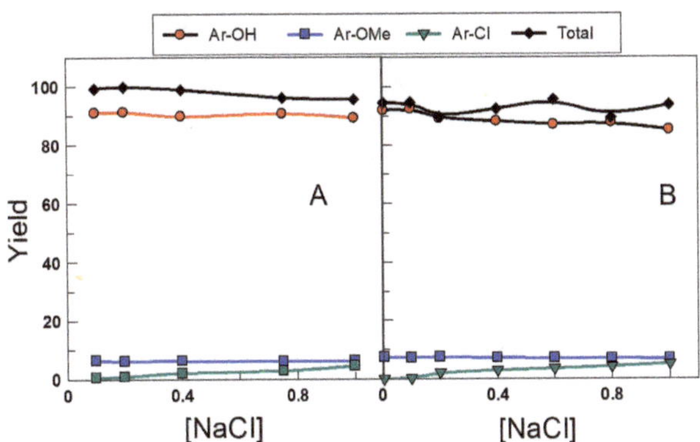

Figure 6. Effects of NaCl on dediazoniation product distribution in 20% MeOH/H_2O mixtures as determined by HPLC. (**A**) 3MBD, (**B**) 4MBD. Experimental conditions: $[3MBD_0] = [4MBD_0] = 10^{-4}$ M, $[HCl] = 10^{-2}$ M, $T = 35$ °C (3MBD), $T = 60$ °C (4MBD).

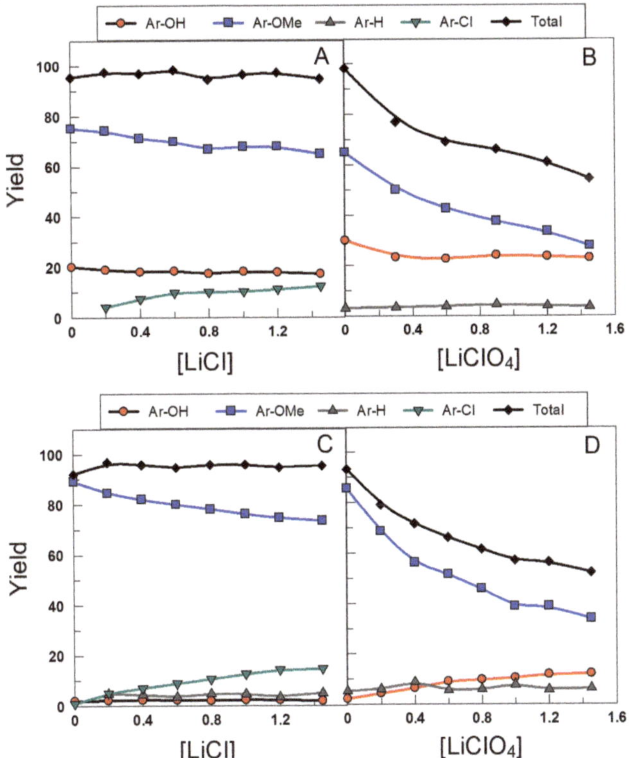

Figure 7. Effects of added electrolytes on the percentage of formation of dediazoniation products in 99.5% MeOH/H_2O mixtures. (**A,B**) 3MBD (**C,D**) 4MBD. Note the formation of the ArCl derivative at the expense of ArOMe (**A,C**). In the presence of $LiClO_4$ the ArCl derivative is not formed but the yield of ArOMe decreases substantially upon increasing $[LiClO_4]$) (**B,D**), attributed to the formation of the arylperchlorate derivative.

At 99.5% MeOH, in the presence of LiCl, the Ar-Cl derivative was detected, in addition to Ar-OH and Ar-OMe, and an increase in [LiCl] of up to 1.5 M had a negligible effect on the product distribution, as seen in Figure 7A,C, so that the quantitative conversion to products was obtained. As expected, when using LiClO$_4$ instead of LiCl, the Ar-Cl derivative was not detected but, surprisingly, a significant reduction in the yields of Ar-OMe was detected, as seen in Figure 7B,D. We note that the total yield decreases because of the presence of an unidentified dediazoniation product, whose yield could not be computed because we made no attempts to fully identify the new product, mainly for safety reasons (see Section 2).

4. Discussion

Solvolytic dediazoniations have long been studied, and there is now a substantial body of knowledge on the reactions. For instance, the sensitivity of dediazoniations towards solvent polarity has been analyzed by employing the Winstein–Grunwald equation [26,27], yielding slopes for 2MBD, 3MBD, and 4MBD of $m = -0.026$, -0.018, and -0.058, respectively. These values are quite low and reflect the astonishing insensitivity of dediazoniations to the nature of the solvent [1]. The low values are interpreted on the grounds of similarities in the structure and the charge allocation between the parent aryldiazonium ion and the corresponding aryl cation [14,50,51].

The selectivity values for a number of aryldiazonium ions have also been obtained, assuming that the nucleophilic attack on the carbocation is under kinetic control. In aqueous solution, aryldiazonium ions have a small selectivity against reacting nucleophiles, compared to water, typically ranging from 0.4 to 6 [46], following the order $Br^- > Cl^- > SO_4^{2-} > H_2O$. Such values are orders of magnitude lower than those observed for nucleophiles competing with water in reactions with stabilized carbocations [14,52,53], and are essentially constant with solvent composition. Table 1 shows the selectivity values for 2-, 3-, and 4-MBD. Such low values are consistent the formation of highly reactive aryl cations and with the preassociation of the stepwise mechanism, as shown in Scheme 1A.

Table 1. Selectivity values for 2-, 3-, and 4MBD. Data from references [14,22,54].

Methylbenzenediazonium Ions	$S_W^{Cl^-}$	S_W^{MeOH}
2MBD	2.7	0.40
3MBD	2.6	0.60
4MBD	2.7	0.74

Figure 5 shows that, at about 80% MeOH/H$_2$O, the yields of ArOH and ArOMe are equal to each other. That is, equal yields of dediazoniation products are obtained when the concentrations of nucleophiles in solution are approximately 8.3 M for [H$_2$O] and approximately 14.6 M for [MeOH], indicating that ArN$_2^+$ ions show favored solvation with H$_2$O molecules. This preferential solvation is fully consistent with the moderate variation in k_{obs} on changing the MeOH content of the solvent, Figure 2. Moreover, barriers for the capture of carbocations by the solvent [55] are sufficiently small, so that the product distribution reflects the solvent distribution in the first solvation shell of the reactant, suggesting, therefore, that the solvation of the ground state ArN$_2^+$ is essentially the same as that in the proximity of the transition state. Thus, at low MeOH content, the presumed aryl cations are mostly solvated by water molecules, and a relatively modest increase in the concentrations of NaCl, LiCl, or LiClO$_4$ (in comparison to that of water) does not have significant effects on both the dediazoniation rate constants (Figure 3), and on product distributions (Figure 6).

The sensitivity of dediazoniations to changes in the solvent polarity can be illustrated by means of a Winstein–Gruwald plot, as shown in Figure 8, where a linear relationship between $\log(k/k_w)$ with the Y parameter is found, with slopes of $m = -0.032 \pm 0.001$ (2MBD), $m = -0.037 \pm 0.001$ (3MBD), and $m = -0.072 \pm 0.003$ (4MBD), consistent with the low selectivity of aryldiazonium ions to solvent effects [52,56]. This low sensitivity can be attributed to the similarity in the structure and charge distribution of the parent aryldiazonium ion, compared to that of the corresponding aryl cation.

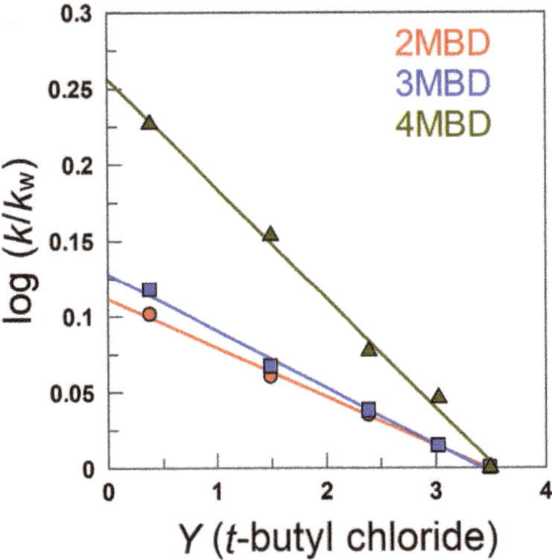

Figure 8. Winstein–Grunwald plots for 2MBD, 3MBD, and 4MBD. Y values were collected from Leffer et al. [57].

Surprisingly, when dediazoniations are carried out in solutions with high concentrations of MeOH, a different behavior is observed. The addition of NaCl or LiCl does not modify k_obs values, as seen in Figure 3D,E, and the analysis of the product distribution confirms the expected formation of the chlorobenzene derivatives. However, upon the addition of $LiClO_4$, both k_obs (Figure 3F) and the yields of the corresponding methyl phenyl ethers (ArOMe, Figure 7B,D) decrease. We note that chromatograms do not show the formation of the reduction products (toluenes, Ar-H) in a large extent and, thus, we can safely presume that the mechanism of the reaction does not change in the presence of $LiClO_4$. Thus, the variations in the yields of Ar-OMe can be attributed to the replacement of MeOH molecules in the solvation shell by ClO_4^- ions, presumably leading to the formation of the potentially explosive (in the solid state) aryldiazonium perchlorates.

Our hypothesis regarding perchlorate ions modifying the solvation shell of aryldiazonium ions seems to be supported by the results of Cruz et al. [58], who carried out a molecular dynamic study of dediazoniations and calculated the local concentrations of solvents around diazonium ions, concluding that, in the absence of salts, the local solvent composition varied linearly on the molar fraction of water in methanol/water mixtures, although they found non-linear variations for other solvent mixtures. The same authors also found that the calculated number of solvent molecules around the aryldiazonium cation can predict the product distribution when preferential solvation is considered. However, results must be taken with some caution because of the failure to identify the new product formed in the presence of $LiClO_4$.

The rates of dediazoniations with uncharged nucleophiles (such as MeOH or H_2O) are hardly affected by the nature of the solvent, as shown in Figure 2, because the electric charges are neither destroyed nor created in the rate-determining step. In fact, the rates of heterolytic dediazoniations have been reported to vary only by a factor of 9 when carried out in 19 solvents [1]. Thus, the low variation in k_obs with [MeOH] suggests that the solvation of the ground-state is similar to the proximity of the transition state; otherwise, unimolecular reactions, where the nucleophilic attack of the solvent is rate-determining, would lead to a strong dependence of k_obs on nucleophile concentration, which is not observed. The rates of the nucleophilic attack on aryl cations have been reported to be close to the diffusion control limit ($\sim10^9$ M$^{-1}\cdot$s^{-1}) and the preassociation of the nucleophile with the aryl cation does not account for much of the trapping [1,2,46].

A different situation is found when charged nucleophiles are present in the solvation shell, as they may distort the charge distribution, compared to that of neutral nucleophiles. The Mayr's reactivity scale is certainly useful to predict whether an electrophile/nucleophile reaction can be expected to take place at room temperature, or to predict the selectivity of competing nucleophiles. The rate constants of electrophile–nucleophile combination reactions are mostly dependent on four factors: nucleophilicity, electrophilicity, temperature, and solvent. Strictly, it is not possible to separate these four factors, mainly because the values of the electrophilicity and nucleophilicity parameters are usually linked to the type of solvent, whose physical properties may change with changing temperatures. Nucleophilicity scales may also be affected by changes in temperature [33,59]. However, to our knowledge, an all-embracing theory of electrophile–nucleophile reactions is still not in sight, and further work on the effects of electrolytes on ionic reactions is needed.

Supplementary Materials: The following supporting information can be downloaded at: https://www.mdpi.com/article/10.3390/compounds2010005/s1, Figure S1: The chromatogram in Figure 1.

Author Contributions: Conceptualization C.B.-D.; investigation, S.L.-B. and C.B.-D.; writing—original draft preparation, C.B.-D.; writing—review and editing, S.L.-B., C.B.-D.; visualization. S.L.-B. and C.B.-D. All authors have read and agreed to the published version of the manuscript.

Funding: This research received no external funding.

Data Availability Statement: Not applicable.

Acknowledgments: The authors thank University of Vigo (Dep. C11) for financial support during the sabbatical leave of C.B.-D.

Conflicts of Interest: The authors declare no conflict of interest.

References

1. Zollinger, H. *Diazo Chemistry I: Aromatic and Heteroaromatic Compounds*; VCH: Weinheim, Germany, 1994; Volume 107, p. 1917.
2. Bravo Díaz, C. Diazohydroxides, diazoethers and related species. In *The Chemistry of Hydroxylamines, Oximes and Hydroxamic Acids*; Rappoport, Z., Liebman, J.F., Eds.; J. Wiley & Sons: Chichester, UK, 2011; Volume 2, p. 853.
3. Firth, J.D.; Fairlamb, I.J.S. A Need for Caution in the Preparation and Application of Synthetically Versatile Aryl Diazonium Tetrafluoroborate Salts. *Org. Lett.* **2020**, *22*, 7057–7059. [CrossRef] [PubMed]
4. Trusova, M.E.; Kutonova, K.V.; Kurtukov, V.V.; Filimonov, V.D.; Postnikov, P.S. Arenediazonium salts transformations in water media: Coming round to origins. *Resour.-Effic. Technol.* **2016**, *2*, 36–42. [CrossRef]
5. Sengupta, S.; Chandrasekaran, S. Modifications of amino acids using arenediazonium salts. *Org. Biomol. Chem.* **2019**, *17*, 8308–8329. [CrossRef] [PubMed]
6. Habraken, E.R.M.; van Leest, N.P.; Hooijschuur, P.; de Bruin, B.; Ehlers, A.W.; Lutz, M.; Slootweg, J.C. Aryldiazonium Salts as Nitrogen-Based Lewis Acids: Facile Synthesis of Tuneable Azophosphonium Salts. *Angew. Chem. Int. Ed.* **2018**, *57*, 11929–11933. [CrossRef]
7. Roglans, A.; Pla-Quintana, A.; Moreno-Mañas, M. Diazonium Salts as Substrates in Palladium-Catalyzed Cross-Coupling Reactions. *Chem. Rev.* **2006**, *106*, 4622–4643. [CrossRef]
8. Mo, F.; Dong, G.; Zhanga, Y.; Wang, J. Recent applications of arene diazonium salts in organic synthesis. *Org. Biomol. Chem.* **2013**, *11*, 1582–1593. [CrossRef]
9. Venkatesh, R.; Singh, A.K.; Lee, Y.R.; Kandasamy, J. Palladium-catalyzed synthesis of α-aryl acetophenones from styryl ethers and aryl diazonium salts via regioselective Heck arylation at room temperature. *Org. Biomol. Chem.* **2021**, *19*, 7832–7837. [CrossRef]
10. Riemer, N.; Shipman, M.; Wessig, P.; Schmidt, B. Iterative Arylation of Itaconimides with Diazonium Salts through Electrophilic Palladium Catalysis: Divergent β-H-Elimination Pathways in Repetitive Matsuda–Heck Reactions. *J. Org. Chem.* **2019**, *84*, 5732–5746. [CrossRef]
11. Habraken, E.R.M.; Jupp, A.R.; Slootweg, J.C. Diazonium Salts as Nitrogen-Based Lewis Acids. *Synlett* **2019**, *30*, 875–884. [CrossRef]
12. Koziakov, D.; Wu, G.; Jacobi von Wangelin, A. Aromatic substitutions of arenediazonium salts via metal catalysis, single electron transfer, and weak base mediation. *Org. Biomol. Chem.* **2018**, *16*, 4942–4953. [CrossRef]
13. Zollinger, H. *Color Chemistry*; VCH: Weinheim, Germany, 1991.
14. Pazo-LLorente, R.; Rodriguez-Sarabia, M.J.; Gonzalez-Romero, E.; Bravo-Díaz, C. Hydroxy- and Chloro-Dediazoniation of 2- and 3-Methylbenzenediazonium Tetrafluoroborate in Aqueous Solution. *Int. J. Chem. Kinet.* **1999**, *31*, 73. [CrossRef]

15. Delamar, M.; Hitmi, R.; Pinson, J.; Saveant, J.M. Covalent modification of carbon surfaces by grafting of functionalized aryl radicals produced from electrochemical reduction of diazonium salts. *J. Am. Chem. Soc.* **1992**, *114*, 5883–5884. [CrossRef]

16. Pinson, J. Attachment of Organic Layers to Materials Surfaces by Reduction of Diazonium Salts. In *Aryl Diazonium Salts*; Chehimi, M.M., Ed.; Wiley-VCH Verlag and Co.: Weinheim, Germany, 2012.

17. Mohamed, A.A.; Salmi, Z.; Dahoumane, S.A.; Mekki, A.; Carbonnier, B.; Chehimi, M.M. Functionalization of nanomaterials with aryldiazonium salts. *Adv. Colloid Interface Sci.* **2015**, *225*, 16–36. [CrossRef] [PubMed]

18. Pinson, J.; Podvorika, F. Attachment of organic layers to conductive or semiconductive surfaces by reduction of diazonium salts. *Chem. Soc. Rev.* **2005**, *34*, 429. [CrossRef]

19. Chehimi, M.M. *Aryl Diazonium Salts: New Coupling Agents in Polymer and Surface Science*; Chehimi, M.M., Ed.; Wiley-VCH: Weinheim, Germany, 2012.

20. Hetemi, D.; Noël, V.; Pinson, J. Grafting of Diazonium Salts on Surfaces: Application to Biosensors. *Biosensors* **2020**, *10*, 4. [CrossRef]

21. Saunders, K.H.; Allen, R.L.M. *Aromatic Diazo Compounds*, 3rd ed.; Edward Arnold: Baltimore, MD, USA, 1985.

22. Pazo-Llorente, R.; Sarabia-Rodriguez, M.J.; Gonzalez-Romero, E.; Bravo-Díaz, C. Solvolysis of o-methylbenzenediazonium Tetrafluoroborate in acidic methanol-water mixtures. Further evidence for nucleophilic attack on a solvent separated aryl cation. *Int. J. Chem. Kinet.* **1999**, *31*, 531. [CrossRef]

23. Pazo-Llorente, R.; Bravo-Díaz, C.; González-Romero, E. Solvolyisis of some arenediazonium salts in binary EtOH/H$_2$O mixtures under acidic conditions. *Eur. J. Org. Chem.* **2003**, *2003*, 3421. [CrossRef]

24. Pazo-Llorente, R.; Bravo-Díaz, C.; González-Romero, E. pH Effects on Ethanolysis of Some Arenediazonium Ions: Evidence for Homolytic Dediazoniation Proceeding through Formation of Transient Diazo Ethers. *Eur. J. Org. Chem.* **2004**, *2004*, 3221. [CrossRef]

25. Pazo-Llorente, R.; Bravo-Díaz, C.; González-Romero, E. Monitoring Micelle Breakdown by Chemical Trapping. *Langmuir* **2003**, *19*, 9142. [CrossRef]

26. Pazo-Llorente, R.; Maskill, H.; Bravo-Díaz, C.; González-Romero, E. Dediazoniation of 4-nitrobencenediazonium ions in acidic MeOH/H$_2$O mixtures: Role of acidity and MeOH concentration on the formation of transient diazo ethers that initiate homolytic dediazoniation. *Eur. J. Org. Chem.* **2006**, *2006*, 2201. [CrossRef]

27. Canning, S.J.; McCruden, K.; Maskill, H.; Sexton, B. Dediazoniation reactions of arenediazonium ions under solvolytic conditions: Fluoride anion abstraction from trifluoroethanol and α-hydrogen atom abstraction from ethanol. *Chem. Commun.* **1998**, *18*, 1971–1972. [CrossRef]

28. Canning, P.S.J.; McCrudden, K.; Maskill, H.; Sexton, B. Rates and mechanisms of the thermal solvolytic decomposition of arendiazonium ions. *J. Chem. Soc. Perkin Trans. 2* **1999**, *12*, 2735. [CrossRef]

29. Canning, P.S.J.; Maskill, H.; Mcrudden, K.; Sexton, B. A product analytical study of the thermal and photolytic decomposition of some arenediazonium ions in solution. *Bull. Chem. Soc. Jpn.* **2002**, *75*, 789. [CrossRef]

30. Mayer, R.J.; Ofial, A.R. Nucleophilic Reactivities of Bleach Reagents. *Org. Lett.* **2018**, *20*, 2816–2820. [CrossRef]

31. Isborn, C.; Hrovat, D.A.; Borden, W.T.; Mayer, J.M.; Carpenter, B.K. Factors Controlling the Barriers to Degenerate Hydrogen Atom Transfers. *J. Am. Chem. Soc.* **2005**, *127*, 5794. [CrossRef]

32. Minegishi, S.; Kobayashi, S.; Mayr, H. Solvent Nucleophilicity. *J. Am. Chem. Soc.* **2004**, *126*, 5174–5181. [CrossRef]

33. Mayr's Database of Reactivity Parameters. Available online: https://www.cup.lmu.de/oc/mayr/reaktionsdatenbantk2/ (accessed on 10 January 2022).

34. Grunwald, E.; Winstein, S.J. The Correlation of Solvolysis Rates. *Am. Chem. Soc.* **1948**, *70*, 846. [CrossRef]

35. Maskill, H. *The Physical Basis of Organic Chemistry*; Oxford University Press: Oxford, UK, 1995.

36. Carey, F.A.; Sundberg, R.J. *Structure and Mechanism, Part A*; Plenum Press: New York, NY, USA, 1993.

37. Sengupta, S.; Bhattacharya, S. Heck Reaction of Arenediazonium Salts: A PAlladium-Catalysed Reaction in an Aqueous Medium. *J. Chem. Soc. Perkin Trans. 1* **1993**, 1943–1944. [CrossRef]

38. Sheng, M.; Frurip, D.; Gorman, D. Reactive chemical hazards of diazonium salts. *J. Loss Prev. Process Ind.* **2015**, *38*, 114–118. [CrossRef]

39. Ullrich, R.; Grewer, T. Decomposition of aromatic diazonium compounds. *Thermochim. Acta* **1993**, *225*, 201–211. [CrossRef]

40. Urbansky, E.T. Perchlorate Chemistry: Implications for Analysis and Remediation. *Bioremediat. J.* **1998**, *2*, 81–95. [CrossRef]

41. Clark, J.J.J. Toxicology of Perchlorate. In *Perchlorate in the Environment*; Urbansky, E.T., Ed.; Springer: Boston, MA, USA, 2000; Volume 57.

42. Garcia-Meijide, M.C.; Bravo-Diaz, C.; Romsted, L.S. A Novel Method for Monitoring Dediazoniations: Simultaneous Monitoring of Rates and Product Distributions of 4-methylbenzenediazonium Tetrafluoroborate. *Int. J. Chem. Kinet.* **1998**, *30*, 31–39. [CrossRef]

43. Bravo-Diaz, C.; Soengas-Fernandez, M.; Rodriguez-Sarabia, M.J.; Gonzalez-Romero, E. Effects of Monovalent and Divalent Anionic Dodecylsulfate Surfactants on the Dediazoniation Reaction of 2-, 3-, and 4-Methylbenzenediazonium Tetrafluorborate. *Langmuir* **1998**, *14*, 5098. [CrossRef]

44. Yamamoto, H.; Ichikawa, K.; Tokunaga, J. Solubility of helium in methanol + water, ethanol + water, 1-propanol + water, and 2-propanol + water solutions at 25 degree.C. *J. Chem. Eng. Data* **1994**, *39*, 155–157. [CrossRef]

45. Fernández-Alonso, A.; Bravo-Diaz, C. Methanolysis of 4-bromobenzenediazonium ions. Effects of acidity, [MeOH] and temperature on the formation and decomposition of diazo ethers that initiate homolytic dediazoniation. *Org. Biomol. Chem.* **2008**, *6*, 4004–4011. [CrossRef] [PubMed]

46. Cuccovia, I.M.; da Silva, M.A.; Ferraz, H.M.C.; Pliego, J.R.; Riveros, J.M.; Chaimovich, H. Revisiting the reactions of nucleophiles with arenediazonium ions. *J. Chem. Soc. Perkin Tans. 2* **2000**, *9*, 1896–1907. [CrossRef]

47. Lewis, E.S.; Hartung, L.D.; McKay, B.M. Reaction of diazonium salts with nucleophiles. XIII. Identity of the rate- and product-determining steps. *J. Am. Chem. Soc.* **1969**, *91*, 419–425. [CrossRef]

48. Bravo-Díaz, C.; González-Romero, E. Monitoring dediazoniation product formation by high-performance liquid chromatography after derivatization. *J. Chromatog. A* **2003**, *989*, 221–229. [CrossRef]

49. Crossley, M.L.; Kienle, R.H.; Benbrook, C.H. Chemical Constitution and Reactivity. I. Phenyldiazonium Chloride and its Mono Substituted Derivatives. *J. Am. Chem. Soc.* **1940**, *62*, 1400–1404. [CrossRef]

50. Glaser, R.; Horan, C.J. Benzenediazonium Ion. Generality, Consistency and Preferability of the Electron Density Based Dative Bonding Model. *J. Org. Chem.* **1995**, *60*, 7518–7528. [CrossRef]

51. Glaser, R.; Horan, C.J.; Lewis, M.; Zollinger, H. σ-Dative and π-Backdative Phenyl Cation−Dinitrogen Interactions and Opposing Sign Reaction Constants in Dual Substituent Parameter Relations. *J. Org. Chem.* **1999**, *64*, 902–913. [CrossRef] [PubMed]

52. Lowry, T.H.; Richardson, K.S. *Mechanism and Theory in Organic Chemistry*, 3rd ed.; Harper-Collins Pub.: New York, NY, USA, 1987.

53. Ritchie, C.D. *Physical Organic Chemistry: The Fundamental Concepts*, 2nd ed.; Marcel Dekker: New York, NY, USA, 1990.

54. Romsted, L.S. Interfacial Compositions of Surfactant Assemblies by Chemical Trapping with Arenediazonium Ions: Method and Applications. In *Reactions and Synthesis in Surfactant Systems*; Texter, J., Ed.; Marcel-Dekker: New York, NY, USA, 2001.

55. Finnenman, J.I.; Fishbein, J.C. Mechanisms of benzyl group transfer in the decay of e-arylmethanediazoates in aqueous solution. *J. Am. Chem. Soc.* **1995**, *117*, 4228. [CrossRef]

56. Pross, A. *Theoretical & Physical Principles of Organic Reactivity*; J. Wiley & Sons: New York, NY, USA, 1995.

57. Leffler, J.E.; Grunwald, E. *Rates and Equilibria of Organic Reactions*; Dover: New York, NY, USA, 1989.

58. Cruz, G.N.; Lima, F.S.; Dias, L.G.; El Seoud, O.A.; Horinek, D.; Chaimovich, H.; Cuccovia, I.M. Molecular Dynamics Simulations of the Initial-State Predict Product Distributions of Dediazoniation of Aryldiazonium in Binary Solvents. *J. Org. Chem.* **2015**, *80*, 8637–8642. [CrossRef]

59. Mayr, H.; Patz, M. Scales of Nucleophilicity and Electrophilicity: A System for Ordering Polar Organic and Organometallic Reactions. *Angew. Chem. Int. Ed. Engl.* **1994**, *33*, 938–957. [CrossRef]

Review

Electrophilic Iodination of Organic Compounds Using Elemental Iodine or Iodides: Recent Advances 2008–2021: Part I

Njomza Ajvazi [1,*] and Stojan Stavber [1,2]

1   Jožef Stefan International Postgraduate School, Jamova 39, 1000 Ljubljana, Slovenia; stojan.stavber@ijs.si
2   Jožef Stefan Institute, Jamova 39, 1000 Ljubljana, Slovenia
*   Correspondence: njomza.ajvazi@rezonanca-rks.com; Tel.: +386-44-258-553

Abstract: The iodination of organic compounds is of great importance in synthetic organic chemistry. It opens comprehensive approaches for the synthesis of various biologically active compounds. The recent advances in iodination of organic compounds using elemental iodine or iodides, covering the last thirteen years, are the objective of the present review.

Keywords: iodination; alkanes; alkenes; alkynes; alkyl carbonyls; elemental iodine; iodides

Citation: Ajvazi, N.; Stavber, S. Electrophilic Iodination of Organic Compounds Using Elemental Iodine or Iodides: Recent Advances 2008–2021: Part I. *Compounds* **2022**, *2*, 3–24. https://doi.org/10.3390/compounds2010002

Academic Editor: Juan Mejuto

Received: 21 November 2021
Accepted: 15 December 2021
Published: 5 January 2022

Publisher's Note: MDPI stays neutral with regard to jurisdictional claims in published maps and institutional affiliations.

1. Introduction

The introduction of iodine as an available, inexpensive, environmentally friendly element into organic molecules has attracted a significant interest providing versatile building blocks in synthetic organic chemistry [1]. Iodine compounds have widespread use in organic chemistry. Iodinated compounds are frequently used as reagents in organic synthesis [2]. The best choice for the iodination of organic compounds is the use of molecular iodine or the iodide anion in combination with environmentally friendly and atom-efficient oxidants in the presence of desirable solvents or under solvent-free protocols, thus enhancing the green chemical profile of the iodination process [3]. Molecular iodine is extensively employed for α-iodination of alkyl carbonyl compounds [4]. It could act as a catalyst promoting enolisation and as a reagent reacting with enol to afford α-iodocarbonyl compounds [5]. Comprehensive synthetic protocols for electrophilic iodination of organic compounds employing I_2 or I^- were reviewed by Stavber and co-workers in 2008 [6]. The reports covered by this review [6] mainly carried low green chemical profiles, thus challenging greener-related protocols. Thus, in the present review, as part I of the matter, related protocols have been elaborated to highlight recent advances in the iodination of organic compounds, including alkanes, alkenes, alkynes, and alkyl carbonyls using elemental iodine or iodides, covering the period from 2008–2021.

2. Iodination of Alkanes

In 2008, Sudalai and co-workers [7] presented $NaIO_4$/KI/NaN_3 an efficient system for mono- and 1,2-difunctionalization of hydrocarbons through activation of C-H bond, providing vicinal azido- and acetoxy iodinations of cyclic hydrocarbons in high yields (Scheme 1). This protocol is successful for acyclic and cyclic alkanes.

Scheme 1. C-H activation of cyclohexane mediated by $NaIO_4$.

Homolytic cleavage of I-N$_3$ gives an azide radical, which removes a proton from cyclic alkane to provide an alkyl radical. The reaction of an alkyl radical with I$_2$ provides alkyl iodide followed by oxidative elimination giving alkene, which undergoes addition of either I-N$_3$ or I-OAc across the double bond (Scheme 2).

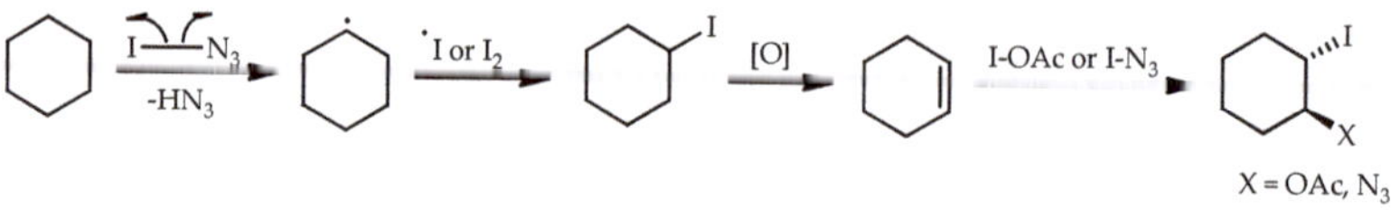

Scheme 2. Plausible reaction pathway.

In 2012, Yu and co-workers [8] reported diastereoselective C-H iodination with *i*-Pr- and *t*-Bu-substituted oxazoline auxiliaries catalyzed by palladium (II) (Scheme 3).

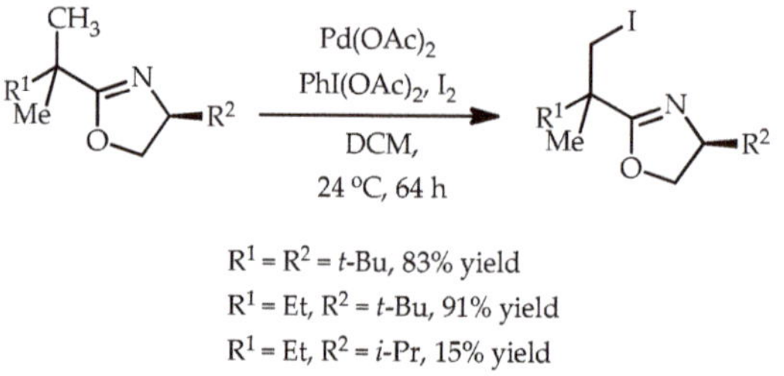

R^1 = R^2 = *t*-Bu, 83% yield

R^1 = Et, R^2 = *t*-Bu, 91% yield

R^1 = Et, R^2 = *i*-Pr, 15% yield

Scheme 3. Diastereoselective C-H iodination *i*-Pr- and *t*-Bu-substituted oxazoline auxiliaries catalyzed by Pd (II).

Zhu and co-workers [9] have presented quinoline-based ligand-enabled palladium (II)-catalyzed iodination of various α-hydrogencontaining carboxylic acid and amino acids (Scheme 4).

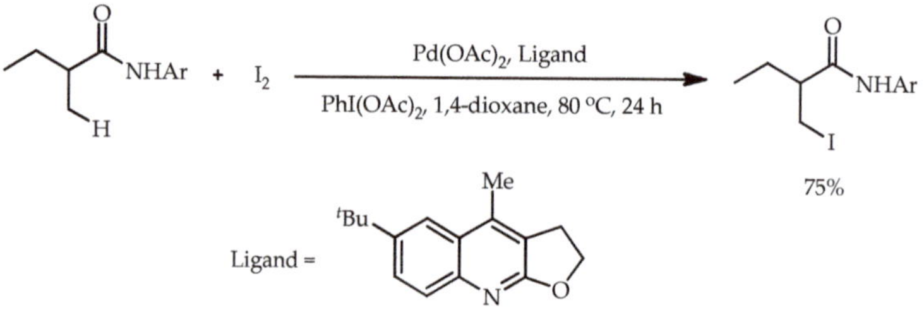

Scheme 4. C(sp^3)-H Iodination of carboxylic acid derivatives.

3. Iodination of Alkenes and Alkynes

The considerably breakthrough in the field of electrophilic iodination of alkenes was achieved already in 2005 [10], where, 30% aqueous H$_2$O$_2$ was used as the oxidant for the iodotransformation. However, later in 2008, Stavber and co-workers [11] even improved the green chemical profile of the reaction establishing an environmentally friendly methodology for aerobic oxidative iodination of alkenes using potassium iodide as iodine source catalyzed by sodium nitrite in acidic media providing the corresponding products in good to quantitative yields (Scheme 5). The authors have reported that the alkene was added as the last reactant to avoid polymerization. Moreover, in the case of phenyl-substituted alkynes (phenylethyne) using MeCN as the solvent, the formation of (*E*)-1,2-diiodo-1-phenyl-1-ethene was observed while, in the case of phenyl-1-propyne, a small amount of (Z)-1,2diiodophenylethene was also obtained (Scheme 6). The presence of an external nucleophile was observed to be beneficial for efficient and selective reactions.

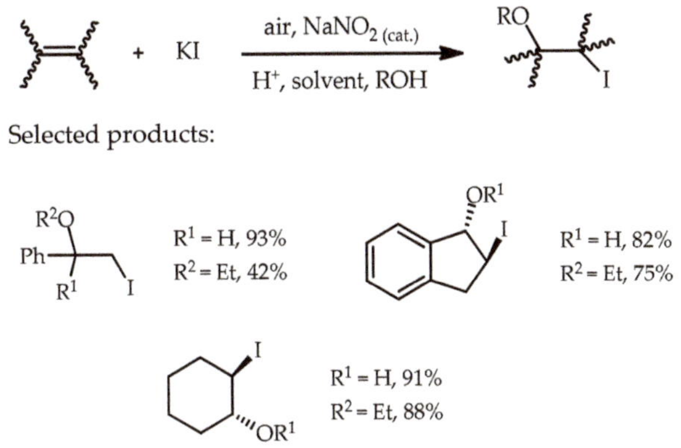
air, NaNO2 (cat.)
+ KI
H+, solvent, ROH
RO
I
Selected products:
R1 = H, 93%
R2 = Et, 42%
R1 = H, 82%
R2 = Et, 75%
R1 = H, 91%
R2 = Et, 88%

Scheme 5. Aerobic oxidative iodination of alkenes.

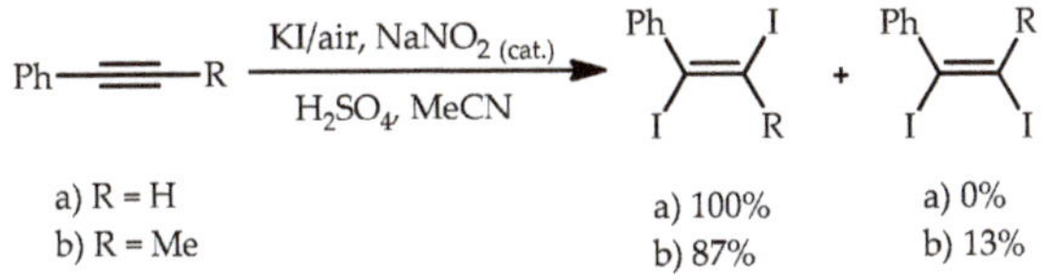
KI/air, NaNO2 (cat.)
H2SO4, MeCN
a) R = H
b) R = Me
a) 100%
b) 87%
a) 0%
b) 13%

Scheme 6. Aerobic oxidative iodination of phenyl-substituted alkynes.

Huang and Yang [12] have reported a novel and efficient method for synthesizing iodocyclopropylmethanol and 3-iodobut-3-en-1-ol derivatives via the iodohydroxylation of alkylidenecyclopropanes with I_2/H_2O system affording ring-opening or ring-keeping products in moderate to excellent yields (Scheme 7).

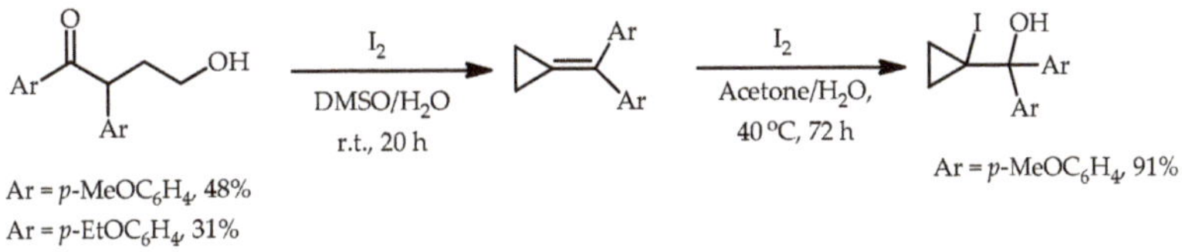
I2
DMSO/H2O
r.t., 20 h
I2
Acetone/H2O,
40 °C, 72 h
Ar = p-MeOC6H4, 48%
Ar = p-EtOC6H4, 31%
Ar = p-MeOC6H4, 91%

Scheme 7. Iodohydroxylation of alkylidenecyclopropanes with substituted aromatic rings.

A convenient and eco-friendly method for the synthesis of α-iodoketones from alkenes and alkynes has been reported by Yadav and co-workers [13] using 2-iodoxybenzoic acid (IBX)/I_2 as the effective reagent system in water providing the corresponding products in good to high yields (Scheme 8).

The authors reported that the interaction between alkene and iodine could generate a cyclic iodonium ion, which reacts with water to provide an iodohydrin. The iodohydrin reacts with IBX to afford α-iodoketone (Scheme 9).

An efficient and facile protocol for the iodomethoxylation and iodohydroxylation of alkenes and alkynes in the presence of *m*-iodosylbenzoic acid as a recyclable reagent has been achieved by Chi and co-workers [14] (Scheme 10).

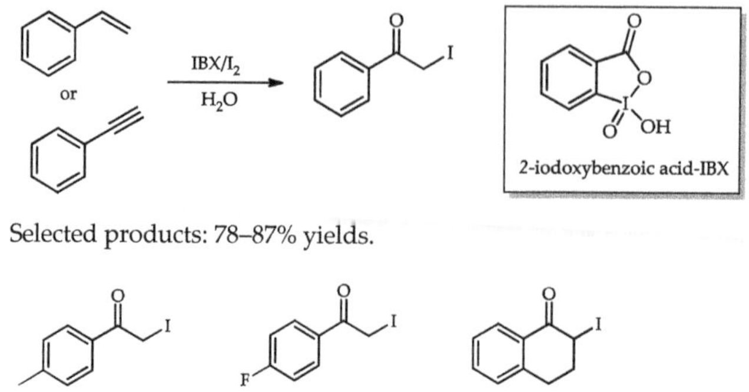

Selected products: 78–87% yields.

Scheme 8. Iodination of alkenes and alkynes mediated by IBX/I$_2$.

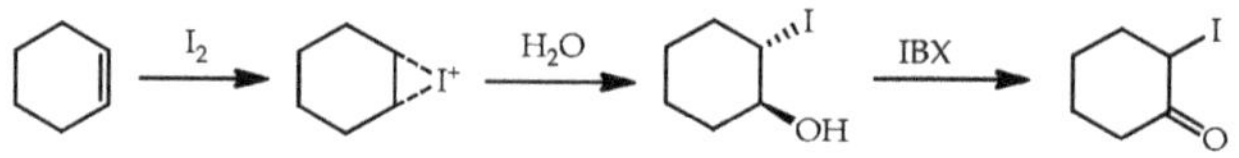

Scheme 9. Possible reaction pathway.

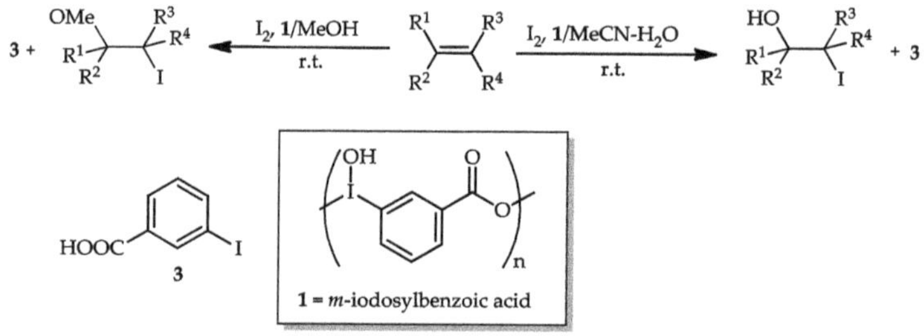

Scheme 10. Iodomethoxylation and iodohydroxylation of alkenes.

In the case of performing the reactions between alkynes and I$_2$/*m*-iodosylbenzoic acid/MeOH diiododimethoxylation products were obtained (Scheme 11).

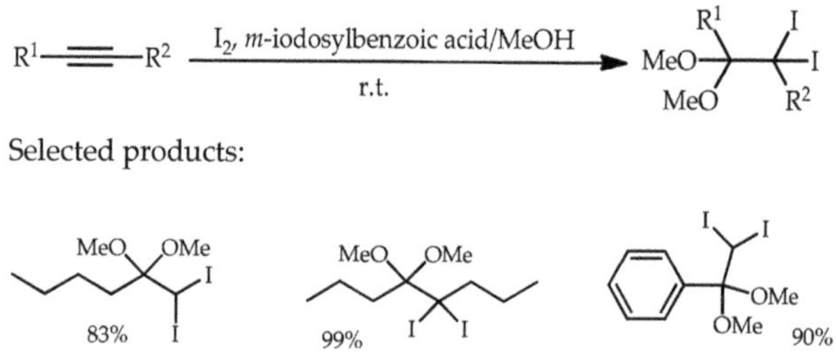

Selected products:

Scheme 11. Diiododimethoxylation of alkynes.

By treatment of the reaction mixture with an anionic exchange resin, pure iodo-functionalized products were provided. Unreacted *m*-iodosylbenzoic acid and reduced *m*-iodobenzoic acid are regenerated from the resin by treatment with HCl.

A simple, efficient, and practical method for the iodination of alkynyl enolates has been described using allenoates as the starting material through an alkynyl enolate as the intermediate. The reaction of the silyl ether of alkynyl enolate with iodine provides iodoallenoate in good yield [15]. Grigg and co-workers [16] developed a protocol for the synthesis of 1-C-(tetra-O-acetyl-β-D-galactopyranosyl)-2,3-diiodo-1-propene using β-C-galactose allene as a starting material with iodine in the presence of ethanol as a solvent (Scheme 12).

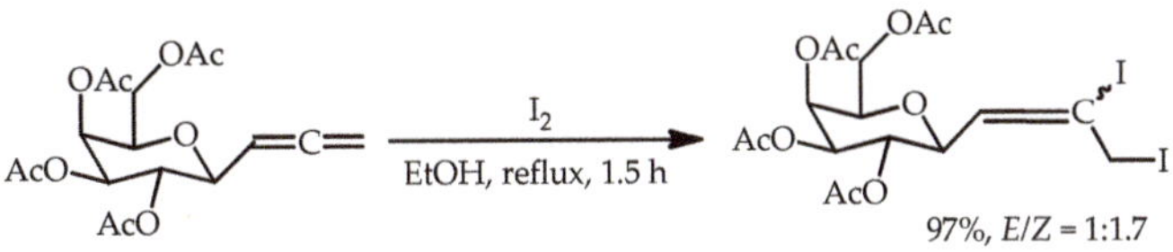

Scheme 12. The reaction of β-C-galactose allene with I_2.

A convenient method for synthesizing of vicinal halohydrins, haloacetates, and halo methyl ethers from olefins with 2:1 I^-/IO_3^- has been described [17]. Iodo reagent was found to be better for reaction with linear alkenes and the elimination of diiodo impurity (Scheme 13). I^-/IO_3^- was not successful for the vicinal functionalization of chalcones and stilbene.

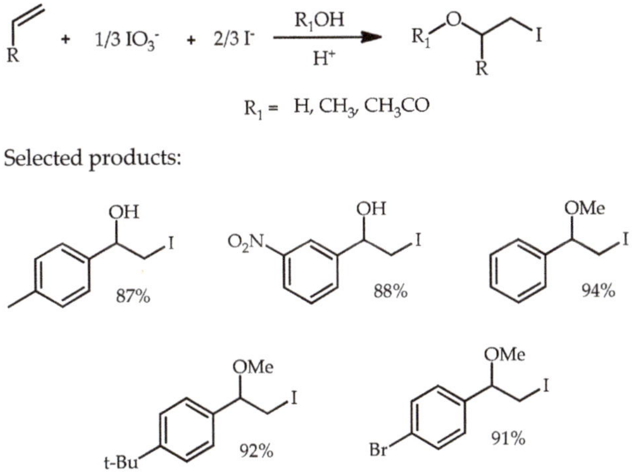

$$R_1 = \text{H, CH}_3\text{, CH}_3\text{CO}$$

Selected products:

Scheme 13. Vicinal functionalization of alkenes with 2:1 I^-/IO_3^- reagents.

Iodofluorination of electron-deficient olefins such as α,β-unsaturated esters, phosphonate, and amides with iodonium cation species generated by the anodic oxidation of iodide anion in Et_3N-5HF/$MeNO_2$ has been reported [18], providing the corresponding iodofluorinated products in good to moderate yields (Scheme 14).

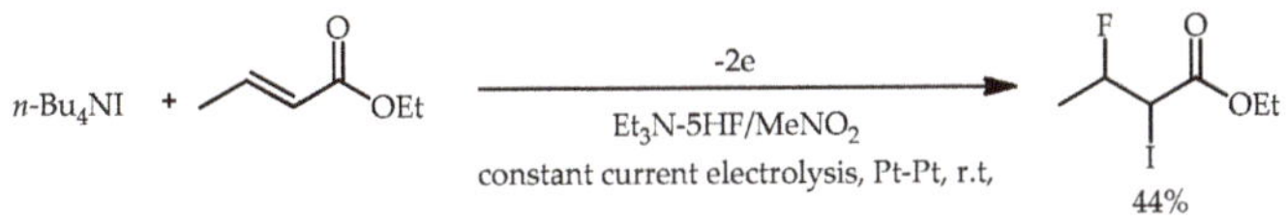

Scheme 14. Electrochemical iodofluorination of α,β-unsaturated esters.

An efficient route for the aerobic photo-oxidative synthesis of phenacyl iodides from styrenes, H_2O and I_2 has been reported by Itoh and co-workers [19], providing the corresponding products in moderate to high yields (Scheme 15).

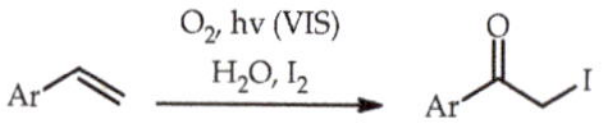

Scheme 15. Photo-oxidative synthesis of phenacyl iodides using I_2.

A simple and efficient method for azidoiodination of alkenes has been reported by Sudalai and co-workers [20] using $NaIO_4/KI/NaN_3$ combination. Through an anti-Markovnikov fashion, the regiospecific 1,2-azidoiodination proceeds to give β-iodoazides in quantitative yields (Scheme 16).

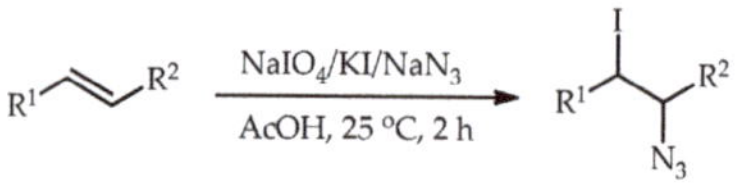

Scheme 16. Azidoiodination of alkenes mediated by $NaIO_4/KI/NaN_3$ system.

$NaIO_4$ oxidizes both KI and NaN_3 to release I_2 and an azide radical, a combination of which gives IN_3. Homolysis of IN_3 affords an azide radical, which reacts with alkenes to form alkyl radical species. The combination of an alkyl radical with either molecular iodine or iodine radical affords β-iodoazides (Scheme 17).

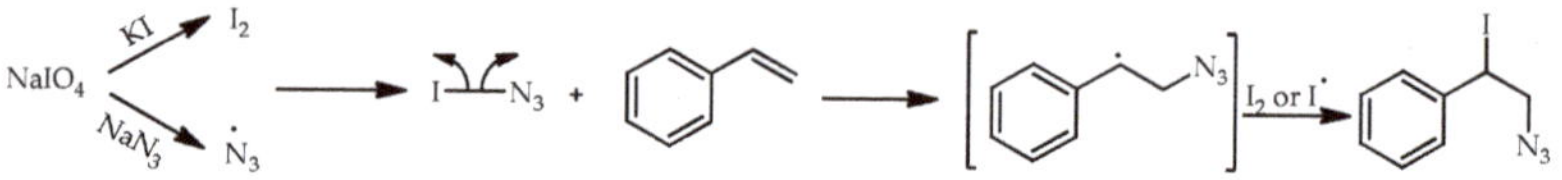

Scheme 17. Plausible reaction mechanism.

Hanessian and co-workers [21] developed a method for the total synthesis of Jerangolid A (shows antifungal activity) where I_2 was used for iodination of lactone in the presence of pyridine in DMF underwent further steps. Kuhakarn and co-workers [22] have described the method for the direct synthesis of β-keto sulfones between the reaction of sodium arensulfinates with alkenes, including styrene derivatives, and aliphatic alkenes mediated by *o*-iodoxybenzoic acid/iodine (IBX/I_2) (Scheme 18).

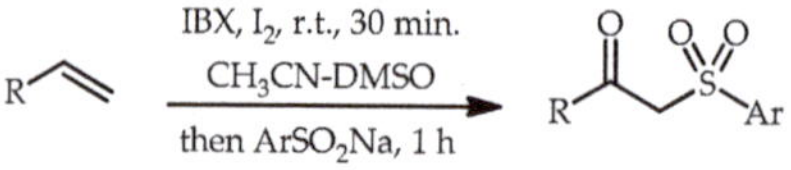

Scheme 18. Synthesis of Keto Sulfones mediated by IBX/I_2.

A new route for synthesizing of 5-hydroxypyrrolin-2-one derivatives from the modified Morita-Baylis-Hillman (MBH) adducts through CuI- mediated aerobic oxidation, allylic iodination, hydration of nitrile, and lactamization has been reported [23].

Krasutsky and co-workers [24] have reported the method for electrophilic monoiodination of terminal alkenes (Scheme 19).

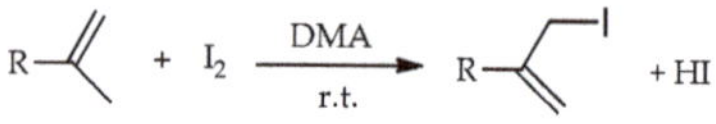

Scheme 19. Monoiodination of terminal alkenes.

The use of the oxidative system (*t*-BuOCl + NaI) as an efficient oxidant and N-iodinating reagent with triflamide and cyclic dienes in acetonitrile for providing 1,1,1-trifluoro-N-((1R,5R)-5-iodocyclopent-2-en-1-yl)methanesulfonamide in low yield has been reported by Shainyan and co-workers [25]. A co-iodination method for alkenes with (diacetoxyiodo)benzene (DAIB) and I_2 combination and different nucleophilic sources (MeCN-*nucleophile*) providing the corresponding products in moderate to high yields have been reported [26] (Scheme 20).

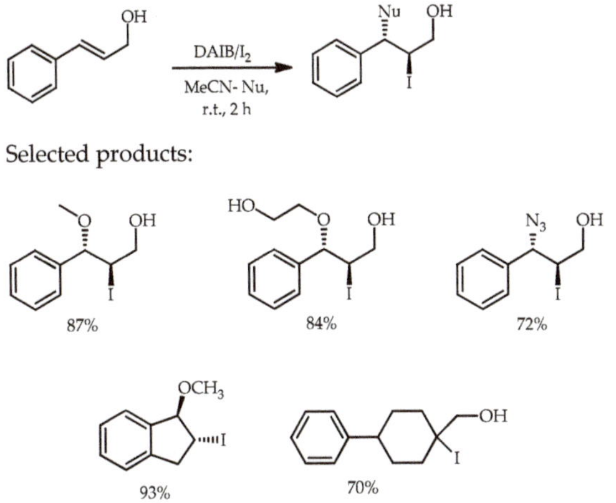

Selected products:

87%

84%

72%

93%

70%

Scheme 20. Co-Iodination of cinnamyl alcohol derivatives and alkenes.

Ma and co-workers [27] established stereoselective iodohydroxylation of 1,2-allenylic sulfoxides using iodine and benzyl thiol, providing 3-hydroxy-2-iodo-2(*E*)-alkenyl sulfides in the presence of MeCN/H_2O as a solvent.

[Bis(trifluoroacetoxy)iodo]benzene (PIFA) was used as a mediator for ethoxyiodination of enamides with potassium iodide, providing the corresponding products in good to quantitative yields [28]. A convenient method for the synthesis of vicinal iodohydrins and iodoesters from olefins has been reported by Narender and co-workers [29] via NH_4I and oxone system in MeCN/H_2O and DMF/DMA, under catalyst-free at room temperature, providing the corresponding products in good to quantitative yields. This protocol is realizable to various olefins, such as a terminal, symmetrical, and 1,2-disubstituted unsymmetrical olefins. Additionally, 1,2-disubstituted olefins provided excellent diastereoselectivity (Scheme 21).

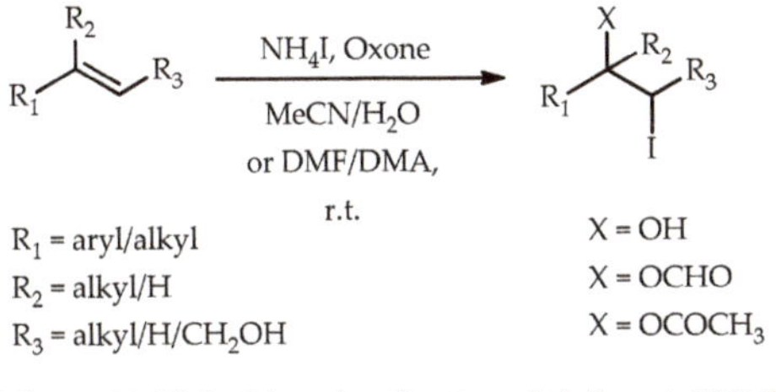

R_1 = aryl/alkyl
R_2 = alkyl/H
R_3 = alkyl/H/CH_2OH

X = OH
X = OCHO
X = $OCOCH_3$

Scheme 21. Vicinal functionalization of olefins via NH_4I/oxone system.

The protocol for the synthesis of iodovinylnaphthols using molecular iodine in the presence of MeCN as the solvent has been developed by Kumar and co-workers (Scheme 22) [30].

Since halohydrins are essential building blocks in organic synthesis and could be transformed to other organic intermediates such as amino-, azidoalcohols, and epoxides, Ning and co-workers [31] established an efficient protocol of iodohydroxylation of olefins with DMSO (dimethylsulfoxide) as an oxidant, an oxygen source, a solvent and HI generated in situ. It was reported that DMSO could oxidize haloanions to halocations under acidic conditions (Scheme 23).

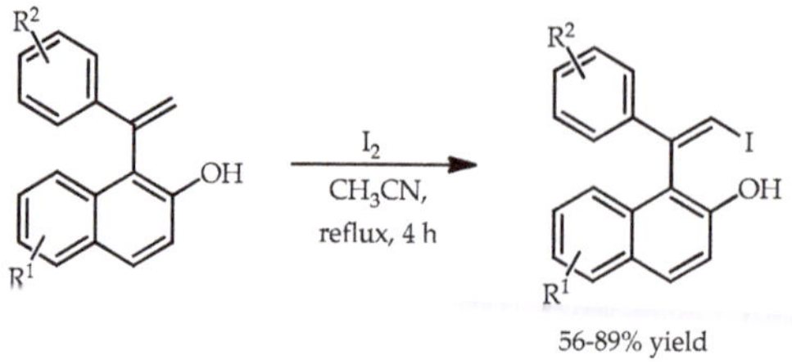

Scheme 22. Vinylic iodination of *ortho*-vinylnaphthols with I_2.

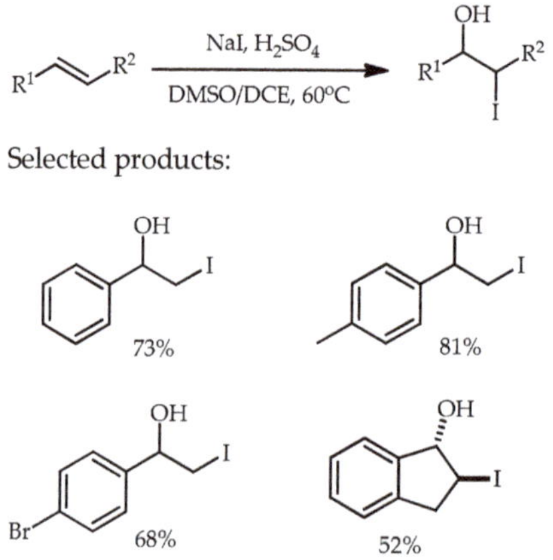

Selected products:

Scheme 23. Iodohydroxylation of olefins with NaI.

Zhu and co-workers [32] developed a one-pot and non-metal strategy for the direct vicinal difunctionalization of alkenes using iodine and *tert*-butyl hydroperoxide (TBHP) to synthesize 1-(*tert*-butylperoxy)-2-iodoethanes in good to high yields (Scheme 24). The method is realizable in the fields of organic synthesis, pharmacology, and medicinal chemistry.

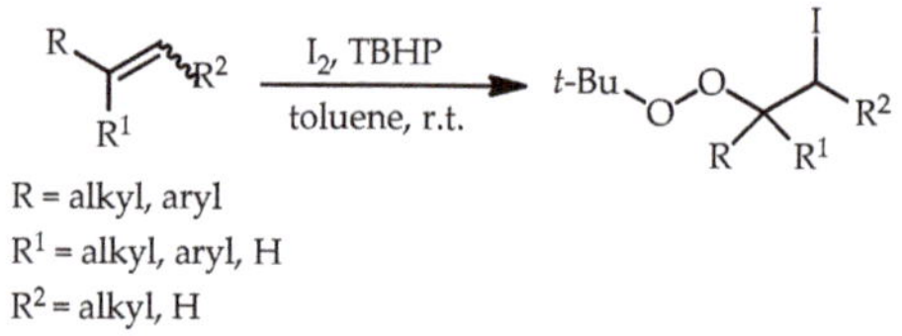

R = alkyl, aryl
R^1 = alkyl, aryl, H
R^2 = alkyl, H

Scheme 24. Synthesis of 1-(*tert*-butylperoxy)-2-iodoethanes with I_2/TBHP.

Iodine monoacetate as an effective reagent was used for the regio- and diastereoselective iodoacetoxylation of alkenes and alkynes. An inexpensive, non-metal, and environmentally friendly protocol for synthesizing iodine monoacetate was presented using iodine and oxone in acetic anhydride and acetic acid combination. It was shown that the reactions with styrene derivatives were more successful than allylic and aliphatic olefins, where regioisomeric mixtures in lower yields were observed (Scheme 25). Additionally, alkynes provided mono- and diiodinations in comparable yields [33].

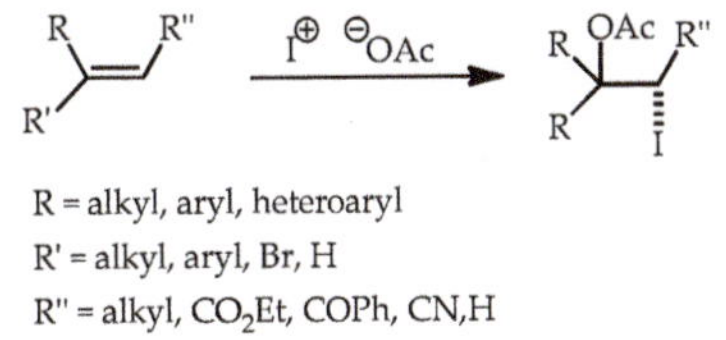

R = alkyl, aryl, heteroaryl
R' = alkyl, aryl, Br, H
R" = alkyl, CO$_2$Et, COPh, CN,H

Scheme 25. Oxyiodinations of alkenes with iodine monoacetate.

Liu and co-workers [34] developed the environmentally friendly method for the iodofunctionalization of olefins using I$_2$O$_5$ as the inorganic oxidant and LiI as the iodine source, in H$_2$O/acetone as the green solvent, providing the iodinated products in good to excellent yields (Scheme 26). The synthesis of different medicinal and agrochemical products could be realized through this method.

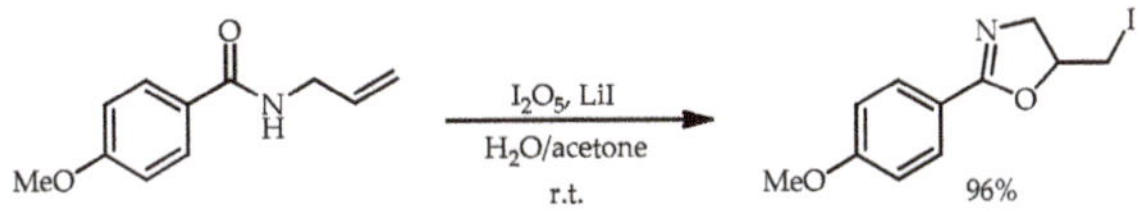

Scheme 26. Iodofunctionalization of olefins using I$_2$O$_5$/LiI.

It is assumed that I$_2$O$_5$ oxidizes LiI to form HOI, which acts as a reactive electrophilic intermediate to cause the electrophile-mediated cyclization. If olefin is tethered with a nucleophilic center, intramolecular cyclization will occur. In contrast, when a substrate without a nucleophilic site was used, iodohydrin adducts were provided *via* the intermolecular nucleophilic attack of water onto the iodonium ions.

Shakhmaev and co-workers [35] have developed an efficient protocol for the synthesis of ethyl 5-phenylpent-2-en-4-ynoate by olefination-dehydrohalogenation of 2-iodo-3-phenylprop-2-enal obtained by the reaction between cinnamaldehyde and molecular iodine in the presence of 4-(dimethylamino)pyridine (DMAP) as the catalyst.

Meng and co-workers [36] have discovered a convenient and efficient method for iodination of arylacetylenes using I$_2$ and DMAP (4-dimethylaminopyridine) (Scheme 27) and the iodination of aryl acetylenic ketones by employing K$_2$CO$_3$ as a base providing the corresponding products in good to excellent yields (Scheme 28).

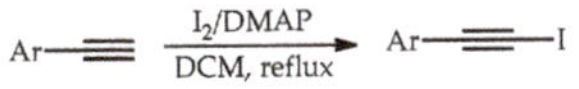

Scheme 27. Iodination of terminal acetylenes with I$_2$/DMAP.

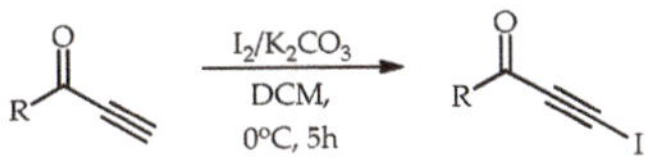

Scheme 28. Iodination of acetylene ketones with I$_2$/K$_2$CO$_3$.

Tsai and co-workers [37] have established a simple and environmentally friendly method for iodination of terminal alkynes catalyzed by CuI/TBAB (tetrabutylammonium bromide) under air in water providing the corresponding products in good to high yields (Scheme 29).

The efficient method for iodoarylation of arylalkynes with I$_2$ and PhI(OCOPh)$_2$ was developed [38], providing the corresponding products in moderate to good yields (Scheme 30).

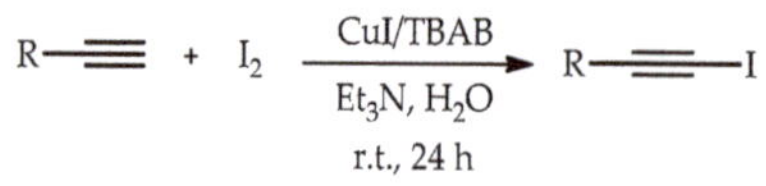

Scheme 29. Iodination of terminal alkynes catalyzed by CuI/TBAB.

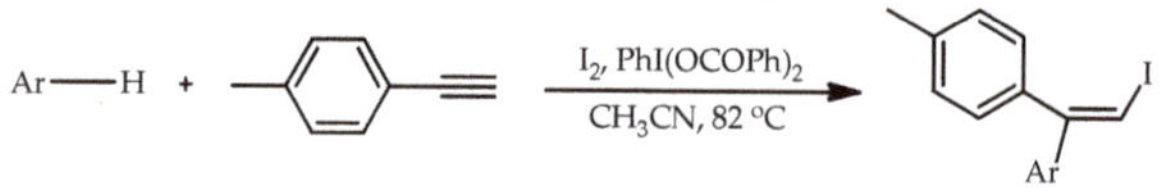

Scheme 30. Iodoarylation of arylalkynes with I_2 and $PhI(OCOPh)_2$.

Reddy and co-workers [39] developed an efficient protocol for oxy-iodination of alkynes using potassium iodide (KI) and *tert*-butyl hydroperoxide (TBHP), in the presence of methanol as a solvent, at room temperature affording the corresponding products in moderate to quantitative yields (Scheme 31).

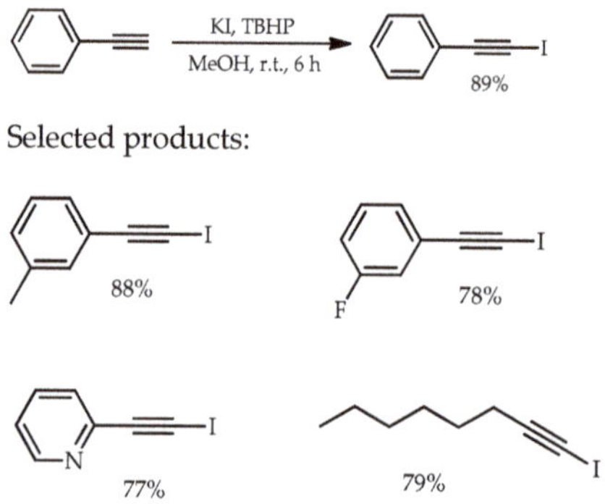

Selected products:

Scheme 31. Synthesis of 1-iodoalkynes employing KI/TBHP.

A novel and convenient protocol for the hydroiodination of alkynes has been established by Ogawa and Kawaguchi [40], employing I_2/hydrophosphine binary system affording the corresponding Markovnikov-type adducts in good yield. It was reported that this method could be applied for iodinations of OH and COOH groups. Chobanyan and co-workers have reported the method for the synthesis of hydroalumantion-iodination of alkyne-1,4-diols [41].

Guo and co-workers [42] have reported a new approach for the synthesis of (E)-diiodoalkenes using alkyne as the starting material, ammonium persulfate as an oxidant, iodide as iodine source, and water as the green solvent, providing the corresponding products in good to quantitative yields (Scheme 32).

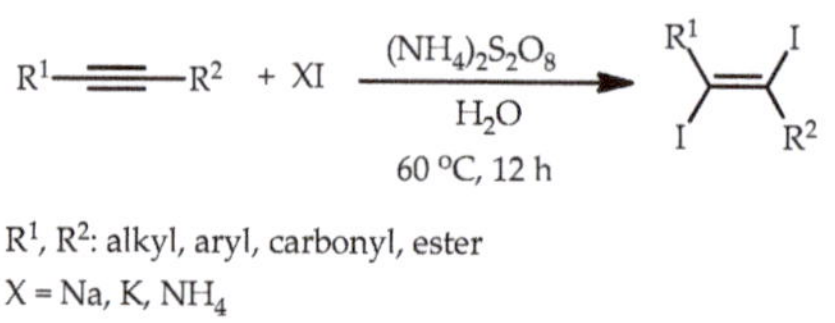

R^1, R^2: alkyl, aryl, carbonyl, ester
$X = Na, K, NH_4$

Scheme 32. Diiodination of alkynes with iodide in water mediated by $(NH_4)_2S_2O_8$.

It is assumed that oxidation of the iodide ion by the persulfate ion generates I_2, which further undergoes electrophilic *anti*-addition into the alkyne to afford the corresponding (E)-1,2-diiodoalkene (Scheme 33).

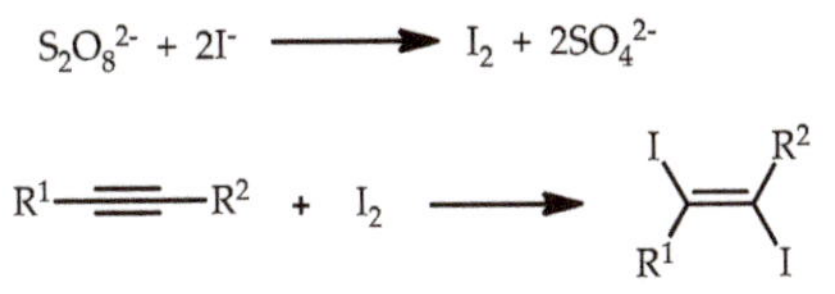

$$S_2O_8^{2-} + 2I^- \longrightarrow I_2 + 2SO_4^{2-}$$

$$R^1\!-\!\!\equiv\!\!-R^2 + I_2 \longrightarrow$$

Scheme 33. Plausible reaction mechanism.

A convenient protocol for the iodination of *N*-propargyltriflamide between trifluo-romethanesulfonamide and trifluoro-*N*-(prop-2-yn-1-yl)methanesulfonamide in the system *t*-BuOCl–NaI provided N-[(2*E*)-2,3-diiodoprop-2-en-1-yl]trifluoromethanesulfonamide has been described by Shainyan and co-workers [43]. Oxidative halogenation of terminal alkynes has been reported by Lui and co-workers [44] mediated by chloroamine salt as the oxidant and KI as the halogen source providing 1-iodooalkynes in good to quantitative yields (Scheme 34).

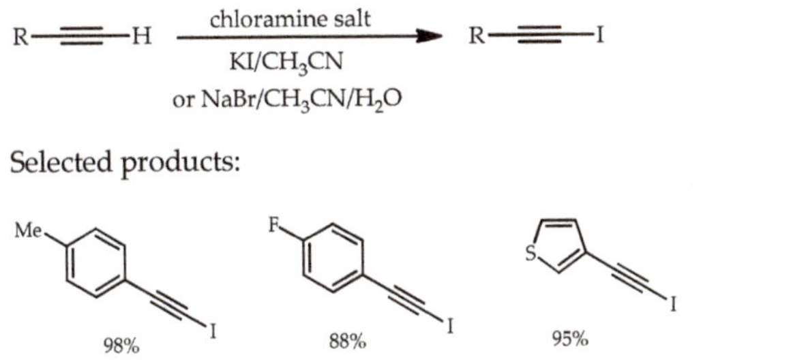

$$R\!-\!\!\equiv\!\!-H \xrightarrow[\substack{\text{KI/CH}_3\text{CN} \\ \text{or NaBr/CH}_3\text{CN/H}_2\text{O}}]{\text{chloramine salt}} R\!-\!\!\equiv\!\!-I$$

Selected products:

98%   88%   95%

Scheme 34. Synthesis of 1-iodoalkynes mediated by chloramine salt.

Inexpensive and non-toxic reagents NH_4I (iodide source) and oxone (oxidant) were used for stereospecific oxidative (*E*)-diiodination of various alkynes such as aliphatic, aromatic, and heteroaromatic alkynes at room temperature in the presence of water as a green solvent [45] (Scheme 35).

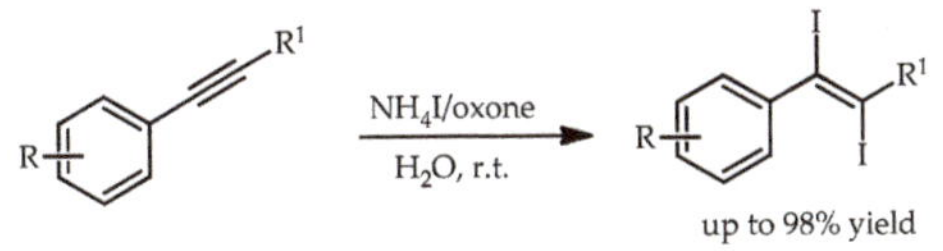

$$\xrightarrow[\text{H}_2\text{O, r.t.}]{\text{NH}_4\text{I/oxone}}$$

up to 98% yield

Scheme 35. Oxidative (*E*)-diiodination of alkynes using NH_4I/oxone system.

It is reported that oxone oxidizes the I^- (NH_4I) to form I^+ (HOI). This reactive species may react with alkyne to give a transient cyclic iodonium species, which further undergoes nucleophilic attack by I^- in situ from the opposite side of the cyclic iodonium ion to provide trans-diiodo alkene.

Ferris and co-workers have reported an efficient method for the iodination of terminal alkynes [46], employing a stoichiometric amount of KI and $CuSO_4$ in a mix of MeCN and sodium acetate buffer solution (pH 5).

Bathophenanthrolinedisulfonic acid (BPDS) was used to solubilize copper species in the solution (Scheme 36).

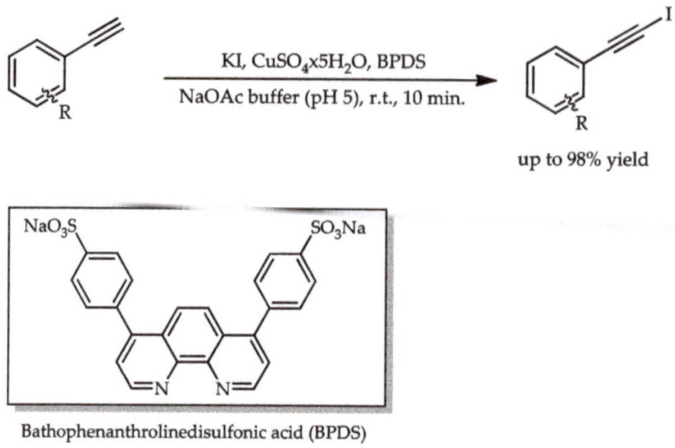
KI, $CuSO_4x5H_2O$, BPDS
NaOAc buffer (pH 5), r.t., 10 min.
R
I
up to 98% yield
NaO_3S
SO_3Na
N
N
Bathophenanthrolinedisulfonic acid (BPDS)

Scheme 36. Iodination of terminal alkynes mediated by $KI/CuSO_4$.

Han and Xiao [47] have developed an efficient and simple procedure for double-iodination of terminal alkynes using I_2 in the presence of water as the green solvent at room temperature. Moreover, by employing I_2/H_3PO_3 system, the selective hydroiodination of different alkenes and alkynes were obtained in good yields (Scheme 37).

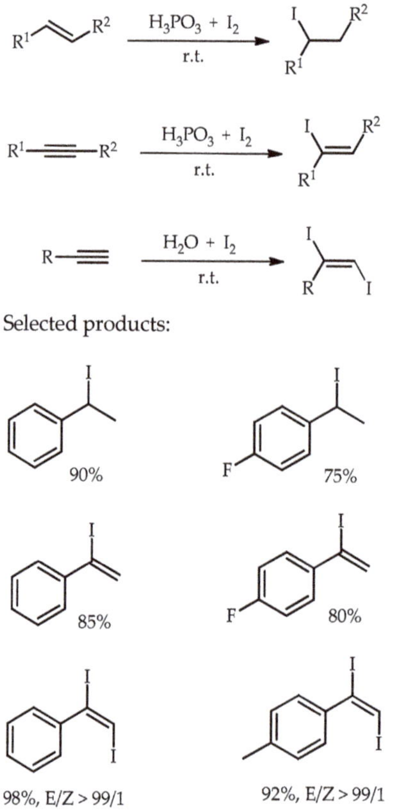
H_3PO_3 + I_2
r.t.
H_2O + I_2
Selected products:
90%
75%
85%
80%
98%, E/Z > 99/1
92%, E/Z > 99/1

Scheme 37. Hydriodination and double-iodination of alkenes and alkynes using I_2.

Regarding the mechanism, it was reported that H_3PO_3 could react with molecular iodine to give hydrogen iodide HI, and then following the Markovnikov rule, provides the corresponding hydroiodination. For the double-iodination, I_2 undergoes electronic *anti*-addition to alkynes via a cyclic iodonium to provide the corresponding diiodoalkene (Scheme 38).

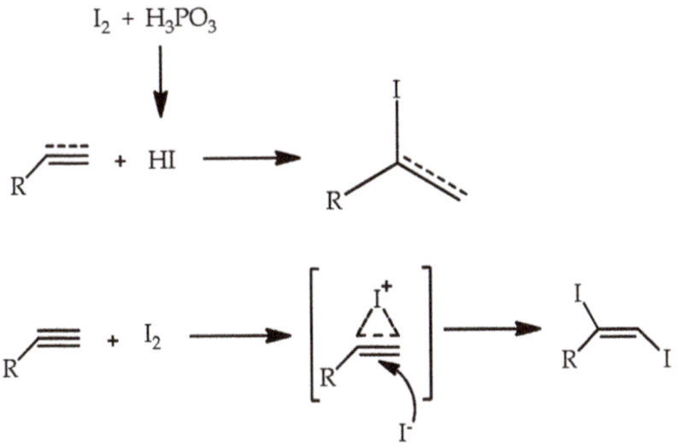

Scheme 38. Possible reaction pathway.

In 2020, Ghosh and co-workers [48] reported a metal- and oxidant-free method for synthesizing 1,1,2-triiodostyrenes by decarboxylative iodination of propiolic acids using I_2/NaOAc providing the corresponding products in good yields (Scheme 39). Moreover, β,β-diarylacrylic acids undergo decarboxylative mono-iodination under the same reaction conditions, affording 1,1-diaryl-2-iodoalkenes.

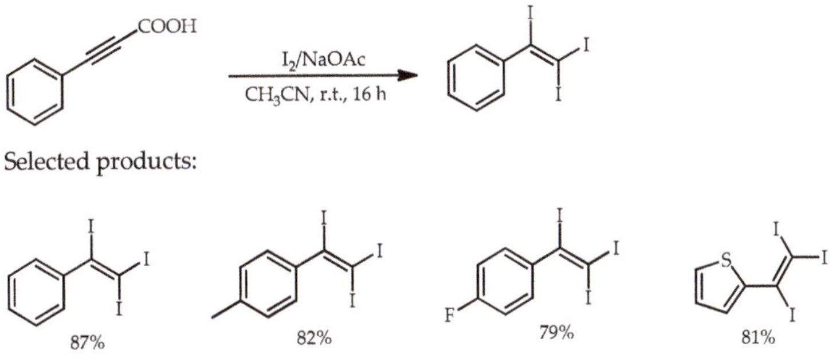

Selected products:

87%　　82%　　79%　　81%

Scheme 39. Synthesis of 1,1,2-triiodostyrenes mediated by I_2/NaOAc.

In the same year, Lingling and co-workers [49] have published a convenient protocol for diiodination of alkynes employing sodium iodide (as iodine source) and air (as an oxidant) under the visible light, providing the corresponding products in moderate to high yields (Scheme 40).

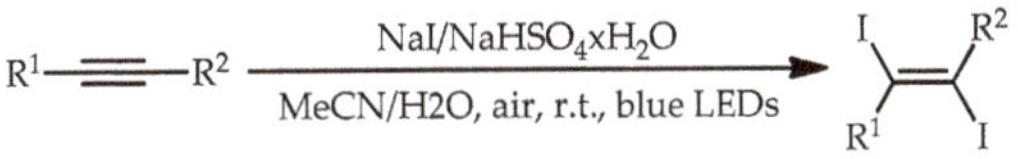

Scheme 40. Diiodination of alkynes mediated by NaI under the visible light.

Recently, an environmentally benign method for the aerobic oxidative iodination of terminal alkynes mediated by sodium sulfinate/KI was presented by Zhuo and co-workers [50] using ethanol as the green solvent at room temperature. Moreover, the synthesis of symmetrical 1,3-diynes was presented via the iodination/homocoupling of terminal alkynes (Scheme 41).

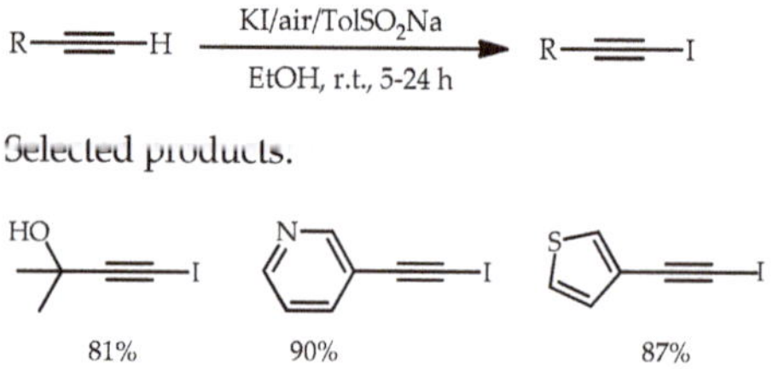

Selected products.

81%   90%   87%

Scheme 41. Synthesis of 1-iodoalkynes mediated by $TolSO_2Na/KI$.

4. Iodination of Alkyl Carbonyls Compounds to α-Iodo Alkyl Carbonyl Derivatives

In 2008, Stavber and co-workers [11] had established an environmentally friendly methodology for aerobic oxidative α-iodination of carbonyl compounds using potassium iodide as iodine source catalyzed by sodium nitrite in acidic media providing the corresponding products in good to quantitative yields. In the case of aryl methyl ketones using MeCN as the solvent, iodination on the aromatic ring was occurred, while in the presence of aqueous EtOH as the solvent, the methyl group was iodinated (Scheme 42).

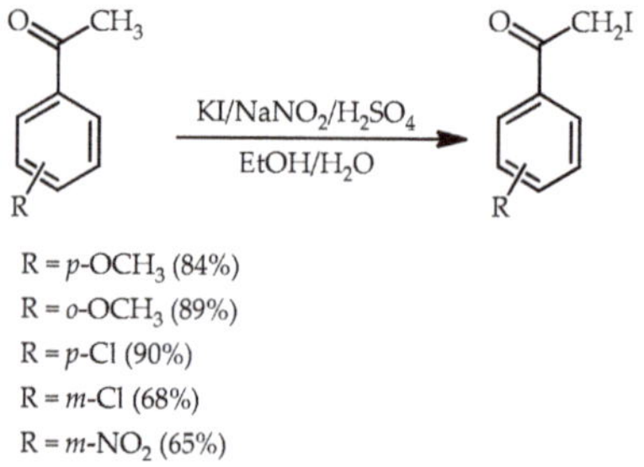

$R = p\text{-}OCH_3$ (84%)
$R = o\text{-}OCH_3$ (89%)
$R = p\text{-}Cl$ (90%)
$R = m\text{-}Cl$ (68%)
$R = m\text{-}NO_2$ (65%)

Scheme 42. Iodination of aryl alkyl ketones mediated by $KI/NaNO_2/H_2SO_4$ in aqueous EtOH.

Pavlinac and co-workers [51] have described an efficient methodology for the iodination of dimethoxy- and trimethoxy benzenes, aryl alkyl ketones and cyclic ketones by employing I_2/UHP (urea-H_2O_2) or $I_2/30\%$ aq. H_2O_2 in the water miscible ionic liquid (IL) 1-butyl-3-methyl imidazolium tetrafluoroborate (bmimBF$_4$) or in water immiscible IL, 1-butyl-3-methyl imidazolium hexafluorophosphate(bmimPF$_6$), providing the corresponding products in excellent yields. In terms of efficiency, 30% aq. H_2O_2 was superior to UHP as the mediator of iodination in both ILs for iodine introduction at methoxy substituted benzenes and alkyl site next to a carbonyl group (Scheme 43).

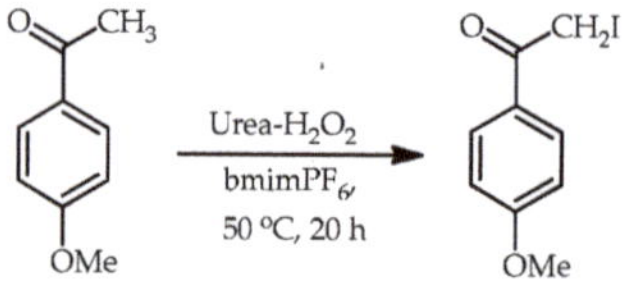

Scheme 43. Iodination of aryl alkyl ketones mediated by bmimPF$_6$/Urea-H_2O_2.

The same group of authors have developed [52] the green methodology for iodination of aryl methoxy substituted 1-indanone, 1-tetralone, and acetophenone using $I_2/30\%$ aq.

H_2O_2 as oxidant under solvent- and catalyst-free reaction conditions (SFRC). In the case of dimethoxy- and trimethoxy benzenes, iodination on the aromatic ring has occurred, while in the case of aryl alkyl ketones, iodination took place at the alkyl position next to a carbonyl group (Scheme 44).

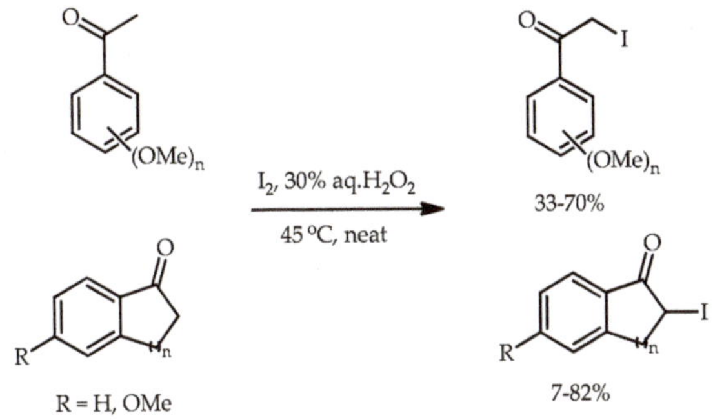

Scheme 44. Iodination of aryl alkyl ketones and cyclic ketones in the presence of I_2/30% aq. H_2O_2 under SFRC.

Furthermore, Iskra and co-workers [53] have reported an efficient, selective, and metal-free protocol for the iodination of aldehydes, alkyl ketones, and aromatics using I_2/$NaNO_2$/air/silica-supported H_2SO_4 in MeCN at room temperature. Air was used as the oxidant for the regeneration of I_2 from eluted HI with 100% iodine atom economy (Scheme 45).

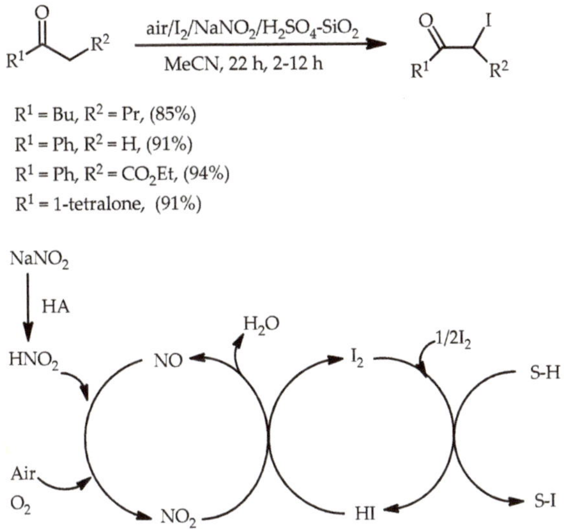

R^1 = Bu, R^2 = Pr, (85%)
R^1 = Ph, R^2 = H, (91%)
R^1 = Ph, R^2 = CO_2Et, (94%)
R^1 = 1-tetralone, (91%)

Scheme 45. Aerobic iodination of alkyl carbonyls using air/I_2/$NaNO_2$/acid system.

Yadav and co-workers [54] have reported a new and efficient method for the synthesis of α-iodo ketones and α-iodo dimethyl ketals in good to high yields, starting from acetophenones in the presence of I_2 and TMOF (trimethylorthoformate), (Scheme 46).

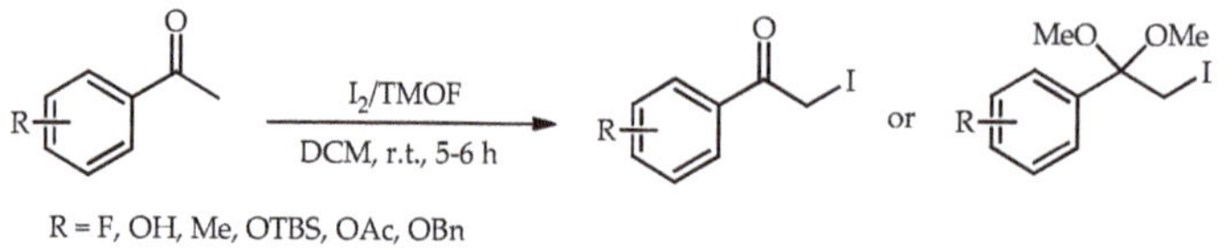

R = F, OH, Me, OTBS, OAc, OBn

Scheme 46. α-iodination of ketones and dimethyl ketals with I_2/TMOF.

A convenient method for iodination of α,β-unsaturated ketones using copper (II) oxide/iodine in the presence of *i*-PrOH as the solvent has been reported by Wang and co-workers [55], providing the corresponding products in good to high yields (Scheme 47).

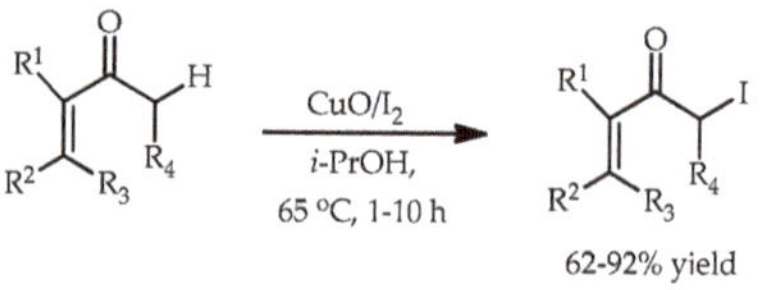

62-92% yield

Scheme 47. Iodination of α,β-unsaturated ketones using CuO/I_2.

Terent'ev and co-workers [56] have reported the convenient method for synthesizing 2-iodo-1-methoxy hydroperoxides and their deperoxidation and demethoxylation to 2-iodo ketones. The reactions have occurred between enol ethers and the I_2-H_2O_2 system, providing the corresponding products in moderate to quantitative yields.

Stavber and co-workers [3] have established a novel and green methodology for iodination of ketones in an aqueous micellar system, in the presence of I_2, as the iodine source, air (terminal oxidant), $NaNO_2$ (catalyst), and H_2SO_4 (activator). The use of the aqueous solution of anionic amphiphile SDS (sodium dodecyl sulfate) was observed to be an excellent promoter than the use of water alone, improving the efficiency of the reactions (Scheme 48).

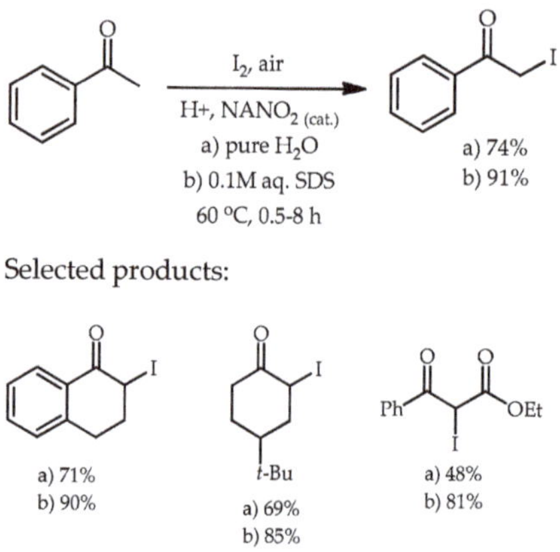

Selected products:

a) 71%
b) 90%

t-Bu
a) 69%
b) 85%

a) 48%
b) 81%

Scheme 48. Aerobic oxidative iodination of ketones.

Lee and co-workers have developed an efficient method for iodination of aryl alkyl ketones using I_2/HTIB [hydroxyl(tosyloxy)iodo]benzene or MeI/HTIB in [bmim]BF_4 ionic liquid, providing the corresponding products in good to excellent yields [57].

An efficient protocol for the synthesis of α-iodo ketones by oxidative iodination of ketones in the presence of iodine and *m*-iodosylbenzoic acid as a recyclable oxidant has been presented. The corresponding iodinated products are separated from side products by treatment with anionic exchange resin Amberlite IRA 900 HCO$_3^-$, *m*-iodosylbenzoic acid can be recovered from Amberlite resin by treatment with HCl [58].

A convenient and selective synthetic protocol for iodination of 1,3-dicarbonyl derivative substrates has been reported by Khan and Ali [59] using vanadyl acetylacetonate, hydrogen peroxide, and sodium iodide at ice-bath temperature, providing the iodinated products in good yields (Scheme 49).

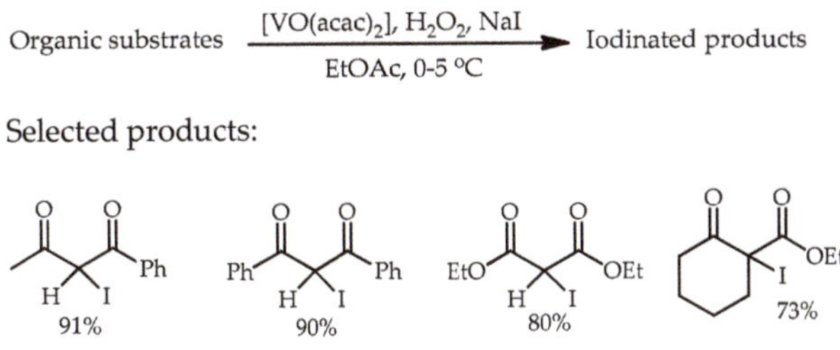

$$\text{Organic substrates} \xrightarrow[\text{EtOAc, 0-5 °C}]{\text{[VO(acac)}_2\text{], H}_2\text{O}_2\text{, NaI}} \text{Iodinated products}$$

Selected products:

91%	90%	80%	73%

Scheme 49. Iodination of 1,3-dicarbonyls using [VO(acac)$_2$]/H$_2$O$_2$/NaI.

Moriya and co-workers [60] have reported a convenient method for reductive iodination of carboxylic acids to alkyl iodides using 1,1,3,3-tetramethyldisiloxane (TMDS) and I$_2$ catalyzed by InBr$_3$ in the presence of CHCl$_3$ as the solvent.

Prebil and co-workers [61] have developed air/NH$_4$NO$_{3\text{(cat.)}}$/I$_2$/H$_2$SO$_{4\text{(cat.)}}$ reaction system in the presence of MeCN as the solvent, for the α-iodination of aryl, heteroaryl, alkyl, and cycloalkyl methyl ketones. In the case of strongly activated aryl methyl ketones iodination took place regioselectively on the aromatic ring, (Scheme 50).

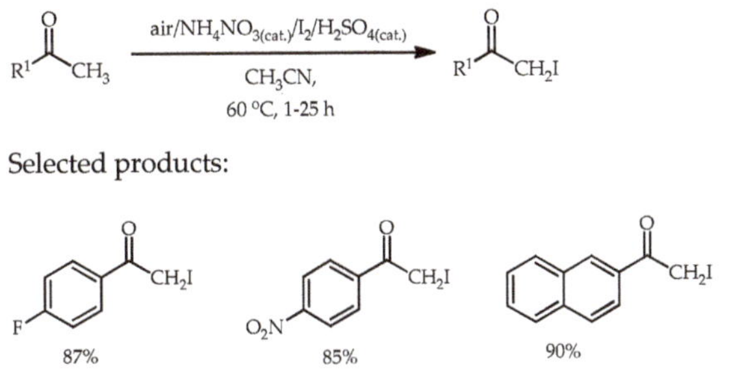

$$R^1\text{-CO-CH}_3 \xrightarrow[\substack{\text{CH}_3\text{CN,}\\60\,°\text{C, 1-25 h}}]{\text{air/NH}_4\text{NO}_{3\text{(cat.)}}/\text{I}_2/\text{H}_2\text{SO}_{4\text{(cat.)}}} R^1\text{-CO-CH}_2\text{I}$$

Selected products:

87%	85%	90%

Scheme 50. Aerobic oxidative α-iodination of alkyl methyl ketones using air/ NH$_4$NO$_{3\text{(cat.)}}$/I$_2$/H$_2$SO$_{4\text{(cat.)}}$ reaction system.

Regarding the mechanism, in cycle I, iodination enol form of the ketone at α-position using I$_2$ has occurred, and I$_2$ has been reduced to HI. The re-oxidation of iodide to I$_2$ by NO$_2$ has been presented as cycle II. NO$_2$ has been reduced to NO when it completes the oxidation of iodide, while the oxidation of NO to NO$_2$ is accomplished with aerial oxygen. Acidic conditions have two leading roles: the first is to transform NH$_4$NO$_3$ to HNO$_3$, which is thermally supported decomposing equilibrium with NO$_2$, and the second is in tuning the reactivity by increasing enolization of the ketone (Scheme 51).

In 2014, Terent'ev and co-workers [62] reported mono-and bicyclic enol ethers reactions with I$_2$/H$_2$O$_2$, I$_2$–Bu^{t}OOH, and I$_2$–tetrahydropyranyl hydroperoxide combinations. The authors have presented that the reaction pathway depends on the nature of peroxide and the ring size. The reaction between 2,3-dihydrofuran and 3,4-dihydro-2*H*-pyran with the I$_2$–hydroperoxide system provides iodoperoxides, α-iodolactones, and α-iodohemiacetals. Bicyclic enol ethers were converted into vicinal iodoperoxides only in the reaction with the I$_2$–H$_2$O$_2$ system (Scheme 52), while I$_2$–Bu^{t}OOH provides the hydroperoxidation product.

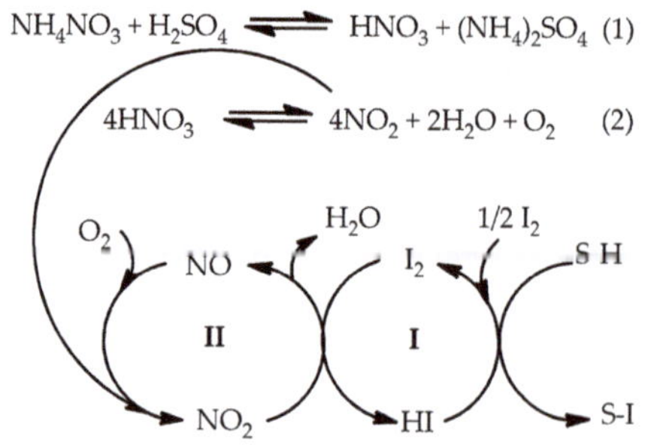

$$NH_4NO_3 + H_2SO_4 \rightleftharpoons HNO_3 + (NH_4)_2SO_4 \quad (1)$$

$$4HNO_3 \rightleftharpoons 4NO_2 + 2H_2O + O_2 \quad (2)$$

Scheme 51. The reaction pathway for the aerobic oxidative α-iodination employing air / $NH_4NO_{3(cat.)}$ / I_2 / $H_2SO_{4(cat.)}$.

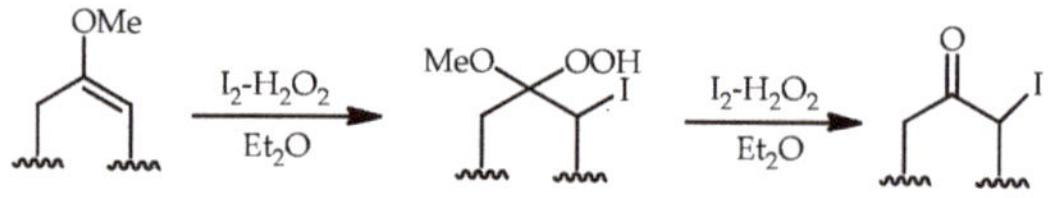

Scheme 52. Synthesis of 2-iodo-1-methoxyhydroperoxides and 2-iodo ketones using I_2–H_2O_2 system.

In 2011, Marri and co-workers [63] had reported a simple and efficient methodology for the α-monoiodination of carbonyl compounds employing NH_4I and oxone in methanol, providing the corresponding products in moderate to excellent yields (Scheme 53).

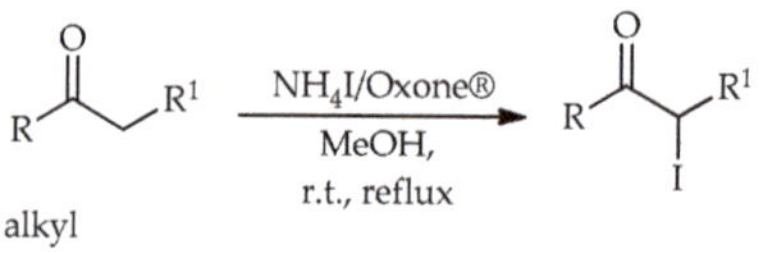

R = aryl, alkyl
R^1 = alkyl, aryl, H

Scheme 53. α-Iodination of alkyl ketones using NH_4I/oxone$^{®}$ system.

It is reported that Oxone$^{®}$ oxidizes the I^- (NH_4I) to I^+ (HOI) which reacts with enol form of carbonyl compound to provide the corresponding α-iodo product (Scheme 54).

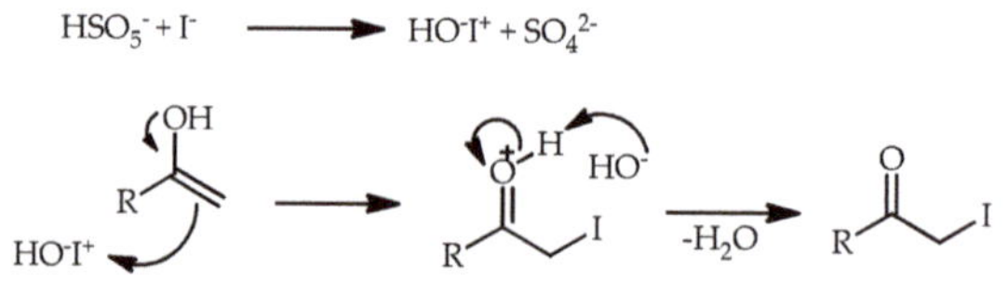

$$HSO_5^- + I^- \longrightarrow HO\text{-}I^+ + SO_4^{2-}$$

Scheme 54. Proposed reaction mechanism.

In 2015, Reddy and co-workers [64] had established a convenient and environmentally friendly protocol for the synthesis of α-iodo alkyl ketones starting from secondary alcohols, including benzylic and aliphatic alcohols (cyclic and acyclic) using ammonium iodide and oxone in aqueous media (Scheme 55).

Zhu and co-workers have reported the convenient method for the β-C (sp^3)-H iodination of ketones in the presence of palladium (II) as the catalyst employing aminooxyacetic acid auxiliary [65]. Sanz-Marco and co-workers [66] have developed an efficient and one-pot methodology for the synthesis of α-iodo alkyl ketones (as single constitutional isomers) starting from allylic alcohols and elemental iodine in combination with $NaNO_2$ as an oxidation catalyst and oxygen as the terminal oxidant. The protocol combines a 1,3-hydrogen shift mediated by Ir(III) complex (Scheme 56).

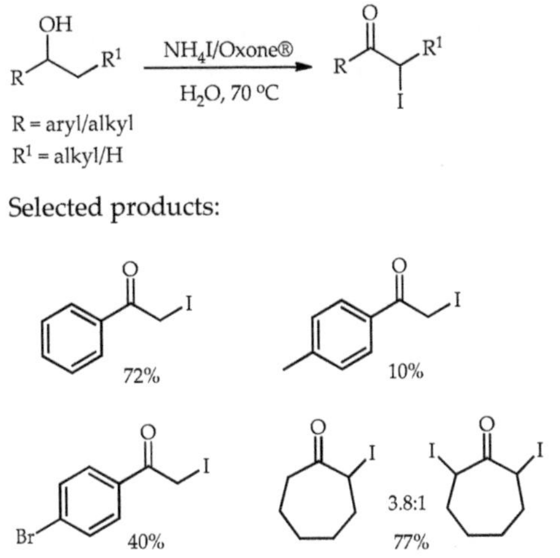

Selected products:

Scheme 55. Synthesis of α-iodo alkyl ketones using NH₄I/oxone® system.

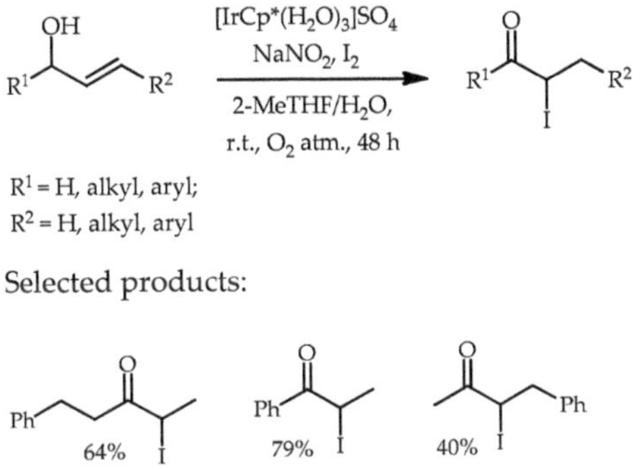

Selected products:

Scheme 56. Synthesis of α-iodoketones through aerobic oxidative iodination.

5. Conclusions

In summary, this review presents the progress of various methods for the iodination of organic compounds, including alkanes, alkenes, alkynes and alkyl carbonyls using elemental iodine or iodides. Aerobic oxidative and non-metal iodination strategies are also established. It should be emphasized that convenient methods have been developed in this field. Still, investigating and developing environmentally friendlier protocols in aqueous reaction media or under solvent-free can be considered an exciting research subject.

Author Contributions: Conceptualization, S.S.; methodology, formal analysis, investigation, writing—original draft preparation, writing—review and editing: N.A. and S.S. All authors have read and agreed to the published version of the manuscript.

Funding: This research was funded by Slovene Human Resources Development and Scholarship Fund (contract: 11011-9/2011), Slovenian Research Agency (contract: Programme P1-0134).

Data Availability Statement: Not applicable.

Conflicts of Interest: The authors declare no conflict of interest.

References

1. Togo, H.; Iida, S. Synthetic Use of Molecular Iodine for Organic Synthesis. *Synlett* **2006**, *2006*, 2159–2175. [CrossRef]
2. Küpper, F.C.; Feiters, M.C.; Olofsson, B.; Kaiho, T.; Yanagida, S.; Zimmermann, M.B.; Carpenter, L.J.; Luther, G.W., III; Lu, Z.; Jonsson, M.; et al. Commemorating Two Centuries of Iodine Research: An Interdisciplinary Overview of Current Research. *Angew. Chem. Int. Ed.* **2011**, *50*, 11598–11620. [CrossRef] [PubMed]
3. Stavber, G.; Iskra, J.; Zupan, M.; Stavber, S. Aerobic oxidative iodination of ketones catalysed by sodium nitrite "on water" or in a micelle-based aqueous system. *Green Chem.* **2009**, *11*, 1262–1267. [CrossRef]
4. Vekariya, R.H.; Balar, C.R.; Sharma, V.S.; Prajapati, N.P.; Vekariya, M.K.; Sharma, A.S. Preparation of α-Iodocarbonyl Compounds: An Overall Development. *ChemistrySelect* **2018**, *3*, 9189–9203. [CrossRef]
5. Mphahlele, M.J. Molecular Iodine-Mediated α-Iodination of Carbonyl Compounds. *J. Chem. Res.* **2010**, *34*, 121–126. [CrossRef]
6. Stavber, S.; Jereb, M.; Zupan, M. Electrophilic Iodination of Organic Compounds Using Elemental Iodine or Iodides. *Synthesis* **2008**, *2008*, 1487–1513. [CrossRef]
7. Chouthaiwale, P.V.; Suryavanshi, G.; Sudalai, A. $NaIO_4$–KI–NaN_3 as a new reagent system for C–H functionalization in hydrocarbons. *Tetrahedron Lett.* **2008**, *49*, 6401–6403. [CrossRef]
8. Giri, R.; Lan, Y.; Liu, P.; Houk, K.N.; Yu, J.-Q. Understanding Reactivity and Stereoselectivity in Palladium-Catalyzed Diastereoselective sp3 C–H Bond Activation: Intermediate Characterization and Computational Studies. *J. Am. Chem. Soc.* **2012**, *134*, 14118–14126. [CrossRef]
9. Zhu, R.-Y.; Saint-Denis, T.G.; Shao, Y.; He, J.; Sieber, J.D.; Senanayake, C.H.; Yu, J.-Q. Ligand-Enabled Pd(II)-Catalyzed Bromination and Iodination of C(sp3)–H Bonds. *J. Am. Chem. Soc.* **2017**, *139*, 5724–5727. [CrossRef] [PubMed]
10. Jereb, M.; Zupan, M.; Stavber, S. Hydrogen peroxide induced iodine transfer into alkenes. *Green Chem.* **2005**, *7*, 100–104. [CrossRef]
11. Stavber, G.; Iskra, J.; Zupan, M.; Stavber, S. Aerobic Oxidative Iodination of Organic Compounds with Iodide Catalyzed by Sodium Nitrite. *Adv. Synth. Catal.* **2008**, *350*, 2921–2929. [CrossRef]
12. Yang, Y.; Huang, X. Iodohydroxylation of Alkylidenecyclopropanes. An Efficient Synthesis of Iodocyclopropylmethanol and 3-Iodobut-3-en-1-ol Derivatives. *J. Org. Chem.* **2008**, *73*, 4702–4704. [CrossRef]
13. Yadav, J.S.; Subba Reddy, B.V.; Singh, A.P.; Basak, A.K. IBX/I2-mediated oxidation of alkenes and alkynes in water: A facile synthesis of α-iodoketones. *Tetrahedron Lett.* **2008**, *49*, 5880–5882. [CrossRef]
14. Yusubov, M.S.; Yusubova, R.Y.; Kirschning, A.; Park, J.Y.; Chi, K.-W. m-Iodosylbenzoic acid, a tagged hypervalent iodine reagent for the iodo-functionalization of alkenes and alkynes. *Tetrahedron Lett.* **2008**, *49*, 1506–1509. [CrossRef]
15. Yang, H.; Xu, B.; Hammond, G.B. Highly Regioselective Fluorination and Iodination of Alkynyl Enolates. *Org. Lett.* **2008**, *10*, 5589–5591. [CrossRef] [PubMed]
16. Sakee, U.; Nasuk, C.; Grigg, R. Synthesis of 1-C-(tetra-O-acetyl-β-d-galactopyranosyl)-2,3-diiodo-1-propene and its reaction with primary amines. *Carbohydr. Res.* **2009**, *344*, 2096–2099. [CrossRef]
17. Agrawal, M.K.; Adimurthy, S.; Ganguly, B.; Ghosh, P.K. Comparative study of the vicinal functionalization of olefins with 2:1 bromide/bromate and iodide/iodate reagents. *Tetrahedron* **2009**, *65*, 2791–2797. [CrossRef]
18. Nagura, H.; Kuribayashi, S.; Ishiguro, Y.; Inagi, S.; Fuchigami, T. Electrochemical iodofluorination of electron-deficient olefins. *Tetrahedron* **2010**, *66*, 183–186. [CrossRef]
19. Nobuta, T.; Hirashima, S.-I.; Tada, N.; Miura, T.; Itoh, A. Facile Aerobic Photo-Oxidative Synthesis of Phenacyl Iodides and Bromides from Styrenes Using I2 or Aqueous HBr. *Synlett* **2010**, *2010*, 2335–2339.
20. Chouthaiwale, P.V.; Karabal, P.U.; Suryavanshi, G.; Sudalai, A. Regiospecific Azidoiodination of Alkenes with Sodium Periodate, Potassium Iodide, and Sodium Azide: A High-Yield Synthesis of β-Iodoazides. *Synthesis* **2010**, *2010*, 3879–3882. [CrossRef]
21. Hanessian, S.; Focken, T.; Oza, R. Total Synthesis of Jerangolid A. *Org. Lett.* **2010**, *12*, 3172–3175. [CrossRef]
22. Samakkanad, N.; Katrun, P.; Techajaroonjit, T.; Hlekhlai, S.; Pohmakotr, M.; Reutrakul, V.; Jaipetch, T.; Soorukram, D.; Kuhakarn, C. IBX/I2-Mediated Reaction of Sodium Arenesulfinates with Alkenes: Facile Synthesis of β-Keto Sulfones. *ChemInform* **2012**, *44*, 1693–1699. [CrossRef]
23. Kim, S.-H.; Kim, S.-H.; Lee, H.-J.; Kim, J.-N. One-Pot Synthesis of 5-Hydroxypyrrolin-2-one Derivatives from Modified Morita-Baylis-Hillman Adducts via a Consecutive CuI-Mediated Aerobic Oxidation, Allylic Iodination, Hydration of Nitrile, and Lactamization. *Bull. Korean Chem. Soc.* **2012**, *33*, 2079–2082. [CrossRef]
24. Yemets, S.V.; Shubina, T.E.; Krasutsky, P.A. Electrophilic monoiodination of terminal alkenes. *Org. Biomol. Chem.* **2013**, *11*, 2891–2897. [CrossRef] [PubMed]
25. Moskalik, M.Y.; Shainyan, B.A.; Astakhova, V.V.; Schilde, U. Oxidative addition of trifluoromethanesulfonamide to cycloalkadienes. *Tetrahedron* **2013**, *69*, 705–711. [CrossRef]
26. Gottam, H.; Vinod, T.K. Versatile and Iodine Atom-Economic Co-Iodination of Alkenes. *J. Org. Chem.* **2011**, *76*, 974–977. [CrossRef] [PubMed]
27. Wang, M.; Fu, C.; Ma, S. Highly Regio- and Stereoselective Iodohydroxylation of 1,2-Allenylic Sulfoxides in the Presence of Benzyl Thiol. *Adv. Synth. Catal.* **2011**, *353*, 1775–1786. [CrossRef]
28. Beltran, R.; Nocquet-Thibault, S.; Blanchard, F.; Dodd, R.H.; Cariou, K. PIFA-mediated ethoxyiodination of enamides with potassium iodide. *Org. Biomol. Chem.* **2016**, *14*, 8448–8451. [CrossRef] [PubMed]
29. Durgaiah, C.; Naresh, M.; Arun Kumar, M.; Swamy, P.; Reddy, M.M.; Srujana, K.; Narender, N. Regio- and stereoselective co-iodination of olefins using NH4I and Oxone. *Synth. Commun.* **2016**, *46*, 1133–1144. [CrossRef]

30. Kaswan, P.; Shelke, G.M.; Rao, V.K.; Kumar, A. Hydroxy-Group-Facilitated Vinylic Iodination of ortho-Vinylnaphthols Using Molecular Iodine. *Synlett* **2016**, *27*, 2553–2556.

31. Xinwei, L.; Song, S.; Ning, J. Oxidative Iodohydroxylation of Olefins with DMSO. *Acta Chim. Sinica* **2017**, *75*, 1202–1206.

32. Wang, H.; Chen, C.; Liu, W.; Zhu, Z. Difunctionalization of alkenes with iodine and tert-butyl hydroperoxide (TBHP) at room temperature for the synthesis of 1-(tert-butylperoxy)-2-iodoethanes. *Beilstein J. Org. Chem.* **2017**, *13*, 2023–2027. [CrossRef]

33. Hokamp, T.; Storm, A.T.; Yusubov, M.; Wirth, T. Iodine Monoacetate for Efficient Oxyiodinations of Alkenes and Alkynes. *Synlett* **2018**, *29*, 415–418.

34. Yi, W.; Wang, P.-F.; Lu, M.; Liu, Q.-Q.; Bai, X.; Chen, K.-D.; Zhang, J.-W.; Liu, G.-Q. Environmentally Friendly Protocol for the Oxidative Iodofunctionalization of Olefins in a Green Solvent. *ACS Sustain. Chem. Eng.* **2019**, *7*, 16777–16785. [CrossRef]

35. Shakhmaev, R.N.; Sunagatullina, A.S.; Ignatishina, M.G.; Yunusova, E.Y.; Zorin, V.V. Synthesis of Ethyl (2E)-5-Phenylpent-2-en-4-ynoate. *Russ. J. Org. Chem.* **2019**, *55*, 897–899. [CrossRef]

36. Meng, L.G.; Cai, P.J.; Guo, Q.X.; Xue, S. Direct Iodination of Monosubstituted Aryl Acetylenes and Acetylenic Ketones. *Synth. Commun.* **2008**, *38*, 225–231. [CrossRef]

37. Chen, S.-N.; Hung, T.-T.; Lin, T.-C.; Tsai, F.-Y. Reusable and Efficient CuI/TBAB-Catalyzed Iodination of Terminal Alkynes in Water under Air. *J. Chin. Chem. Soc.* **2009**, *56*, 1078–1081. [CrossRef]

38. Rahman, M.A.; Kitamura, T. Iodoarylation of Arylalkynes with Molecular Iodine in the Presence of Hypervalent Iodine Reagents. *Molecules* **2009**, *14*, 3132–3141. [CrossRef] [PubMed]

39. Rajender Reddy, K.; Venkateshwar, M.; Uma Maheswari, C.; Santhosh Kumar, P. Mild and efficient oxy-iodination of alkynes and phenols with potassium iodide and tert-butyl hydroperoxide. *Tetrahedron Lett.* **2010**, *51*, 2170–2173. [CrossRef]

40. Kawaguchi, S.-I.; Ogawa, A. Highly Selective Hydroiodation of Alkynes Using an Iodine−Hydrophosphine Binary System. *Org. Lett.* **2010**, *12*, 1893–1895. [CrossRef] [PubMed]

41. Gharibyan, H.A.; Makaryan, G.M.; Hovhannisyan, M.R.; Kinoyan, F.S.; Chobanyan, Z.A. Some special features of hydroalumination-iodination of alkyne-1,4-diols. *Russ. J. Gen. Chem.* **2014**, *84*, 457–464. [CrossRef]

42. Jiang, Q.; Wang, J.-Y.; Guo, C.-C. (NH4)2S2O8-Mediated Diiodination of Alkynes with Iodide in Water: Stereospecific Synthesis of (E)-Diiodoalkenes. *Synthesis* **2015**, *47*, 2081–2087. [CrossRef]

43. Astakhova, V.V.; Ushakov, I.A.; Shainyan, B.A. Oxidative iodination of N-propargyltriflamide. *Russ. J. Org. Chem.* **2017**, *53*, 953–954. [CrossRef]

44. Liu, X.; Chen, G.; Li, C.; Liu, P. Chloramine Salt Mediated Oxidative Halogenation of Terminal Alkynes with KI or NaBr: Practical Synthesis of 1-Bromoalkynes and 1-Iodoalkynes. *Synlett* **2018**, *29*, 2051–2055. [CrossRef]

45. Banothu, R.; Peraka, S.; Kodumuri, S.; Chevella, D.; Gajula, K.S.; Amrutham, V.; Yennamaneni, D.R.; Nama, N. An aqueous medium-controlled stereospecific oxidative iodination of alkynes: Efficient access to (E)-diiodoalkene derivatives. *New J. Chem.* **2018**, *42*, 17879–17883. [CrossRef]

46. Ferris, T.; Carroll, L.; Mease, R.C.; Spivey, A.C.; Aboagye, E.O. Iodination of terminal alkynes using KI/CuSO4–A facile method with potential for radio-iodination. *Tetrahedron Lett.* **2019**, *60*, 936–939. [CrossRef] [PubMed]

47. Xiao, J.; Han, L.-B. Ready access to organoiodides: Practical hydroiodination and double-iodination of carbon-carbon unsaturated bonds with I2. *Tetrahedron* **2019**, *75*, 3510–3515. [CrossRef]

48. Ghosh, S.; Ghosh, R.; Chattopadhyay, S.K. Oxidant- and additive-free simple synthesis of 1,1,2-triiodostyrenes by one-pot decarboxylative iodination of propiolic acids. *Tetrahedron Lett.* **2020**, *61*, 152378. [CrossRef]

49. Lingling, L.; Yiming, L.; Xuefeng, J. Visible-Light-Promoted Diiodination of Alkynes Using Sodium Iodide. *Chinese J. Org. Chem.* **2020**, *40*, 3354–3361.

50. Zhou, P.; Feng, S.; Qiu, H.; Zhang, J. Sodiump-Toluenesulfinate/KI-Mediated Aerobic Oxidative Iodination of Terminal Alkynes for Synthesis of 1-Iodoalkynes and 1,3-Diynes. *Chinese J. Org. Chem.* **2021**, *41*, 394–399. [CrossRef]

51. Pavlinac, J.; Laali, K.K.; Zupan, M.; Stavber, S. Iodination of Organic Compounds with Elemental Iodine in the Presence of Hydrogen Peroxide in Ionic Liquid Media. *Aust. J. Chem.* **2008**, *61*, 946–955. [CrossRef]

52. Pavlinac, J.; Zupan, M.; Stavber, S. Iodination of Organic Compounds Using the Reagent System I2–30% aq. H2O2 under Organic Solvent-free Reaction Conditions. *Acta Chim. Slov.* **2008**, *55*, 841–849.

53. Iskra, J.; Stavber, S.; Zupan, M. Aerobic oxidative iodination of organic molecules activated by sodium nitrite. *Tetrahedron Lett.* **2008**, *49*, 893–895. [CrossRef]

54. Yadav, J.S.; Kondaji, G.; Shiva Ram Reddy, M.; Srihari, P. Facile synthesis of α-iodo carbonyl compounds and α-iodo dimethyl ketals using molecular iodine and trimethylorthoformate. *Tetrahedron Lett.* **2008**, *49*, 3810–3813. [CrossRef]

55. Wang, Z.; Yin, G.; Qin, J.; Gao, M.; Cao, L.; Wu, A. An Efficient Method for the Selective Iodination of α,β-Unsaturated Ketones. *Synthesis* **2008**, *2008*, 3675–3681.

56. Terent'ev, A.O.; Borisov, A.M.; Platonov, M.M.; Starikova, Z.A.; Chernyshev, V.V.; Nikishin, G.I. Reaction of Enol Ethers with the I2-H2O2 System: Synthesis of 2-Iodo-1-methoxy Hydroperoxides and Their Deperoxidation and Demethoxylation to 2-Iodo Ketones. *Synthesis* **2009**, *2009*, 4159–4166. [CrossRef]

57. Lee, J.-C.; Kim, J.-M.; Park, H.-J.; Kwag, B.-M.; Lee, S.-B. Direct Metal-free α-Iodination of Arylketones Induced by Iodine or Iodomethane with HTIB in Ionic Liquid. *Bull. Korean Chem. Soc.* **2010**, *31*, 1385–1386. [CrossRef]

58. Yusubov, M.S.; Yusubova, R.Y.; Funk, T.V.; Chi, K.-W.; Kirschning, A.; Zhdankin, V.V. m-Iodosylbenzoic Acid as a Convenient Recyclable Hypervalent Iodine Oxidant for the Synthesis of α-Iodo Ketones by Oxidative Iodination of Ketones. *Synthesis* **2010**, *2010*, 3681–3685. [CrossRef]

59. Khan, A.T.; Ali, S. A Useful and Convenient Synthetic Protocol for Iodination of Organic Substrates Using a Combination of Vanadyl Acetylacetonate, Hydrogen Peroxide, and Sodium Iodide. *Bull. Chem. Soc. Jpn.* **2012**, *85*, 1239–1243. [CrossRef]

60. Moriya, T.; Yoneda, S.; Kawana, K.; Ikeda, R.; Konakahara, T.; Sakai, N. Indium(III)-Catalyzed Reductive Bromination and Iodination of Carboxylic Acids to Alkyl Bromides and Iodides: Scope, Mechanism, and One-Pot Transformation to Alkyl Halides and Amine Derivatives. *J. Org. Chem.* **2013**, *78*, 10642–10650. [CrossRef]

61. Prebil, R.; Stavber, S. Aerobic oxidative α-iodination of carbonyl compounds using molecular iodine activated by a nitrate-based catalytic system. *Tetrahedron Lett.* **2014**, *55*, 5643–5647. [CrossRef]

62. Terent'ev, A.O.; Zdvizhkov, A.T.; Kulakova, A.N.; Novikov, R.A.; Arzumanyan, A.V.; Nikishin, G.I. Reactions of mono- and bicyclic enol ethers with the I2–hydroperoxide system. *RSC Adv.* **2014**, *4*, 7579–7587. [CrossRef]

63. Marri, M.R.; Macharla, A.K.; Peraka, S.; Nama, N. Oxidative iodination of carbonyl compounds using ammonium iodide and oxone®. *Tetrahedron Lett.* **2011**, *52*, 6554–6559. [CrossRef]

64. Reddy, M.M.; Swamy, P.; Naresh, M.; Srujana, K.; Durgaiah, C.; Rao, T.V.; Narender, N. One-pot synthesis of α-iodoketones from alcohols using ammonium iodide and Oxone® in water. *RSC Adv.* **2015**, *5*, 12186–12190. [CrossRef]

65. Zhu, R.-Y.; Liu, L.-Y.; Yu, J.-Q. Highly Versatile β-C(sp3)–H Iodination of Ketones Using a Practical Auxiliary. *J. Am. Chem. Soc.* **2017**, *139*, 12394–12397. [CrossRef]

66. Sanz-Marco, A.; Možina, Š.; Martinez-Erro, S.; Iskra, J.; Martín-Matute, B. Synthesis of α-Iodoketones from Allylic Alcohols through Aerobic Oxidative Iodination. *Adv. Synth. Catal.* **2018**, *360*, 3884–3888. [CrossRef]

Review

Modern Software for Computer Modeling in Quantum Chemistry and Molecular Dynamics

Marina V. Malyshkina [1,2] **and Alexander S. Novikov** [2,*]

1   Presidential Physics and Mathematics Lyceum No. 239, Kirochnaya St., 8, Bldg. A, 191028 Saint Petersburg, Russia; mar_mal05@mail.ru
2   Infochemistry Scientific Center, ITMO University, Kronverksky Pr., 49, Bldg. A, 197101 Saint Petersburg, Russia
*   Correspondence: novikov@itmo.ru

Abstract: The most popular modern programs for quantum chemical and molecular dynamics (classical, ab initio, and QM/MM) calculations, which are relevant for the investigation of nature and various properties of different molecules and periodic chemical systems such as nanotubes, surfaces and films, polymers, and crystalline solids, are highlighted and briefly discussed.

Keywords: quantum chemistry; computational chemistry; molecular dynamics; modeling; open-source software; proprietary software

Citation: Malyshkina, M.V.; Novikov, A.S. Modern Software for Computer Modeling in Quantum Chemistry and Molecular Dynamics. *Compounds* **2021**, *1*, 134–144. https://doi.org/10.3390/compounds1030012

Academic Editor: Juan Mejuto

Received: 31 August 2021
Accepted: 2 November 2021
Published: 16 November 2021

In the modern world, computer modeling can quite successfully compete with experimental methods (and compliment them) in many areas of science and technology. The quantum and computational chemistry as well as molecular dynamics are no exception, and currently there are many open-source and proprietary program solutions for the investigation of nature and various properties of different molecules and periodic chemical systems such as nanotubes, surfaces and films, polymers, and crystalline solids. In this mini-review, we would like to highlight and briefly discuss some of such programs for computer modeling in chemistry and materials science. A summary of the programming language, type of license, and possibility to calculate NMR, as well as vibrational spectra and thermochemistry, for all discussed software for computer modeling in quantum chemistry and molecular dynamics is given in Table 1.

ABINIT [1] is an open-source suite that was developed for programs of physics, chemistry, materials science, and others. ABINIT realizes implementation of density functional theory (DFT) by performing the solution of the Kohn–Sham equations. The equations describe the electrons in a chemical system, the presentation of which is implemented in a plane wave basis set. The energy minimum is determined through usage of a self-consistent conjugate gradient method. Computational efficiency and high speed are achieved by fast Fourier transforms, and effective core potentials (aka pseudopotentials) to describe core (non-valence) electrons. ABINIT implements an alternative to standard norm-conserving pseudopotentials. The application of the projector augmented-wave method is possible. In addition to total electronic energy, stresses and forces are calculated; moreover, geometry optimizations and ab initio molecular dynamics are capable of being performed and calculated.

Aces II and III [2] are computational chemistry packages that were elaborated for implementing high-level ab initio quantum chemical calculations. Approaching many-body techniques, for instance, cluster techniques, in particular, to treat electron correlation in details and many-body perturbation theory, allows the precise prediction of atomic and molecular energies in addition to properties.

BerkeleyGW [3] is a code, specialized on excited states, for many-body perturbation theory. The BerkeleyGW package implements the solution for optical properties of materials and quasiparticle excitations by performing the GW plus Bethe–Salpeter equation

---

(GW-BSE) method and the GW method. The latest BerkeleyGW can be applied to study systems up to a few thousand atoms.

Table 1. Summary of the programming language, type of license, and possibility to calculate NMR, as well as vibrational spectra and thermochemistry, for all discussed software for computer modeling in quantum chemistry and molecular dynamics.

Package	Language	License	NMR	Vib. Spec.	Thermo.
1 ABINIT	Fortran	Free, GPL	+	+	+
2 ACES	Fortran, C++	Free, GPL	+	+	+
3 BerkeleyGW	Fortran	Free, GPL	+	+	+
4 BigDFT	Fortran	Free, GPL	-	+	+
5 BrianQC	C++, CUDA	Commercial	+	+	+
6 CASTEP	Fortran 95, Fortran 2003	Academic, commercial	+	+	+
7 COLUMBUS	Fortran	Academic	-	+	+
8 CP2K	Fortran 95	Free, GPL	+	+	+
9 CPMD	Fortran	Academic	+	+	+
10 CRYSTAL	Fortran	Academic (UK), Commercial (IT)	-	+	+
11 DACAPO	Fortran	Free, GPL	-	+	-
12 Dalton	Fortran	Free, LGPL	+	+	+
13 deMon2k	Fortran	Academic, commercial	+	+	+
14 DMol3	Fortran 90	Commercial	+	+	+
15 DP	Fortran 90, C	Free, LGPL	-	-	-
16 ErgoSCF	C++	Free, GPL	-	-	-
17 Exabyte.io	Python	Cloud, Free Tier	-	+	+
18 FHI-aims	Fortran	Academic, commercial	+	+	+
19 FPLO	Fortran 95, C++, Perl	Commercial	-	-	-
20 GAMESS (UK)	Fortran	Academic UK, Commercial	+	+	+
21 GAMESS (US)	Fortran	Academic	+	+	+
22 Gaussian	Fortran	Commercial	+	+	+
23 GPAW	Python, C	Free, GPL	+	+	+
24 HORTON	Python, C++	Free, GPL	-	+	+
25 HyperChem	C++	Commercial	+	+	+
26 JDFTx	C++	Free, GPL	+	+	+
27 Maple	Maple, C, Fortran, Python	Commercial	-	-	-
28 MOLCAS	Fortran, C, C++, Python, Perl	Academic, commercial	+	+	+
29 MOLGW	Fortran	Free, GPL	-	-	-
30 MOLPRO	Fortran	Commercial	+	+	+
31 MOPAC	Fortran	Academic, commercial	-	+	+
32 MPQC	C++	Free, LGPL	+	+	+
33 MRCC	Fortran	Academic	+	+	+
34 NTChem	Unknown	Commercial	+	+	+
35 NWChem	Fortran 77, C	Free, ECL v2	+	+	+
36 ONETEP	Fortran 2003	Academic (UK), Commercial	-	+	+
37 OpenAtom	Charm++ (C++)	Academic	-	+	+
38 OpenMx	C	Free, GPL	+	+	+
39 ORCA	C++	Academic, commercial	+	+	+
40 PARSEC	Fortran	Free, GPL	-	-	-
41 PSI	C, C++, Python	Free, GPL	-	+	+
42 PyQuante	Python	Free, BSD	-	+	+
43 PySCF	Python	Free, BSD	+	+	+
44 QMCPACK (QMC)	C++	Free, U. Illinois Open Source	-	-	-
45 Quantum ESPRESSO	Fortan	Free, GPL	+	+	+
46 RMG	C, C++	Free, GPL	-	+	+
47 SAMSON	C++, Python	Free	-	-	-
48 SIESTA	Fortran 2003	Free, GPL	-	-	-
49 VOTCA-XTP	C++	Free, Apache License	-	-	-
50 Yambo	Fortran	Free, GPL	-	-	-

BigDFT [4] is an open-source software package, developed for physicists and chemists for innovative research of materials and macro-molecular systems at the nanoscale. In this software package implementation, DFT is realized by solution of the Kohn–Sham equations, a self-consistent direct minimization, and Davidson diagonalization methods. The Kohn–Sham equations describe the electrons in a material, expanded in a Daubechies wavelet basis set, and Davidson diagonalization methods determine the energy minimum on the potential energy surface. Computational efficiency is achieved with fast short convolutions and pseudopotentials to describe core (non-valent) electrons. BigDFT was among the first massively parallel density functional theory codes that benefited from graphics processing units (GPUs) with the usage of CUDA and then OpenCL languages.

BrianQC is a GPU module for Q-Chem [5]. As a Q-Chem software, it is able to calculate high angular momentum orbitals, and is highly efficient for simulating large molecules and quantum systems. BrianQC makes Q-Chem run on GPUs and speeds up DFT and Hartree–Fock (HF) single point, geometry optimization, and frequency calculations, as well as many other methods.

CASTEP [6] is a fully functional materials modelling code. It is based on a first-principles quantum mechanical description of electrons and nuclei (including usage of path-integral molecular dynamics). The CASTEP package is capable of simulation of an extensive range of materials properties such as vibrational properties, energetics, and electronic response properties and structure at the atomic level using density functional theory. Moreover, it has a great list of spectroscopic features, such as IR and Raman spectroscopies, core level spectra, and NMR, which are related directly to the experiment.

COLUMBUS [7] is a set of programs for high-level ab initio molecular electronic structure calculations. Specially, calculations on electronic ground and excited states of atoms and molecules were implemented by performing distinctive methods such as multi-reference average quadratic coupled-cluster (MR-AQCC), multireference averaged coupled-pair-functional (MR-ACPF), multiconfiguration self-consistent field (MCSCF), multireference configuration interaction with all single and double excitations (MR-CISD), and so on. COLUMBUS's unity of programs is available to select reference configurations in addition to typical classes of reference wave functions (RAS or CAS).

CP2K [8] is a package developed as a software for calculations that are relevant for solid state physics and quantum chemistry. Implementation of atomistic simulations for a list of systems, such as crystalline solids, liquid materials, and biological systems, is realized in the CR2K package. The framework of CP2K implies modeling methods such as Gaussian/plane wave (GPW) basis-set and density functional theory (DFT), along with methods that present support for the general structure of package. For CR2K, a list of such methods is accomplished; for instance, it includes semi-empirical methods, random phase approximation (RPA), the Møller–Plesset perturbation theory (MP2), the classical force fields, the density functional based tight binding (DFTB) method, the local-density approximations (LDAs), the generalized gradient approximation (GGA), and so on. Runs of simulations are also available in THE CR2K package, specifically, CR2K carries out simulations of Monte Carlo, molecular dynamics, metadynamics, and Ehrenfest dynamics; moreover, the package implies energy minimization, transition state optimization, core level spectroscopy, vibrational analysis, and so on.

CPMD [9] is a code that performs the implementation of density functional theory by using extremely well parallelized plane/pseudopotential wave for ab initio molecular dynamics. Using the Car–Parrinello molecular dynamics scheme, the CPMD code executes quantum molecular dynamics simulations limited to systems of a few hundred atoms. Moreover, the CPMD code is capable of a hybrid quantum mechanical and molecular mechanics interface, retaining practices from the GROMOS96 molecular dynamics code, in order to expand its domain of applicability to larger biologically relevant systems. It can also be used for the simulation of chemical reactions, particularly in the condensed phase. Moreover, with the restricted open-shell Kohn–Sham excited state gradient, it allows the molecular dynamics simulation of photoreactions.

CRYSTAL [10] is a purpose program for the exploration of crystalline solids. The program carries out calculations on the electronic structure of Hartree–Fock periodic systems, various hybrid approximations, or density functional theory. In addition, the CRYSTAL program implements linear combinations of atom-centered Gaussian functions to expand the Bloch functions of the periodic systems. The calculations on restricted (closed shell) and unrestricted (spin-polarized) are feasible with all-electron and valence-only basis sets with effective core pseudo-potentials.

DACAPO [11] is a program designed for the implementation of calculations on total energy's system. The performance of calculations is realized by DACAPO through the usage of the camp atomic simulation environment (ASE). The base of the program is density functional theory (DFT). For the valence electronic states, DACAPO implies a plane wave basis; in addition, the description of the interactions of core–electron is realized by usage of Vanderbilt ultrasoft pseudo-potentials. The calculations on various generalized gradient approximation (GGA) exchange-correlation potentials and local density approximation (LDA) are carried out through program the accomplishment of state-of-the-art iterative algorithms. The solutions to the Schrodinger equations along with density functional theory allow implementing contemporaneously structural relaxation and molecular dynamics. The program implies compilation for parallel along with serial execution.

DALTON [12] is an ab initio chemistry program suitable for quantum calculations on distinct molecules properties. The realization of the program is carried out through the implementation of MCSCF, Hartree–Fock perturbation theory, MP2, and coupled cluster theories. The DALTON package is capable of automatic determination of a huge number of molecular properties based on a coupled cluster, density functional theory (DFT), Møller–Plesset perturbation theory (MP2), the Hartree–Fock (HF) method, or multi-configurational self-consistent field (MCSCF) reference wave function.

deMon2k [13] is a software package elaborated for calculations on DFT. At the core of the package, it is presented as an approach, based on the linear combination of Gaussian-type orbital (LCGTO), to the self-consistent solution of the Kohn–Sham (KS) DFT equations. The deMon2k program approaches the introduction of an auxiliary function basis for the variation fitting of the Coulomb potential. This allows avoiding the calculation of the four-center electron repulsion integrals.

DMol3 [14] is a software package that was developed based on density functional theory (DFT) in conjunction with a numerical radial function basis set. The prime focus of the package is presented as a conjunction of calculations on the surfaces, clusters, molecules, and solids properties of electronic structures. The package implies the availability of an application of lower-dimensional periodicity simulations or 3D periodic boundary conditions for crystalline solid materials or boundary conditions of gas phase; in addition, DMol3 applies, for quantum simulations of recently wetted surfaces and solvated molecules, the implementation of the COSMO solvation model as a conductor screening model. Geometry optimization and the search for saddle point along with realization of the calculation of electronic configuration properties, including or not the geometry constraints.

DP [15] is an open-source software package. It is specified mainly for physicists' implementation of ab initio time-dependent density functional theory (TD-DFT) with a linear-response on a plane wave basis set and frequency-reciprocal space. It allows calculating both dielectric spectra, such as inelastic electron energy-loss spectroscopy, X-ray scattering spectroscopy, and coherent inelastic X-ray scattering spectroscopy, and optical spectra, e.g., optical absorption, refraction index, and reflectivity. The range of systems lies in the gap from crystalline or periodic solids, to clusters, surfaces, atoms, and molecules. A nice feature of this software package is the implementation of adiabatic local-density approximation and non-local approximations (even neglecting or including local-field effects), as well as the random phase approximation.

ErgoSCF [16] is a quantum chemistry program. Large-scale self-consistent field calculations can be implemented by ErgoSCF. Linear scaling is achieved by modern techniques such as hierarchic sparse matrix algebra, efficient integral screening, fast multipole methods, and density matrix purification.

Exabyte.io [17] is a cloud-based digital platform for content research and development. The platform allows designing and building molecular models and running the stimulations.

FHI-aims package [18] (Fritz Haber Institute ab initio molecular simulations package) is a code package designed to perform computational molecular and materials science. FHI-aims maintains density functional theory and many-body perturbation theory for the realization of semi-local and hybrid exchange-correlation functionals. The FHI-aims is capable of calculations of an amount near to thousands of atoms and provides efficient usage for (ten) thousands of cores.

The FPLO [19] package is code that was developed for finite systems to solve the Kohn–Sham equations with free boundary or on a conditions regular lattice. Orbital polarization correction, the full-featured linearized augmented plane wave (LAPW) method implementations, generalized gradient approximation (GGA) functional, and the local spin density approximation (LSDA) are available and realized by the FPLO package, which is programmed for a local-orbital full-potential minimum-basis. Relativistic effects either are considered in scalar-relativistic or in full four-component formalism. A level comparable to the advanced level of numerical precision was achieved. Because of this high accuracy, implementation calculations on system's entire potential unit cells include up to 300 atoms; moreover, calculations can be carried out on uniprocessor machines.

GAMESS-UK [20] and GAMESS-US [21] (General Atomic and Molecular Electronic Structure System) are general-purpose computer software, designed for the computational chemistry program. GAMESS-UK and GAMESS (US) implement basic general calculations on computational chemistry, such as multiconfigurational self-consistent field, generalized valence bond, density functional theory, and the Hartree–Fock method. Estimation of correlation corrections after the self-consistent field calculations is realized by the approaching of second-order Møller–Plesset perturbation theory, coupled cluster theory, and configuration interaction. Calculations on relativistic corrections are estimated through the application of third-order Douglas–Kroll scalar terms. The solvent effect is treated by the usage of molecular mechanics and quantum mechanics through discrete effective fragment potentials or continuum models, such as the polarizable continuum model (PCM).

The Gaussian [22] package is a general-purpose software, developed for computational chemistry. The Gaussian package makes available electronic structure modeling with state-of-the-art capabilities. The implementation of Gaussian orbitals instead of Slater-type orbitals allows accelerating calculations on molecular electronic structure. The prediction of molecules' properties, such as Raman and IR spectra, molecular energies, vibrational frequencies, atomic charges, structures of transition states, NMR shielding, and magnetic susceptibilities, among others, is implied by Gaussian. Scientific and modeling feature are contained in the Gaussian package; moreover, calculations are not entangled with any artificial limitations, except for your computing resources and time.

GPAW [23] is a Python code for density-functional theory (DFT). GPAW is based on the atomic simulation environment (ASE) and the projector-augmented wave (PAW) method. Realization of the description of wave functions is done with real-space uniform grids, multigrid methods and the finite-difference approximation (FD), atom-centered basis-functions (LCAO), and plane-waves (PW). GPAW calculations are controlled through scripts written in the programming language Python. GPAW relies on the atomic simulation environment (ASE) that handles molecular dynamics, analysis, visualization, geometry optimization, and more.

HORTON [24] is a modular quantum chemistry program that allows to carry out various ab initio and DFT calculations following the post-processing options such as atoms-in-molecule analysis including Hirshfeld, extended Hirshfeld, iterative Hirshfeld, Becke, and iterative stockholder; orbital entanglement analysis; and electrostatic potential fitting of atomic charges.

HyperChem [25] is an environment that was developed for molecular modeling. The base features of HyperChem are presented in the program as a conjunction of 3D visualization and 3D animation with molecular dynamics, molecular mechanics, and quantum calculations on chemical systems. The program HyperChem is available on computers with Windows as well as Linux operative systems. The kinetics, spectra, and thermodynamics are implemented; moreover, geometry optimization, calculations in search of transition state, and application of modeling surfaces' potential energy are also realized by HyperChem. In addition, Langevin dynamics and metropolis Monte Carlo simulations are run in the environment. Calculations of free energies, heat capacities, and entropies are accomplished through treatment of vibrational partition, rotational, and translational functions.

JDFTx [26] is a code, developed for plane-wave density-functional theory (DFT), and specifically the code that supports of all the standard functionalities presented in electronic DFT software and several semi-local, meta-GGA, and EXX-hybrid exchange-correlation functions, which are available with additional options by linking to the library of exchange-correlation functionals (LibXC). DFT + U is capable of approaching localized electrons. Including van der Waals interactions in calculations is possible in pair potential dispersion corrections. The usage of analytically continued energy functionals allows JDFTx to implement total energy minimization. Before performing the first electronic solution, following straight after the end of initialization, JDFTx carries prints out that present a list for optional features of current relevant citations of the code. JDFTx is based on C++ code, developed with an approach of highly templated and object-oriented functions. This was applied for the expression of all the physics that appear in the DFT++ algebraic framework; moreover, it allows to contemporaneously support a range of hardware architectures (such as graphics processing unit (GPU) using computed unified device architecture (CUDA)) and to maintain a small memory footprint, requiring any implementations for each architecture, especially hand-optimized.

The Maple Quantum Chemistry (MQC) package [27] is a toolbox, developed for the predictions, research, and design of novel molecules in an available and easy-to-build encirclement. The Maple Quantum Chemistry package presents a combination of up-to-date quantum chemistry software, including advanced reduced density matrix (RDM) techniques, density functional theory (DFT), and wave function methods, with the mathematical calculations and approachability of the MQC package allowing to provide a complex, available for usage encirclement for the parallel computation of electronic energies and molecules' properties. Moreover, the MQC package uses a database that includes nearly 96 million molecules to define molecules; in addition, the package runs quantum computations with an approach of electronic structure methods, analyzes molecular energies, and carries out high-quality 2D and 3D plots and animations.

Molcas [28] is a quantum chemistry ab initio software, which presents itself as a package developed on a stable version of OpenMolcas. As the basic idea of code is an ability to consider general electronic structures, the focus of the package is primarily turned to multiconfigurational methods; therefore, the number of approaches is typically connected to the consideration of highly degenerate states.

MOLGW [29] is a Gaussian-type orbital (GTO) code designed to implement a self-consistent mean-field calculation for finite systems (clusters, molecules, atoms) with a sequent application of a many-body perturbation theory on the system. The MOLGW code performs the approximation for the optical excitations applying the Bethe–Salpeter equation and, for the self-energy, the code accomplishes the density-functional theory (DFT) code for preparation of the subsequent many-body perturbation theory (MBPT) runs. MOLGW is capable of semi-local and standard local approximations of density-functional theory (DFT) along with several range-separated hybrid and hybrid functionals; moreover, the usage of a Gaussian-type orbitals basis set allows for the entire standard quantum chemistry tools to be reused. MOLGW is available for direct calculations on systems containing near to 100 atoms; larger calculations are feasible, but require computers that are more powerful.

Molpro [30] is a complex system composed of ab initio programs designed for calculations on molecular electronic structures. Molpro comprises programs that were well connected in parallel. Standard computational chemistry approaches include the multireference wave function methods, state-of-the-art high-level coupled-cluster, density functional theory (DFT) with a large choice of functionals, and so on. Molpro allows to implement electronically excited states by consideration of the complete active space self-consistent field (CASSCF)/the multiconfigurational self-consistent field (MCSCF) methods, the complete active space second-order perturbation theory (CASPT2) method, the multireference configuration interaction (MRCI) method, or the full configuration interaction (FCI) methods, or by response methods such as the time-dependent density functional theory (TDDFT), coupled cluster methods (CC2), and equation-of-motion coupled cluster with single and double excitations (EOM-CCSD). Molpro accomplishes the modules that were developed for computing molecular properties such as further wave function analysis, calculations on harmonic as well as inharmonic vibrational frequencies, and geometry optimization. The analytical availability of energy gradients is carried out through the coupled cluster single-double (CCSD) method, the coupled cluster method for single and double excitations with exponential interelectronic distance explicit correlation correction (CCSD-F12), density functional theory (DFT), the Hartree–Fock (HF) method, the Møller–Plesset perturbation theory (MP2), the second-order Møller–Plesset perturbation theory with exponential interelectronic distance explicit correlation correction (MP2-F12), the quadratic configuration interaction with single and double excitations (QCISD), the quadratic configuration interaction with single and double excitations and triple excitations (QCISD(T)), the complete active space self-consistent field (CASSCF) method, and the complete active-space second-order perturbation theory (CASPT2). Approximations of density fitting (the density functional (DF) or the resolution-of-the-identity (RI) approximations) are able to accelerate the density functional theory (DFT) and Møller–Plesset perturbation theory (MP2) calculations that imply large basis sets and are clearly correlated with methods such as MP2-F12, CCSD(T)-F12, CASPT2-F12, and MRCI-F12. This allows to minimize errors that might appear owing to incompleteness of the basis set, thus achieving results near to CBS quality using the triple-zeta basis sets. An application of Molpro to large molecules with high accuracy is implemented by the combination of local approximations and efficient parallelization, high-level methods such as the accurate intermolecular interaction energies using explicitly correlated local coupled-cluster methods (PNO-LCCSD(T)-F12), and the pair natural orbital local second-order Møller–Plesset perturbation theory (PNO-LMP2-F12). Moreover, wave function in DFT (WF-in-DFT) calculation application or quantum mechanics/molecular mechanics (QM/MM) methods are available to be extended in the applicability as ab initio methods applied to systems of a large size that present biochemical and chemical interests.

MOPAC [31] (Molecular Orbital PACkage) is a quantum chemistry program based on Thiel's NDDO and Dewar approximation. The package allowed to use the improved handling of large biomolecules, including the ability to more easily manipulate macromolecules, e.g., to ionize and de-ionize individual atoms and residues, superimposition and calculation of RMSD for pairs of systems, and easier specification of individual atoms.

The Massively Parallel Quantum Chemistry (MPQC) [32] platform is a platform developed as a package for building simulation capable of run ab initio of the electronic structure of periodic solids and other molecules. The focus is directed on methods of many-body electronic structure, such as coupled-cluster (CC2) or the Møller–Plesset perturbation theory (MP2). The current (4th) version of the MPQC4 package presents as the original MPQC platform revision of the conceptual design. The package is able to use the Gaussian integrals library Libint embedding, distribution of task-based programming model, and runtime MADWorld, causing the massively parallel tensor framework TiledArray.

MRCC [33] is a software that presents itself as a suite developed for quantum chemistry programs of ab initio and density functional with high-accuracy calculations on an electronic structure. MRCC's special feature is the performance of automated programming tools that are able to develop routines of tensor manipulation, which are not independent of the quantity of the corresponding tensors indices, thus the application of quantum chemical methods is generally simplified. The approach based on the automated tools of the program allows realizing techniques of high complexity for quantum chemistry models. The applied methods include and compose multi-reference CC approaches, arbitrary coupled-cluster (CC) taken as a reference and configuration interaction (CI) methods, arbitrary perturbative CC approaches, CC and CI energy derivatives, and response functions. In addition, the number of features that the package implies are capable of availability with relativistic Hamiltonians, which allows accurate calculations on heavy element systems. The techniques were developed specifically for cost-reduction on calculations; moreover, approaches with local correlation allow high-precision calculations on large- and medium-sized molecules.

NTChem [34] is a package that implies itself as a high-performance software developed in R-CCS from scratch. It is specified explicitly for the computation for general purpose of the molecular electronic structure calculation. NTChem presents a complex new software of ab initio quantum chemistry; moreover, it contains standard quantum chemistry and original applications.

NWChem [35] is a software package for ab initio computational chemistry, and was designed in a way to include both molecular dynamics and quantum chemical functionality at the same time. The prime focus of NWChem remains to run on parallel supercomputers with high-performance along with conventional workstation clusters. The aims of the package are to be scale both in its usage of available parallel computing resources, and in its ability to consider efficiently problems.

ONETEP [36] (Order-N Electronic Total Energy Package) is a linear-scaling code based on density functional theory (DFT). It was developed explicitly for calculations on quantum-mechanical systems. The prime focus of ONETEP is to improve and apply the overall linear scaling and to control accuracy, keeping it high. The optimization procedures and the density-matrix formulation of density functional theory (DFT) were both included in the method of code; therefore, the description of the local orbitals or non-orthogonal generalized Wannier functions and for the density-kernel is implemented. The approaches of these methods to a great number of systems are able to demonstrate the performance of the realization prime focus of ONETEP.

OpenAtom [37] is an ab initio software for molecular dynamics; being a parallel simulation, it allows to study atomic along with molecular systems that are based on principles of quantum chemical. The main approach of the software is the CPAIMD algorithm that runs calculations on the forces acting independently on each atom as the result of the summation of multiple terms determined by plane-wave density functional solutions. OpenAtom implies research on complex systems, such as atomic systems and electronic physics in semiconductor, metallic, biological, and other molecular systems' implementation; in addition, OpenAtom is programmed on top of Charm++ that is a parallel framework for programming based on an over-decomposition.

OpenMx [38] is an open source software that allows estimation of a wide variety of advanced multivariate statistical models. OpenMx consists of a library of functions and optimizers that allow a quick and flexible definition of an SEM model and estimation of parameters given the observed data.

ORCA [39] is a program package that implies a general purpose tool for quantum chemistry. ORCA contains a number of modern methods for calculations on electronic structure, such methods as the semi-empirical quantum chemistry methods, the multireference methods, the coupled cluster (CC2), the many-body perturbation theory (MBPT), the density functional theory (DFT), and so on. The prime focus of ORCA is directed to the availability of applying the calculations to transition metal complexes and their optical properties, large molecules, and others. As a great amount of attention in the ORCA package is paid to spectroscopic open-shell molecules' properties, the package implies an extensive list of standard quantum chemical methods for this issue. The methods range from single- and multireference correlated ab initio methods and the consideration of ambient and relative effects to semi-empirical methods and density functional theory (DFT).

PARSEC [40] is a package that implements the realization of dentinal functional theory (DFT) calculations on solids and molecules. The code implies non-periodic boundary conditions without the performance of super-cells; moreover, the code accomplishes an equal well handling with periodic boundary and periodic conditions. PARSEC is programmed capable of an easy amenable process for efficient massive parallelization that performs highly effective calculations on far more than large systems.

PSI4 [41] is an ab initio quantum chemistry program that was programmed as a hybrid of C++ and Python. The main focus of the PSI4 program is the performance of calculations on an electronic structure. PSI4 implies realization of the density functional theory (DFT), the coupled cluster (CC) theory, the Hartree–Fock (HF) method, the configuration interaction (CI), the many-body perturbation theory (MBPT), the density cumulant theory (DCT), the symmetry-adapted perturbation theory (SAPT), and so on.

PyQuante [42] is a software that was developed as a suite for programs that imply Gaussian-type orbital (GTO) basis sets for implementation of quantum chemistry methods. The basic code, which is presented in PyQuante, is programmed in a way such that it is easily comprehended and easily modified, despite that it is time-consuming. The software provides a set of tools that can be used by scientists to construct self-developed quantum chemistry programs without the necessity of writing such things as every low-level routine. PyQuante might be suitable for educational purposes for students who are involved in a quantum and computational chemistry development.

PySCF [43] (Python-based Simulations of Chemistry Framework) is a Python/C-based package of a number of open-source compositions of electronic structure module development. PySCF implies a platform that is specified on the calculations on quantum chemistry systems and on the development of methodology. The package is capable of simulation of molecules' and crystals' properties. The PySCF provides Hamiltonians with post-mean-field and mean-field methods. Extensibility is implied and implemented by the fact that almost all of the PySCF features are realized in Python, whereas parts critical to the calculation are programmed in C; therefore, the combination of Python/C allows the package to be complete efficiently and to accomplish C- or FORTRAN-based quantum chemistry programs.

QMCPACK [44] (Quantum Monte Carlo PACKage) is a package designed for ab initio calculations on an electronic structure. QMCPACK implies computations of such systems as model Hamiltonians, atoms, molecules, and insulating and metallic solids. The package contains real space quantum Monte Carlo algorithms with embedded diffusion and reputation. Slater–Jastrow type trial wave functions are implemented as one with a complicated optimizer that is programmed in a way of an opportunity to improve thousands and tens thousands of parameters. QMCPACK carries out the quantum Monte Carlo method for the orbital space auxiliary-field; moreover, approaching cross validation between different highly accurate methods allows for optimization of the realization of calculations. A large number systems of electrons is currently available on high-performance computers; in particular, such calculations are relevant for graphical processing unit systems and multicore central processing units.

Quantum ESPRESSO [45] (opEn-Source Package for Research in Electronic Structure, Simulation, and Optimization) is a suite developed for computer codes to accomplish calculations on an electronic structure and modulations of nanoscale materials. The general framework of Quantum ESPRESSO is based on the density functional theory (DFT), pseudopotentials (ultrasoft and norm conserving), the density functional perturbation theory, and plane waves basic sets.

RMG [46] is an extraordinary example of highly scalable DFT electronic structure code, which is developed for solutions of Kohn–Sham equations right on a 3D real space grid without including the usage of basis set functions.

SAMSON [47] is the platform specified on integrated molecular design. It allows to model nanosystems through their structures, dynamics, interactions, visuals, and properties.

SIESTA [48] is a computer program implementation that allows performing calculations on electronic structures and on molecular dynamics simulations for solids and molecules. The program implies the usage of strictly localized basis sets to improve efficiency; in addition, the realization of algorithms based on linear scaling allows to apply SIESTA to huge systems. The feature observed in the code presents itself with accuracy. The cost of application is placed in a wide tern ranging from simulations with high accuracy, matching the quality of a number of other approaches, to quick exploratory calculations, for instance, all-electron and plane-wave methods.

VOTCA-XTP [49] is an open-source library that was programmed for calculations on the organic material properties of electronic structures, such as excited state properties through the GW plus Bethe–Salpeter equation (GW-BSE) method, spectra usage of the quantum mechanics/molecular mechanics (QM/MM) approach, along with excited state consideration with GW-BSE. Calculations on hole and electron mobilities were realized using electronic couplings and kinetic Monte Carlo. Usage of the Monte Carlo kinetic and electronic couplings allows to perform singlet, electron, triplet, and hole diffusion constants. Energetic disorder for electrons or holes or excitons was implemented along with the quantum mechanics/molecular mechanics (QM/MM) or molecular mechanics multipole methods.

Yambo [50] is a software package designed for the implementation of calculations, based on many-body perturbation theory (MBPT) methods (such as GW and BSE) and time-dependent density functional theory (TDDFT), for molecule systems and crystalline solid materials. Yambo implies calculations on the excited state properties of physical systems; moreover, the prediction of fundamental properties, such as defect quasi-particle energies, band gaps of semiconductors, optics, and so on, is accurate.

Thus, we call on the scientific community to pay attention to this software for computer modeling in chemistry and related areas of natural sciences and to widely use these program packages in daily research work.

Author Contributions: Conceptualization, A.S.N.; methodology, A.S.N.; software, M.V.M. and A.S.N.; validation, M.V.M. and A.S.N.; formal analysis, M.V.M. and A.S.N.; investigation, M.V.M. and A.S.N.; resources, M.V.M. and A.S.N.; data curation, M.V.M. and A.S.N.; writing—original draft preparation, M.V.M. and A.S.N.; writing—review and editing, A.S.N.; visualization, M.V.M. and A.S.N.; supervision, A.S.N.; project administration, A.S.N.; funding acquisition, this review was written without attracting additional external funding from any scientific foundations. All authors have read and agreed to the published version of the manuscript.

Funding: This review was written without attracting additional external funding from any scientific foundations.

Data Availability Statement: Not applicable.

Conflicts of Interest: The authors declare no conflict of interest.

References

1. ABINIT. Available online: https://www.abinit.org/ (accessed on 24 September 2021).
2. ACES. Available online: http://www.qtp.ufl.edu/ACES/ (accessed on 24 September 2021).
3. BerkeleyGW. Available online: https://berkeleygw.org/ (accessed on 24 September 2021).
4. BigDFT. Available online: https://bigdft.org/ (accessed on 24 September 2021).
5. BrianQC and Q-Chem. Available online: https://www.brianqc.com/ and https://www.q-chem.com (accessed on 24 September 2021).
6. CASTEP. Available online: http://www.castep.org/CASTEP/CASTEP (accessed on 24 September 2021).
7. COLUMBUS. Available online: https://www.univie.ac.at/columbus/ (accessed on 24 September 2021).
8. CP2K. Available online: https://www.cp2k.org/ (accessed on 24 September 2021).

9.	CPMD. Available online: https://bioexcel.eu/software/cpmd/ (accessed on 24 September 2021).
10.	CRYSTAL. Available online: https://www.crystal.unito.it/index.php (accessed on 24 September 2021).
11.	Dacapo. Available online: https://wiki.fysik.dtu.dk/dacapo/dacapo (accessed on 24 September 2021).
12.	Dalton and LSDalton. Available online: https://daltonprogram.org/ (accessed on 24 September 2021).
13.	deMon2k. Available online: http://www.demon-software.com/public_html/index.html (accessed on 24 September 2021).
14.	DMol^3. Available online: http://dmol3.web.psi.ch/dmol3.html (accessed on 24 September 2021).
15.	The DP code. Available online: http://www.dp-code.org/ (accessed on 24 September 2021).
16.	ErgoSCF. Available online: http://www.ergoscf.org/ (accessed on 24 September 2021).
17.	EXABYTE.IO. Available online: https://exabyte.io/ (accessed on 24 September 2021).
18.	FHI-aims. Available online: https://aimsclub.fhi-berlin.mpg.de/ (accessed on 24 September 2021).
19.	FPLO. Available online: https://www.fplo.de/ (accessed on 24 September 2021).
20.	GAMESS-UK. Available online: https://computingforscience.com/gamess-uk/index.shtml (accessed on 24 September 2021).
21.	GAMESS. Available online: https://www.msg.chem.iastate.edu/ (accessed on 24 September 2021).
22.	Gaussian. Available online: https://gaussian.com/ (accessed on 24 September 2021).
23.	GPAW. Available online: https://wiki.fysik.dtu.dk/gpaw/ (accessed on 24 September 2021).
24.	HORTON. Available online: https://theochem.github.io/horton/ (accessed on 24 September 2021).
25.	HyperChem. Available online: https://www.chemits.com/en/software/molecular-modeling/hyperchem/ (accessed on 24 September 2021).
26.	JDFTx. Available online: https://jdftx.org/ (accessed on 24 September 2021).
27.	Maple Quantum Chemistry Toolbox. Available online: https://www.maplesoft.com/products/toolboxes/quantumchemistry/ (accessed on 24 September 2021).
28.	Molcas. Available online: https://www.molcas.org/ (accessed on 24 September 2021).
29.	MOLGW. Available online: http://www.molgw.org/ (accessed on 24 September 2021).
30.	Molpro. Available online: https://www.molpro.net/ (accessed on 24 September 2021).
31.	MOPAC. Available online: http://openmopac.net/ (accessed on 24 September 2021).
32.	MPQC. Available online: https://mpqc.org/ (accessed on 24 September 2021).
33.	MRCC. Available online: https://www.mrcc.hu/ (accessed on 24 September 2021).
34.	NTChem. Available online: https://www.r-ccs.riken.jp/software_center/software/ntchem/overview/ (accessed on 24 September 2021).
35.	NWChem. Available online: https://nwchemgit.github.io/ (accessed on 24 September 2021).
36.	ONETEP. Available online: https://www.onetep.org/ (accessed on 24 September 2021).
37.	OpenAtom. Available online: https://charm.cs.illinois.edu/OpenAtom/doxy/html/index.html (accessed on 24 September 2021).
38.	OpenMx. Available online: https://openmx.ssri.psu.edu/ (accessed on 24 September 2021).
39.	ORCA. Available online: https://orcaforum.kofo.mpg.de/app.php/portal (accessed on 24 September 2021).
40.	PARSEC. Available online: https://parsec.oden.utexas.edu/ (accessed on 24 September 2021).
41.	PSI4. Available online: https://psicode.org/ (accessed on 24 September 2021).
42.	PyQuante. Available online: http://pyquante.sourceforge.net/ (accessed on 24 September 2021).
43.	PySCF. Available online: https://pyscf.org/ (accessed on 24 September 2021).
44.	QMCPACK. Available online: https://qmcpack.org/ (accessed on 24 September 2021).
45.	Quantum ESPRESSO. Available online: https://www.quantum-espresso.org/ (accessed on 24 September 2021).
46.	RMG. Available online: http://www.rmgdft.org/ (accessed on 24 September 2021).
47.	SAMSON. Available online: https://www.samson-connect.net/ (accessed on 24 September 2021).
48.	SIESTA. Available online: https://siesta-project.org/siesta/ (accessed on 24 September 2021).
49.	VOTCA-XTP. Available online: https://github.com/votca/xtp (accessed on 24 September 2021).
50.	YAMBO. Available online: http://www.yambo-code.org/ (accessed on 24 September 2021).

MDPI
St. Alban-Anlage 66
4052 Basel
Switzerland
Tel. +41 61 683 77 34
Fax +41 61 302 89 18
www.mdpi.com

Compounds Editorial Office
E-mail: compounds@mdpi.com
www.mdpi.com/journal/compounds